De Novo Peptide Design

Principles and Applications

De Novo Peptide Design

Principles and Applications

Edited by

Vibin Ramakrishnan
Biosciences and Bioengineering, Indian Institute of Technology Guwahati, Guwahati, Assam, India

Kirtikumar Patel
N. B. Mehta Science and Commerce College, Bordi, Palghar, Maharashtra, India

Ruchika Goyal
Department of Biosciences and Bioengineering, Indian Institute of Technology Guwahati, Guwahati, Assam, India

ELSEVIER

ACADEMIC PRESS
An imprint of Elsevier

Academic Press is an imprint of Elsevier
125 London Wall, London EC2Y 5AS, United Kingdom
525 B Street, Suite 1650, San Diego, CA 92101, United States
50 Hampshire Street, 5th Floor, Cambridge, MA 02139, United States
The Boulevard, Langford Lane, Kidlington, Oxford OX5 1GB, United Kingdom

ISBN: 978-0-323-99917-5

For Information on all Academic Press publications
visit our website at https://www.elsevier.com/books-and-journals

Publisher: Andre G. Wolff
Acquisitions Editor: Glyn Jones
Editorial Project Manager: Samantha Allard
Production Project Manager: Sreejith Viswanathan
Cover Designer: Greg Harris

Typeset by MPS Limited, Chennai, India

Dedication

Dedicated to Professor Susheel Durani

Contents

List of contributors

Debika Datta Department of Nanoengineering, University of California San Diego, La Jolla, CA, United States

Bhupesh Goyal School of Chemistry & Biochemistry, Thapar Institute of Engineering & Technology (Deemed to be University), Patiala, Punjab, India

Deepti Goyal Department of Chemistry, DAV College, Chandigarh, India

Ruchika Goyal Department of Biosciences and Bioengineering, Indian Institute of Technology Guwahati, Guwahati, Assam, India

Gaurav Jerath Pepthera Laboratories Pvt. Ltd., Technology Incubation Centre, Indian Institute of Technology Guwahati, Guwahati, Assam, India; Department of Medical Biotechnology, Gujarat Biotechnology University, Gandhinagar, Gujarat, India

Gaurav Pandey Department of Chemistry, University of California Riverside, Riverside, CA, United States; Present address: Scripps Research, Department of Molecular Medicine, La Jolla, CA, United States

Kirtikumar Patel N. B. Mehta Science and Commerce College, Bordi, Palghar, Maharashtra, India

Vivek Prakash Department of Biosciences and Bioengineering, Indian Institute of Technology Guwahati, Guwahati, Assam, India

Aparna Rai Pepthera Laboratories Pvt. Ltd., Technology Incubation Centre, Indian Institute of Technology Guwahati, Guwahati, Assam, India

Vibin Ramakrishnan Department of Biosciences and Bioengineering, Indian Institute of Technology Guwahati, Guwahati, Assam, India

Ranjit Ranbhor Pergament & Cepeda LLP, Florham Park, NJ, United States

Amay Redkar Department of Biosciences and Bioengineering, Indian Institute of Technology Guwahati, Guwahati, Assam, India

Jahnu Saikia Department of Biosciences and Bioengineering, Indian Institute of Technology Guwahati, Guwahati, Assam, India

Mouli Sarkar Department of Biosciences and Bioengineering, IIT Guwahati, Guwahati, Assam, India

Chapter 1

Structural organization of peptides

Kirtikumar Patel*
N. B. Mehta Science and Commerce College, Bordi, Palghar, Maharashtra, India

Nature makes every molecule for a specific purpose, such as nucleic acids to store and transmit genetic information to the next generation, carbohydrates as source of energy and raw materials for various natural products, lipids to build structure of the cell, proteins to carry out reactions efficiently at ambient conditions. Among these, the proteins are by far the most versatile biomolecules. They adapt to perform various roles in the cell such as catalyzing cellular processes, storing of essential metal ions, quenching of harmful intermediates such as superoxide, building defenses for the organism in the form of antibodies, and many more. They are among the fastest and highly specific in their function.

Proteins are polymers of L-α-amino acid in alphabet that encodes the information from its sequence to perform its function as an enzyme (Branden & Tooze, 1998). Although astronomical in possibilities of interactions along their sequences, proteins form a defined three dimensional structure, which according to Anfinsen are sequence-defined minima in free energy (Anfinsen, 1973; Anfinsen & Scheraga, 1975). Anfinsen's experiment with ribonuclease identified amino acid sequence as the sole arbiter of protein folding (Anfinsen, 1973). The studies of Anfinsen have now become the subject of interest to understand the role of sequence in lowering free energy in protein folding and its function.

The three-dimensional structure adopted by a protein is dependent on its environment, the temperature, the pressure, and the solvent or molecular surroundings. Often, the "folded" conformation of a polypeptide is only marginally more stable than the "unfolded" conformation (Daura et al., 2001; Daura, Gunsteren, & Mark, 1999). The combined results of experiment and theory with atomistic models of polypeptides are laying the foundations for the knowledge of protein folding.

---

*Corresponding Author

De Novo Peptide Design. DOI: https://doi.org/10.1016/B978-0-323-99917-5.00008-1

Polypeptides are designed sequentially in amino acid side-chain structures as alphabet for their specific role as receptors, enzymes, antibodies, etc. The method evokes interest as a technology for practical applications, but remains in mystery about its working principles. Researches are undertaken in individual amino acid side-chain structures to clarify their roles in structure of a protein or catalytic activity of an enzyme. The studies are pursued either biochemically or chemically. The biochemical approach involves interchanging the amino acid side-chain positions and studying the effects in protein structure or enzyme function also known as top-down approach. The chemical approach, on the other hand, involves modifying the amino acid side chains in specific structural detail and studying the effects in synthetic polypeptides also known as bottom-up approach. The bottom-up approach offers the advantage that simple polypeptide structures can be used and experiment and theory can be combined to study the principles. In addition, the bottom-up approach offers the advantage that de novo protein design can be extended in scope with additions to the alphabet of nature.

Anfinsen pioneered the top-down approach with his experiment of folding-unfolding of ribonuclease (Anfinsen, 1973). Combined with the site-directed mutagenesis, top-down approaches are insightful of the residue contributions in stability and specificity of conformation. Anfinsen suggested that the folding process might be highly cooperative (Anfinsen, 1973). Cooperativity of protein folding is shown experimentally by Tanuichi. In the experiment, RNAse A was unable to fold when four C-terminal residues were removed (Tanuichi, 1970). Thus showing the importance of sequence in protein folding and also eradicating the possibility of folding during protein synthesis. Similar results have been reported in small peptide models where segments like β-turns have been used as conformational nucleators to enable the peptide to fold into a predefined three-dimensional structure (Venkatraman, Shankaramma, & Balaram, 2001). The top-down approaches have mainly focused on the effects of specific residues either in folding or imparting function to the protein.

The bottom-up approach, pioneered by Pauling (Pauling & Corey, 1951a, 1951b) and Ramachandran (Ramachandran & Sasisekaran, 1968; Ramachandran, Ramakrishnan, & Sasisekaran, 1963), identifies generic properties of the polypeptide structure that might be critical in folding. Pauling clarified the importance of amide planarity and hydrogen bond geometry in folding. Ramachandran demonstrated the role of excluded volume effects in constraining φ,ψ space in polypeptides. Apart from hydrogen bond and excluded volume effects, variety of molecular interactions are operative in native proteins like, electrostatic interactions (Avbelj & Baldwin, 2002; Avbelj & Moult, 1995), dispersion interactions, hydrophobic effect (Chan & Dill, 1991; Dill, Truskett, Vlachy, & Hribar-Lee, 2005; Dill, 1990; Kauzmann, 1959) etc.

The bottom-up approach allows to test the involvement of each interaction in protein folding and in imparting function to the polypeptide chain.

This approach offers an advantage in de novo design of protein folds that may impart novel functions or it may be used to study the effect of certain interactions in protein folding. Recent studies involving interactions within protein structures and with surrounding solvent molecules have begun to give some understanding of the principles involved but much remains to be uncovered. The principles may be analyzed at the level of interactions involving main chain structures, the side-chain structures, and the structures of cofactors and substrates involved in enzyme catalysis.

1.1 Molecular interactions for protein folding

The overall stability of three-dimensional structure of the protein is dependent on the contributions made by several different forces such as hydrogen bonding, intrinsic propensities of sequence, main chain and side-chain electrostatics, main chain conformation and side-chain entropy, and van der Waals forces. These forces between protein side chains form the determining factor in protein folding and are thus determinants of protein tertiary and quaternary structure.

1.1.1 Hydrogen bonds

Hydrogen bond is an electrostatic force of attraction between partial positive charge of hydrogen atom bonded to an electronegative atom or group and electron-rich another electronegative atom possessing lone pair of electrons. The main component of the hydrogen bond is an electrostatic interaction between the dipole of the donor atom with partial negative charge, with the other electronegative atom, the acceptor, bonded to hydrogen atom with partial positive charge. Due to its small size and partial positive charge when covalently bonded to an electronegative atom, hydrogen atom is able to interact strongly with another electronegative atom. Strong hydrogen bond is formed when the donor, hydrogen, and acceptor are collinear, any deviations from this result in decrease in the hydrogen bond energies (Scheiner, Redfern, & Hillenbrand, 1986). The strength of the hydrogen bond is in the order of 2–10 kcal/mole.

In proteins, hydrogen bonding is an electrostatic interaction between the amide dipoles that comprise main chain structure of proteins. Chemical groups in the proteins that most commonly serve as hydrogen bond donors are N – H, O – H and less frequently S – H and C – H. It is natural to assume that hydrogen bonding is important for the conformational preferences of proteins. While the main chain hydrogen bonding is important to protein folding defining the secondary structural elements, hydrogen bonding between the side chains is crucial to protein – protein interactions. The contributions of hydrogen bonds in protein interactions are clearly evident from the burial of polar residues across protein interfaces, in spite of their desolvation cost, guiding the specificity of protein – protein interaction (Zhou, 2018).

1.1.2 CH − π interactions

The traditional hydrogen bond is formed between an electron-rich donor and hydrogen attached to an electron-deficient atom. Recently a number of nonconventional hydrogen bonds are found in chemistry and biology that do participate in the stabilization (Jiang & Lai, 2002; Weiss, Brandl, Suhnel, Pal, & Hilgenfeld, 2001). Though weaker than conventional hydrogen bonds, these kinds of hydrogen bonds participate in specificity for recognition. The CH − π interaction, between electron-deficient CH and an electron-rich aromatic ring, is one such kind of a nonconventional hydrogen bond. Examples include the interaction of the electron-deficient $^{\delta}$CH of proline with aromatic rings (Bhattacharyya & Chakrabarti, 2003). The CH residues of sugars interact favorably with the aromatic residues; the sugar C − H residues have a significant partial positive charge on their hydrogen atoms as a consequence of an inductive effect provided by the neighboring O and OH groups (Sujatha, Sasidhar, & Balaji, 2007).

1.1.3 van der Waals interactions

Atoms attract each other even in the absence of charge due to the induced polarization effects. These interactions between two atoms are called the van der Waals interactions and are basically steric interactions (Dahiyat & Mayo, 1997a,1997b). As two atoms approach each other there is an attraction between atoms, weak and close range, varying as the sixth power of distance between them (Pollard & Earnshaw, 2017). However, as they come near enough for their electron orbitals to overlap there is a strong repulsion. The repulsive energy is often said to increase with the 12th power of distance between the centers of two atoms. The attraction between atoms can be that between two permanent dipoles, a permanent and induced dipole, and those between two mutually induced dipoles, which are also called as London or dispersion forces. Thus these forces are predominantly electrostatic in origin. The dispersion interaction is most commonly described by the Lennard Jones 6−12 potential. The optimal distance for interaction is usually 0.3−0.5 Á greater than the sum of their van der Waals radii.

1.1.4 Hydrophobic interaction

Hydrophobic interaction is one of the most common interactions that contributes toward protein structure, stability, and folding (Kauzmann, 1954, 1959). This involves interactions between nonpolar groups. The magnitude of hydrophobic interaction is usually measured as the free-energy of transfer ΔG_{tr} of a nonpolar molecule in gas, liquid, or solid state to water. According to the hydrophobic mechanism, a protein folds by burial of its nonpolar residues in the core, secluded from the solvent while exposing its polar residues to the solvent.

1.1.5 Electrostatic interactions

All molecular interactions are primarily electrostatic in origin. Unlike all other interaction forces, the electrostatic interactions are long range varying inversely with the second power of the distance as described in the Coulomb equation.

In proteins, electrostatics is contributed by the peptide backbone and the charged amino acid side chains. Both the electrostatic interactions of the peptide as well as specific charged interactions between the ion pairs are vital to protein stability. Protein recognition sites and interfaces are rich in both polar and charged residues where they play important structural and functional roles. The energetic role of salt bridges in protein structure stability has been found to vary (Fersht, 1972; Meyer, Castellano, & Diederich, 2003; Perutz & Raidt, 1975; Sheinerman, Norel, & Honig, 2000; Warshel & Russell, 1984). Solvent-exposed salt bridges contribute only marginally whereas a buried salt bridge contributes up to 5 kcal/mol toward native state stability. Burial of salt bridges in cores of protein interfaces is probably to intensify the electrostatic interactions across subunits.

1.1.6 Aromatic – aromatic ($\pi - \pi$) interactions

Proteins contain aromatic residues like Phenylalanine, Tyrosine, Tryptophan, and Histidine that are generally found in protein cores and are known to be involved in aromatic interactions (Burley & Petsko, 1985; Meyer et al., 2003). These interactions though weaker than hydrogen bonding are found to be quite prevalent in structural studies of proteins (McGaughey, Gagne, & Rappe, 1998). They contribute significantly in biomolecular recognition. Aromatic cluster in the active site of ATP synthetase is found to recognize ATP via $\pi - \pi$ interactions (Masunov & Lazaridis, 2003; Moodie, Mitchell, & Thornton, 1996).

1.1.7 Cation – π interaction

The cation – π interactions can be considered an ion – quadrupole interaction between positively charged group and an electron-rich π-cloud of the aromatic ring (Ma & Dougherty, 1997; Tatko & Waters, 2003). Like conventional interactions, hydrophobicity or hydrogen bonds, even cation-π interactions play an important role in molecular recognition. One of the classical examples is that of binding of acetylcholine to the enzyme acetylcholine esterase. The enzyme plays a crucial role at cholinergic synapses by hydrolyzing acetylcholine, thereby terminating synaptic transmission. The enzyme active site interacts via its aromatic cluster with quaternary ammonium group of acetylcholine.

1.2 Poly-alanine models and the energetics of protein folding

Alanine is the parent prototype for all the protein-coding amino acids, except glycine, which are its derivatives modified in side chain. The side chains are α-carbon substituents joined always in L-configuration. The consequent chirality of alanine and its derivatives renders N − Cα and Cα − CO bonds in residue-level structure rotationally anisotropic. The anisotropy imparts asymmetry to Ramachandran φ,ψ diagram for L-alanine dipeptide (Ac-Ala-NHMe) and its derivatives, contrasted with the symmetry of glycine residue (Ac-Gly-NHMe) being achiral in α-carbon (Ramachandran & Sasisekaran, 1968; Ramachandran et al., 1963). Alanine and its derivatives have sterically permitted access in β, α_L, and α_R conformational zones of Ramachandran diagram. Consequently, being perceived to unfold from "order-to-disorder," unfolded proteins are thought of as accessing the sterically allowed spaces in Ramachandran diagram statistically, being considered as "random coils" (Ramachandran et al., 1963) in so-called protein-folding funnel model represented as its smooth upper edge. Lacking apparent coding with side chains, oligo-alanines may be thought of nonspecific in conformation. However, when greater than the critical threshold of about 20 residues, oligo-L-alanines are α-helical folds showing rudiments of two-state cooperativity, making them legitimate "protein" models (Kumar, Ramakrishnan, Ranbhor, Patel, & Durani, 2009). Based on this observation, alanine is considered as helix-loving in its coding characteristics. The ordering of oligo-alanine helices manifests length-dependent cooperativity in polypeptide structure, as the helices are lost if the chain is shorter than the critical length (Shi, Olson, Rose, Baldwin, & Kallenbach, 2002; Shi, Woody, & Kallenbach, 2002). Thus less than about 20 residue poly-alanines were regarded as the ideal models for "unfolded" protein, which was presumed to be "random coil."

Experimental and theoretical studies with alanine and oligo-alanines have altered the perception radically (Head-Gordon et al., 1992). Instead of sampling the sterically accessible Ramachandran φ,ψ space statistically, alanine dipeptide has been found by experiment and theory to adopt extended conformation in correspondence of the right half of Ramachandran β-conformational zone now called as its polyproline II helix (PPII) subbasin (Shi, Olson, et al., 2002; Zagrovic et al., 2005). The preference for PPII conformation has been observed, in alanine dipeptide by nuclear magnetic resonance spectroscopy (NMR) (Poon, Samulski, Weise, & Weisshaar, 2000), molecular dynamics (MD) (Apostolakis, Ferrera, & Caflisch, 1999; Tobias & Brooks, 1992), and by quantum mechanical calculations (Han, Jalkanen, Elstner, & Suhai, 1998), in tri-alanine peptide by vibrational spectroscopy

(Eker, Griebenov, & Schweitzer-Stenner, 2003; Schweitzer-Stenner, 2002) and MD (Graf, Nguyen, Stock, & Schwalbe, 2007; Mu & Stock, 2002; Tashiro, Kobayashi, & Fujita, 2006), and in longer alanine peptides by UV Raman (Asher, Mikhonin, & Bykov, 2004). NMR and circular dichroism (CD) evidence have shown oligo-alanines as ordered ensembles in PPII conformation (Kumar et al., 2009), while UV-Raman experiments confirm that a 21-residue α-helical alanine peptide melts into predominantly PPII conformation (Mikhonin & Asher, 2006). Quantum mechanical calculations reported by Dannenberg and co-workers have clarified that alanine has its conformation defined with combination of sterics and peptide solvation (Wieczorek & Dannenberg, 2003, 2004, 2005). In absence of methyl side chain, glycine dipeptide and its oligomers have the energy minima in symmetry-related corners of φ,ψ surface in correspondence of $\varphi = +/-180$ and $\psi = +/-180$ (Henzler, Wilson, Lee, & Ramamoorthy, 2003; Munoz, Puiggali, Rodriguez, & Subirana, 1983; Ohnishi, Kamikubo, Onitsuka, & Shortle, 2006) Methyl in L-configuration will break the symmetry-shifting minimum to the left edge of Ramachandran β-basin. Immersion of the structure in water will cause minima on φ,ψ surface to shift as solvent-promoted minima to the right edge of Ramachandran β-basin called its PPII subbasin.

Extended but not to sufficient length for α-helix folding, oligo-alanines were proposed by Kallenbach and co-workers to adopt extended PPII conformation called PPII helix (Shi, Olson, et al., 2002) The studies were performed with the hepta-alanine model peptide XAO of the structure Ac-$X_2Ala_7O_2$-amide with diaminobutyric acid (X) and ornithine (O) residues as solubilizers. Based on NMR and CD the peptide was suggested to be in PPII conformation to at least 90% at 275K, devoid of even nascent type α-helices, and displaying ~12% increase in β-type conformation at 325K (Shi, Olson, et al., 2002). The follow-up studies of this peptide and similar other oligo-alanines have established that while these peptides were macroscopically in PPII helix conformation as reflected with NMR and CD, they were dynamic ensembles of PPII microstates largely folded in β-turn conformation (Nguyen, Marchut, & Hall, 2004; Ramakrishnan, Ranbhor, & Durani, 2004; Soto, Baumketner, & Shea, 2006; Thorpe, Zhou, & Voth, 2008). The basis for the paradoxical extended conformation in macrostate and folded conformation in microstate was observed to be the consequence of homochiral stereochemistry (Kumar, Ramakrishnan, Ranbhor, Patel, & Durani, 2009; Ramakrishnan, Ranbhor, Kumar, & Durani, 2006a, 2006b).

The poly-L stereochemistry was probed for possible role in protein folding using end-protected octa and nona-peptides of eight alanines or six alanines interspersed with three equally spaced Lysine solubilizers (Kumar et al., 2009). When poly-L in structure, the peptides were shown with NMR and CD as the PPII ensembles unfolded as helices but with microcalorimetry and MD as the PPII ensembles folded in β-hairpin conformation (Kumar et al., 2009). The poly-L structure was proven to promote the dynamic equilibrium involving the unfolded state extended as helices and the folded state bent as β-hairpins. The folding – unfolding equilibrium was under solvent control, the high dielectric environment in water promoting folding and the low dielectric environment in methanol promoting unfolding. Spectroscopic evidence with lysine solubilized nona-peptide proved its two-state equilibrium with the appearance of isodichoric point in CD (Kumar et al., 2009). The computational studies have suggested the role of the equilibrium to manifest the interactions among peptide dipoles becoming frustrated due to poly-L structure. This can be cured by modifications to alternating-L, D structure where the chain became impervious to solvent. The frustration providing the likely control principle for chain folding was proven to involve hydrogen bonds and electrostatics of peptide dipoles in mutual conflict as by abolishing the conflict the chain became impervious to solvent. The broad conclusion made was that the two-state nature of protein folding is a result of electrostatic interactions among the dipoles of main chain of poly-L structure in addition to the interactions of the side chains.

1.3 De novo protein design and stereochemical logic of protein folding

De novo design critically tests the folding principles of the protein structure. Designed proteins evoke interest as the aids for possibility of atom-by-atom analysis of folding (Baltzer, Nilsson, & Nilsson, 2001). For the design of receptors and enzymes for biosensing, signal transduction, catalysis, etc. (Dwyer & Hellinga, 2004; Park, Yang, Saven, & Curr, 2004) and of therapeutics against bacterial, fungal, and viral infections and as complements of immune system (Edwards, Cohen, & Bloom, 1999; Lazar, Marshall, Plecsy, Mayo, & Desjarlais, 2003; Lee et al., 2002), the coded design could be a powerful approach. The challenge in developing therapeutic peptides lies in addressing proteolysis and the lack of bioavailability via oral route. The susceptibility of natural proteins to denaturation and proteolysis (Fersht, 1998) motiviates explorations of non-peptides based on isosteric or iso-electronic exchanges or involving β (Abele & Seebach, 2000; Gellman, 1998; Imamura et al., 2009) or γ-peptides (Seebach, Beck, & Bierbaum, 2004), and peptidomimetics (Farrera, Giralt, Royo, & Albericio, 2007) as the design templates novel in structure. While the foldamer design is currently an exciting field of exploration (DeGrado, Wasserman, & Lear, 1989; Kaplan & DeGrado, 2004), de novo design of polypeptide folds harnessing L and D structure as the stereochemical alphabet has also given exciting results (Durani, 2008).

1.3.1 Stereochemical principles in protein design

α-amino acid structure is stereogenic with the possibility of L and D stereochemistry in side-chain attachment. The choice of configuration has fundamental bearing on conformational space in Ramachandran diagram, being asymmetric due to the structural chirality in α-amino acid residue (Ramachandran & Sasisekaran, 1968; Ramachandran et al., 1963). Although there is the choice of enantiomeric structure in every residue, proteins are poly-L due to biological reasons. The repeated stereochemistry over the length of peptide chain defines both α-helix and β-sheet folds. Polypeptides of irregular sequence chirality, that is, heterochiral sequences, can be sequence-defined folds (Durani, 2008), and thus they can be stereochemically deteremined protein shapes. Nature does use the conformational space favored for D-amino acids, but selectively with Glycine as the surrogate D-amino acid specifically in β-turns, which are crucial in their role in protein folding.

For fold design using stereochemistry as the alphabet, Ramachandran diagram is the important starting point. Developed in the early sixties (Ramachandran & Sasisekaran, 1968; Ramachandran et al., 1963), the diagram represents intramolecular atom − atom interactions within a dipeptide unit as a function of the rotational angles φ and ψ, and captures the residue level stereochemical effects in peptide conformation. The effect of R-group size and attachment chirality is directly evident from Ramachandran diagram. Sequence effects manifest the interactions among the R-groups that are topological neighbors in a folded protein. The chemical and streochemical roles of R-group as protein codes are important for protein structure. The realization is the basis for protein design in which side chains are the alphabet in not only their chemical but also the stereochemical structures, being the basis for so-called shape-specific protein design. The shape-specific protein design using L and D alphabet has rich precedent in the understanding of stereochemistry in its role in protein β-turns.

1.3.2 β-Turn as stereochemically diverse conformational nucleators

The most common structural element that brings about the chain reversal in proteins is the so-called β-turn (Gunasekaran, Ramakrishnan, & Balaram, 1997; Milner -White & Poet, 1986; Sibanda, Blundell, & Thornton, 1989; Sibanda & Thornton, 1985). β-turns are frequently found as connecting units of secondary structure such as α-helices and β-sheets. Different kinds of β-turns occur and are differentiated based on the backbone dihedral angles of the central two residues (Rose, Gierasch, & Smith, 1985). The most common turns are types I, II, and III and their inversion symmetry variants I′, II′, and III′. The characteristic φ,ψ angles of these turn types are as shown in Table 1.1.

1.3.2.1 Homochiral turns

The turns in which the defining torsional angles are favored for either LL or DD chiral dipeptide segment characterize the homochiral turns. The types

TABLE 1.1 Torsional angle values of types I, I′, II, II′, III and III′ β–turns for (i + 1) and (i + 2) residues.

Type	$\varphi_{(i+1)}$	$\psi_{(i+1)}$	$\varphi_{(i+2)}$	$\psi_{(i+2)}$
I	−60	−30	−90	0
I′	60	30	90	0
II	−60	120	80	0
II′	60	−120	−80	0
III	−60	−30	−60	−30
III′	60	30	60	30

I and III turns and their inversion symmetry variants types I′ and III′ turns are all homochiral turns. The two turn types share an identical set of torsional angles for the first corner residue and differ in the values of the second corner residue. In the types I and I′ turns the second corner residue occurs in the neck region between the α and β zones of Ramachandran plot. The type I turn is a prototypical chain inverter involved in hairpin formation (Alvarado, Blanco, & Serrano, 1996; Maynard, Sharman, & Searle, 1998). The type III turn, on the other hand, is helix primitive than chain inverter being more commonly associated with helix termini (Nemethy & Scheraga, 1981).

1.3.2.2 Heterochiral turns

The types II and II′ β-turns are with φ,ψ values for the central two residues that are favored in, respectively, L, D and D, L chiral dipeptide segments. The majority of the turn conformations in proteins and peptides are of this fundamental heterochiral type (Haque, Little, & Gellman, 1994; Raghothama, Awasthi, & Balaram, 1998; Sibanda, Blundell, & Thornton, 1989; Struthers, Cheng, & Imperiali, 1996). The ability of LD/DL segment to nucleate turns was theoretically recognized by Ramachandran and co-workers (Ramachandran & Sasisekaran, 1968) and has subsequently been widely exploited for designing synthetic sequences. Proline is the most favored first corner residue in this turn type in proteins. Nature uses wide conformational flexibility of Glycine residue and places it in the second corner position of type II and first corner position of type II′ turns, which are otherwise inaccessible by the naturally occurring L-α-substituted amino acids. Thus nature compensates the lack of D-amino acids by using Glycine as the pseudo D-amino acid in the heterochiral turns. The majority of the turn conformations in proteins and peptides are of this fundamental heterochiral type. The reason for this choice of nature partly comes from the fact that types II and II′ turns form stronger hydrogen bond due to favored alignment of participating NH and CO groups. Because of their heterochiral nature, types II and II′ are particularly attractive targets for de novo design approachable by use of sequence chirality as the design aid.

Turns play a major role in protein folding. The turns are more critically determined by the local sequence than either helix or sheet. Helices and strands of β-sheet are often separated by β-turns, by changing their direction and hence allowing the elements of secondary structure to form tertiary structure (Alvarado et al., 1996; DeAlba, Jimenez, & Rico, 1997). Given that, the segments of α-helices and β-strands are at best marginally stable in aqueous solution, the formation of stabilizing interactions through β-turns could be crucial step in the folding process. The β-turns have been postulated to play an important role in protein folding and have been suggested to act as a type of start/stop code for the formation of helices and sheets. In general, almost one-third of the residues in proteins are involved in turns (Guruprasad & Rajkumar, 2000). The observation of significant population of β-turn conformation in short linear peptides in aqueous solution argues for the possibility that turn may participate actively in protein folding (Rose et al., 1985).

The critical role of β-turns as chain inverters or helix primitives has prompted their use as conformation nucleators in de novo design (Imperiali & Ottensen, 1999; Wetlaufer, 1990). Studies have been carried out by various groups to determine the propensity of amino acids at different positions in these turns. Serrano and co-workers (Alvarado, Blanco, Niemann, & Serrano, 1997) have experimentally determined propensities of amino acids at various positions in the turn. Similarly, Guruprasad and co-workers (Guruprasad & Rajkumar, 2000) have extensively studied statistical propensities of amino acids at various positions for all turn types.

The type II′ β-turn has been successfully exploited as a nucleator in the design of artificial antiparallel β-hairpins and simple β-sheet motifs. Durani and co-workers have demonstrated the use of ^{D}Pro-Gly as a type II′ turn nucleator in the designs of Bracelet (Rana, Kundu, & Durani, 2004), Boat (Rana, Kundu, & Durani, 2005), Canoe (Rana et al., 2007a) and π-cup (Rana, Kundu, & Durani, 2007b). They have also used inverse symmetrical variant ^{L}Pro-Gly for type II turn in Canoe (Rana et al., 2007a). The ^{D}Pro-Ser unit has been reported by Imperiali and co-workers (Struthers et al., 1996) to nucleate a tight β-turn in a designed ββα fold.

Fundamentally, helix nucleation can happen at N-terminus, C-terminus, or in the center of a peptide. Both natural and artificial residues, even turns (Bobde, Beri, & Durani, 1993; Felcy, Pattabhi, Raju, & Durani, 1997), are capable of helix nucleation. The conformationally constrained residue Aib, with access largely confined to the right and left-handed helical regions of Ramachandran space, has been extensively used for stabilizing helical conformation both in aqueous and organic media (Karle & Balaram, 1990; Marshall et al., 1990; Nagaraj, Shamala, & Balaram, 1979; Shamala, Nagaraj, &

Balaram, 1977, 1978; Toniolo & Benedetti, 1991; Yamada et al., 1993). Β-turns—types I, III—are nearly equivalent to a single turn of a 3_{10}-helix, stabilized by a 4→1 hydrogen bond and has been used with Aib as helix nucleating turn (Ravi, Prasad, & Balaram, 1983). Similarly, αβ-dehydrophenylalanine (ΔPhe), a highly conformationally constrained residue, is known to stabilize 3_{10}-helical conformations (Jain, Singh, & Chauhan, 1994; Rajashankar, Ramakumar, Jain, & Chauhan, 1995; Rajashankar, Ramakumar, Mal, Jain, & Chauhan, 1995; Ramagopal, Ramakumar, Joshi, & Chauhan, 1998; Ramagopal, Ramakumar, Sahal, & Chauhan, 2001). Nambiar and co-workers suggest that certain amino acids demonstrate better helix-inducing trend than others at N and C-termini (Forood, Feliciano, & Nambiar, 1993). According to their study, helix inducing effect at N-terminus follows the trend—Asp > Asn > Ser > Glu > Gln > Ala, whereas at C-terminus it is Arg > Lys > Ala. In proteins, Pro is frequently observed at N-terminus of helices. As the NH group of first three residues in helical conformation lack intrahelical hydrogen bonds, ^{L}Pro−$\varphi \sim -60$ degrees makes an obvious choice. Studies on synthetic peptides in organic solvents have indicated that presence of neighboring Pro residue can initiate 3_{10}-helix conformation (Venkatachalapathi & Balaram, 1979). A covalently locked Ac-Pro-Pro analog of Kemp and co-workers is also an excellent helix nucleus (Kemp, Boyd, & Muendel, 1991; Kemp, Curran, Davis, Boyd, & Muendel, 1991).

1.3.3 Design of β-sheet proteins

Being one of the major secondary structures, β-sheets display parallel and antiparallel structures and differ in the hydrogen bond patterns. Β-sheets may twist, curl, and even fold back on themselves to varying degree (Chothia, 1984; Richardson, 1981; Rose et al., 1985). In parallel β-sheets ($\varphi = -119$ degrees, $\psi = 113$ degrees), the backbone hydrogen bonds are evenly spaced and are angled across adjacent chains. In antiparallel β-sheets ($\varphi = -139$ degrees, $\psi = 135$ degrees), the hydrogen bonds are nearly perpendicular to main chain. Natural β-sheets tend to be right-handed twisted due to intrastrand nonbonded interactions, and geometric constraints of interstrand interaction (Chothia, 1973; Chou, Némethy, & Scheraga, 1983). In their tertiary structure interactions, β-sheets exhibit either aligned packing with mutual twist angle ~ -30 degrees or orthogonal packing at $\sim$ 90 degrees angle (Chothia, Levitt, & Richardson, 1977; Chou, Némethy, Rumsey, Tuttle, & Scheraga, 1986). The β-strands in β-hairpin are in mutual $\sim$ 180 degrees angle due to the presence of β-turn (Hutchison & Thornton, 1994; Sibanda, Blundell, & Thornton, 1989; Sibanda & Thornton, 1985).

1.3.3.1 β-hairpin design

Several groups have studied the factors that stabilize the interaction of one β-strand with another through either studying statistical propensities from the natural database of proteins whose structures are known (Branden & Tooze, 1998; Creighton, 1993) or through design of novel β-sheet peptide models (Alba, Jimenez, & Rico, 1997; Karle, Awasthi, & Balaram, 1996; Karle, Gopi, & Balaram, 2001; Sharman, 1997)(Espinosa & Gellman, 2000). Recently Bartlett and co-workers have attempted to quantify amino acid conformational preferences and side-chain–side-chain interactions in β-hairpins of 8-residue length by incorporating an artificial 1,2 dihydro-3(6 H)-pyridinyl unit as a probe for secondary structure (Phillips, Peirsanti, & Bartlett, 2005). Klimov and Thirumalai have studied mechanisms and kinetics of β-hairpin formation and predicted that the mechanism of formation of β-hairpin depends on turn stiffness (Klimov & Thirumalai, 2000). Similarly, Koide and co-workers (Yan et al., 2007) and Jimenez and co-workers (Santiveri, Santoro, Rico, & Jimenez, 2004) have found hydrophobic surface burial as the major stability determinant of a flat, single layer β-sheet. Andersen and co-workers (Olsen, Fesinmeyer, Stewart, & Andersen, 2005) have observed that the rates of β-hairpin folding depends on sequences at the turn region as well as away from the turn. Gai and co-workers (Du, Zhu, Huang, & Gai, 2004) have shown that among four 12-residue β-hairpins, differing only in their turn sequences, the one containing ^{D}Pro-Asn turn segment folds fastest, whereas Gellman and co-workers (Stanger & Gellman, 1998) have argued that ^{D}Pro-Gly turn sequence has a considerably greater tendency to nucleate hairpins than Asn-Gly. Similarly, Cochran and co-workers (Cochran, Skelton, & Starovasnik, 2001) have designed stable Trp-Zip by using ^{D}Pro-Gly turn sequence. Durani and co-workers have successfully used ^{D}Pro-Gly turn sequence in the design of heterochiral β-hairpins resulting in various shapes—Bracelet (Rana et al., 2004), Boat (Rana et al., 2005). Gellman and co-workers (Syud, Stanger, & Gellman, 2001) have reported studies on significance of lateral and diagonal pairing in interstrand side-chain–side-chain interactions in a designed β-hairpin comprising ^{D}Pro-Gly type II′ β-turn. Also, they have studied the effects of types I′ and II′ β-turns in β-hairpin stabilities. Searle and co-workers (Griffiths-Jones, Maynard, & Searle, 1999) have studied by both NMR and MD analysis, the contributions of β-turn and β-strand to the folding of a β-hairpin. Similarly, Balaram and co-workers (Gunasekaran et al., 1997) have extensively studied the protein data bank for propensities of number of residues in turn region and observed preponderance of two-residue turn followed by three-residue turns. Also, they observed higher propensity for $6-8$ residue strand for hairpins with type I′ turn and $4-6$ residue strand for hairpins with type I turn. A representative β-hairpin is shown in Fig. 1.1.

1.3.3.2 Multiple stranded sheets

The answer to the relationship between length and stability of β-sheet secondary structure is more complex than it is for an α-helix due to the fact that the

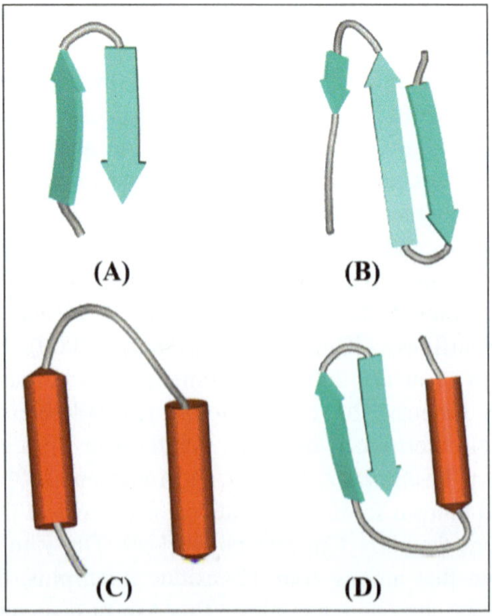

FIGURE 1.1 Representative structures. (A) β-hairpin; (B) multiple stranded protein; (C) all α-helix protein; (D)αβ mixed–protein.

latter has only one dimension along the axis. Two orthogonal dimensions must be considered for β-sheets, along the strand and perpendicular to the strand. Length dependence effects on antiparallel β-sheet stability have been studied using β-hairpin as a model and have observed 5 − 7 residue long strand to be most stable (Gunasekaran et al., 1997). Several groups on the other hand have studied length dependence on the other direction by increasing the number of strands from two in β-hairpin. Sharman and Searle (1997, 1998) have observed that three-stranded β-sheet is more stable than two-stranded in aqueous methanol. Similarly, Serrano and co-workers (Kortemme, Ramirez-Alvarado, & Serrano, 1998) designed three-strand β-sheet, which is 100%, folded in water. Ottesen and Imperiali (2001) redesigned the naturally occurring toxin hand motifs to a discretely folded all β-protein in a 29-residue sequence. Such three-stranded and higher order β-sheets have been applied to the study of the factors governing β-sheet folding and to assay the role of cooperativity along individual hairpins constituting a sheet and across strands of a β-sheet. A representative β-sheet is shown in Fig. 1.1.

1.3.3.3 Mixed αβ-protein design

The mixed αβ structures have also attracted considerable attention. A popular approach in protein design has been to examine the properties of the fold of choice and to attempt an adaptation of such a fold to synthetic strategies.

The design of zinc-finger motif has been attempted by two groups. Imperiali and co-workers (Struthers et al., 1996) followed an iterative procedure of rational design to arrive at a ββα fold with a novel sequence. An alternate approach is the Meccano (Lego) set approach (Balaram, 1992, 1999), wherein individually constructed, rigid, secondary structural modules are assembled to form a novel structure. Balaram and co-workers (Karle, Das, & Balaram, 2000) and DeGrado and co-workers (Lahr et al., 2005) in design of a synthetic peptide containing helical and hairpin segments have utilized this approach. Moe and co-workers (Butcher, Bruch, & Moe, 1995; Butcher & Moe, 1996) also attempted to design a mixed αβ structure wherein a single β-strand packs against an α-helix. Attempts have also been made to test the design rationale by mutation of certain key residues. Mayo and co-workers (Dahiyat & Mayo, 1997a,1997b) redesigned zinc finger to fold it without metal. A representative αβ-mixed protein is shown in Fig. 1.1.

1.3.4 All α-helix protein design

Helix is the other important protein secondary structure in which hydrogen bonding pattern distinguishes α-helix ($\varphi = -57$ degrees, $\psi = -47$ degrees and i to i + 4 hydrogen bonding) from the closely related 3_{10} helix ($\varphi = -50$ degrees, $\psi = -30$ degrees and i to i + 3 hydrogen bonding). Mixed α and 3_{10} helical structures have been documented in both proteins and peptides. The review by Baltzer et al. (2001) and Micklatcher and Chmielewski (1993) provides excellent guidelines for helical protein design. Several groups have attempted rational design of α-helical hairpins and helical bundles. Hodges and co-workers (Lau, Taneja, & Hodges, 1984; Lu & Hodges, 2004) as well as other groups (Burkhard, Meier, & Lustig, 2001) have used heptad repeats for designing coiled-coil structures. Without using heptad repeats Osterhout and co-workers (Fezoui, Weaver, & Osterhout, 1994) reported design of 38 residue helical hairpin. DeGrado and co-workers (Hill, Anderson, Wesson, DeGrado, & Eisenberg, 1990) attempted the design of 16-residue helical peptide, which associates to form four-helix bundle. A representative α-helix is shown in Fig. 1.1.

1.3.5 Metalloprotein design

Metalloproteins are among the most efficient and diverse biocatalysts as they employ metal ions or metal containing cofactors or prosthetic groups such as heme. Several groups have reported progress toward the "de novo" design of specific metal sites using peptide scaffolds. One goal of such studies is to design minimal systems that will reproduce the structural, electronic, and catalytic properties of metalloproteins, learning about cumulative influences of proteins in the process. Dutton and co-workers (Shifman, Gibney, Sharp, & Dutton, 2000) have used "consensus sequence" approach in incorporating

4Fe-4S clusters within 16 amino acid ferredoxin consensus sequence that was further truncated to seven amino acids. Hodgson and co-workers (Musgrave, Laplaza, Holm, Hedman, & Hodgson, 2002) have similarly reported thorough study of the metal binding properties of Cys-Gly-Val-Cys and Cys-Ile-Ala-Cys motifs that chelate Ni. Similarly, DeGrado and co-workers (Lombardi et al., 2000) have reported design of Cys-X-X-Cys containing peptides based on "retrostructural analysis" of the native rubredoxin protein structure. In a yet different study Marino and Regan (1999) have incorporated potential second-shell interactions into a previously designed mononuclear His_3Cys Zn(II)binding site in a small protein.

In a classical work Dutton and co-workers (Gibney & Dutton, 1999) have demonstrated the construction of stable "maquettes" in which histidine ligands support heme groups within four-helix bundles. A recent study by same group incorporates two heme groups A and B into such a maquette, which could be a step toward construction of a peptide model for cytochrome-c-oxidase (Koder & Dutton, 2006). Some groups have examined electron transfer through and between α-helices and in electrode-supported layered assemblies that perform biomimetic catalysis.

Several groups have recently taken an important first step in this direction through the design of zinc-binding sites via the introduction of zinc-binding residues [histidine, cysteine, or an amino acid containing carboxyl side chain (Ruan, Chen, & Hopkins, 1990)] separated by one turn of an α-helix (Handel & DeGrado, 1990), or on adjacent strands of a β-sheet (Roberts et al., 1990). Handel and DeGrado (1990) have prepared derivatives of α_2 and α_4, which bind Zn^{2+} with three histidine residues, two from one helix and a third from a neighboring helix positioned at three corners of a tetrahedron or octahedron. The resulting protein binds Zn^{2+} in a 1:1 stoichiometry with association constants that are much greater than $10^4\ M^{-1}$. Interestingly, the protein loses much of its molten-globule-like character when it binds Zn^{2+} despite the fact that the binding occurs without an appreciable change in secondary structure. Regan and Clarke (1990) have introduced a $(Cys)_2(His)_2$ tetrahedral metal-binding site (similar to those found in zinc fingers) near the ends of two helices in α_4. The resulting protein binds Zn^{2+} and Co^{2+} with dissociation constants of 25 nM and 16 μM, respectively. The optical absorption spectrum of the Co^{2+} form of the protein is consistent with the tetrahedral geometry envisioned in the design.

1.4 Shape-specific design of heterochiral proteins

Although proteins are astronomical in possibilities of interactions along their sequences, they are restricted to only about 1000 topological possibilities (Chothia, 1992). Main chain defines folds due to solvent-mediated screening of electrostatics of interpeptide interactions in conflict with the interpeptide hydrogen bonding due to homochiral structure (Ramakrishnan, Ranbhor, Kumar, & Durani, 2006a, 2006b). Due to the effect of homochirality, α-helix and β-sheet motifs are the building blocks of protein structure, being sequentially selectable options of secondary structure. This limits the scope of morphological possibility in protein tertiary structure to just about 1000 topological forms (Chothia, 1992). Geometric rules of interpeptide interaction under constraint of polypeptide stereochemistry are the defining consideration of protein structure.

Due to poly-L structure, proteins access mainly left-hand space of the φ coordinate of Ramachandran diagram in correspondence of α-helix and β-sheet secondary structure. Due to stereospecificity of conformation, residue stereochemistry can be a powerful tool for design of proteins with L and D structure as the alphabet. The principle is observed in nature in the form of Gramicidin-A, a microbial peptide alternately L,D in structure is a β-helical fold. In the synthetic variants, peptides with even number of alternately L,D residues are ring shaped molecules self-stackable as nanotubes (Horne, Stout, & Ghadiri, 2003).

D-amino acids have recently found place in de novo protein design. The most common usage of D-amino acids is for design of β-turn and interhelical linkers. Nanda and DeGrado (2006) reported putative helices in polypeptide sequences of mixed L,D amino acids. In possibly the first demonstration of design in L and D alphabet, Durani and co-workers reported a hexapeptide in a complex stereochemical fold (Fabiola et al., 1997). More recently, a family of shape-specific heterochiral mini-proteins for chosen sequence plans in L and D structure were reported as Bracelet (Rana et al., 2004), Boat (Rana et al., 2005), Canoe (Rana et al., 2007a), and π-cup (Rana, Kundu, & Durani, 2007b) shaped molecules.

1.5 Enzymes: functional proteins

Enzymes are the workhorses of living organisms. Living organisms are necessitated to countless chemical transformations efficiently and, above all, selectively to sustain critical biological functions. Enzymes perform these chemical transformations either by using efficiently the available 20 amino acids or with the help of cofactors such as metal ions and organic molecules such as NAD etc. as depicted in the Fig. 1.2.

Enzymes are catalysts and they increase the rate of a chemical reaction without undergoing any permanent chemical change themselves. The basic reaction of enzyme catalysis is as follows:

$$S + E \rightarrow ES \rightarrow P + E,$$

where E represents enzyme catalyzing the reaction, S the substrate, P the product of the reaction, and ES is the enzyme − substrate complex, an intermediate in the chemical reaction. The enzymes are thought to increase the

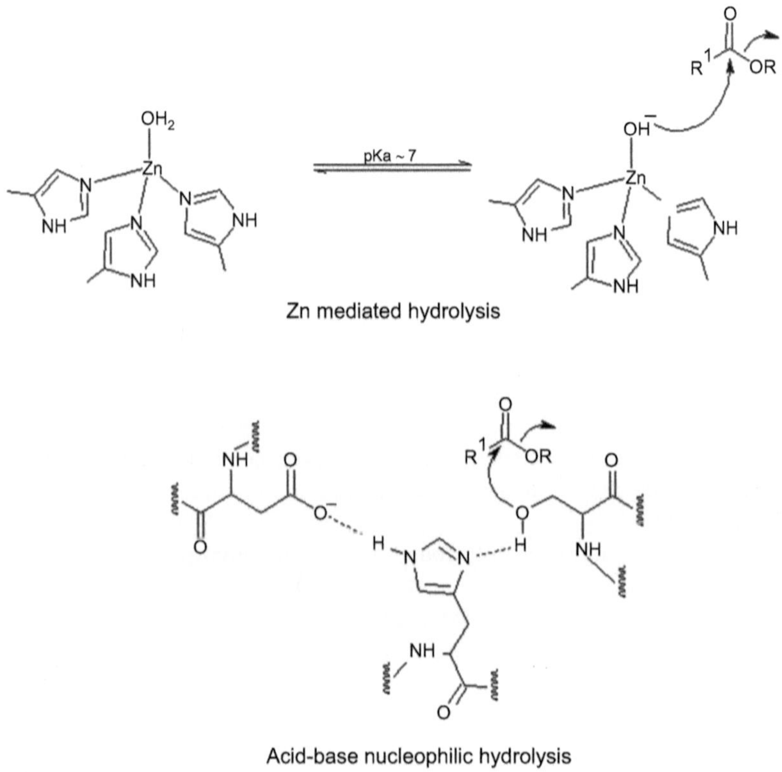

FIGURE 1.2 Schematic representation of the mechanism of zinc-mediated and acid-base-nucleophilic hydrolysis of ester.

rate of a reaction by reducing the activation energy required for the necessary transformation of substrate to a product.

1.5.1 Hydrolase enzymes

Among the six classes of enzymes, the hydrolase class constitutes enzymes performing countless hydrolysis reactions efficiently and, above all, selectively to sustain countless critical biological functions as shown in Fig. 1.2 (Breslow, 2006; Neurberger & Brocklehurst, 1987). Nature has solved this problem by the efficient use of available 20 natural amino acids. It has used the specific side-chain properties—polar, pKa etc.—to catalyze and/or bind the substrate molecules tightly enough and with high specificity (Blow, Birkoft, & Hartley, 1969). The HIV protease catalyzes the cleavage of peptide bond via two aspartyl residues and water molecule (Davies, 1990). Similarly, serine protease uses the assembly of serine, histidine, and aspartate residues where aspartate carboxyl group forms hydrogen-bonding interaction with N^{δ} of histidine, which in turn leads to polarization of serine O − H bond through N^{ε} of histidine that leads to attack on the substrate moiety thus using side-chain polarity to facilitate the necessary chemical transformation (Fersht, 1998).

In many other cases nature has solved this problem very elegantly by placing one or more divalent metal ions close to each other in the active site of an enzyme (Breslow, 2006; Neurberger & Brocklehurst, 1987). The positive charge of metal ions is a key feature in metalloenzymes that lowers the pKa of coordinated water and provides locally high concentration of the otherwise unavailable highly nucleophilic OH^- ions, which then hydrolyzes the substrate molecule under biological conditions, conditions where these reactions normally cannot take place (Breslow, 2006). In a similar manner, the positively charged metal center can also serve as a general Lewis acid for activation of a substrate molecule modulating its reactivity following coordination. The metal ions also assist in stability of the transition state arising from the attack of nucleophile on substrate molecule and orienting certain groups that facilitate the removal of leaving group and product from the active site of the enzyme (Fersht, 1998). These are examples of enzymes being truly co-catalytic where both metal ions and protein molecule work synergistically as a single unit. Through long evolution, nature has optimized the structure of the active site in order to obtain the necessary selectivity for the specific reaction process present in the biological medium.

It is highly challenging to mimic this natural high throughput and very complex machines into small and simple constructs to understand the underlying principles responsible for their activity and specificity. The combination of substrate binding and catalytic activity makes this mimic not only interesting as catalytic species in their own respect but also, possibly, helpful in understanding enzymatic activity. Efforts have been made to design catalysts that mimic the catalytic principles of enzymes. Catalytic antibodies are examples of semisynthetic artificial enzymes (Lerner, Benjovic, & Schultz, 1991; Schultz & Lerner, 1993, 1995; Wentworth & Janda, 2001). Fully synthetic molecules have also been designed as enzyme mimics by using either peptidic (Baumeister, Sakai, & Matile, 2001; Broo, Nilsson, Flodberg, & Baltzer, 1998) or nonpeptidic (Breslow, 1995; Suh, 2001, 2003) molecules.

Baltzer and co-workers (Broo, Nilsson, Flodberg et al., 1998) have used His pair and Arg to catalyze ester hydrolysis. Similarly, Matile and co-workers (Baumeister et al., 2001) have reported p-octiphenyl β-barrels for esterase activity. The studies have been concentrated on reproducing major characteristics of enzymatic action such as complex formation, rate acceleration, and high selectivity. In a different approach, several groups have used synthetic polymers as mimics of enzyme activity. Suh and co-workers (Suh, 2003) have used metal-bound synthetic polymers for hydrolysis of peptide bonds in proteins and phosphodiester bonds in DNA. Cram and co-workers (Cram, 1988a, 1988b) have demonstrated catalytic activity by randomly placing functional groups on polymers.

Many hydrolytic enzymes carry one or more metal ions in their active sites displaying cooperativity and hence increasing the activity of enzyme (Ragsdale, 2006). Because metals are relatively easily approximated in model systems, have been target for designing mimetics of enzymes. Akkaya and co-workers (Ozturk & Akkaya, 2004) have designed novel zinc-bound calixarene derivatives as bifunctional models. Recently, Liskamp and co-workers (Albada & Liskamp, 2008) have reported mimetics of type 3 copper binding sites in protein by the use of combinatorial approach.

Phillips and co-workers (Corey & Phillips, 1994) and Burnier and co-workers (Wells, Fairbrother, Otlewski, & Burnier, 1994) have used peptide-based models for catalysis. Similarly, Tsikaris and co-workers (Stavrakoudis, Demetropoulos, Sakarellos, Sakarellos-Daitsiotis, & Tsikaris, 1997) have reported cyclic peptide scaffolds as models. Baltzer and co-workers (Broo et al., 1998) have reported α-helices containing histidine residues as models for enzyme activity. Miller and co-workers (Lewis et al., 2006), Dudaczek and co-workers (Schmuck & Dudaczek, 2007), Wennemers and co-workers (Revell & Wennemers, 2007), and Liskamp and co-workers (Albada & Liskamp, 2008), among others, have reported combinatorial approaches for the discovery of peptide-based catalysts or enzyme active site mimics. Reymond and co-workers (Darbre & Reymond, 2006) have shown impressive hydrolytic properties of peptidic dendrimers containing the serine protease catalytic triad aspartate-histidine-serine and serine-histidine dyads.

References

Abele, S., & Seebach, D. (2000). *European Journal of Organic Chemistry*, *2000*(1), 1–15.

Alba, D. A., Jimenez, M. A., & Rico, M. (1997). *Journal of the American Chemical Society*, *119*(1), 175–183. Available from https://doi.org/10.1021/ja962325e.

Albada, B. H., & Liskamp, R. M. (2008). *Journal of Combinatorial Chemistry*, *10*(6), 814–824. Available from https://doi.org/10.1021/cc800065a.

Alvarado, M. R., Blanco, F. J., & Serrano, L. (1996). *Nature Structural Biology*, *3*, 604–612. Available from https://doi.org/10.1038/nsb0796-604.

Alvarado, R. M., Blanco, F. J., Niemann, H., & Serrano, L. (1997). *Journal of Molecular Biology*, *273*(4), 898–912. Available from https://doi.org/10.1006/jmbi.1997.1347.

Anfinsen, C. B. (1973). Principles that Govern the Folding of Protein Chains. *Science (New York, N.Y.)*, *181*(4096), 223–230. Available from https://doi.org/10.1126/science.181.4096.223.

Anfinsen, C. B., & Scheraga, H. A. (1975). Experimental and Theoretical Aspects of Protein Folding. *Advances in Protein Chemistry*, *29*, 205–300. Available from https://doi.org/10.1016/S0065-3233(08)60413-1.

Apostolakis, J., Ferrera, P., & Caflisch, A. (1999). Calculation of conformational transitions and barriers in solvated systems: Application to the alanine dipeptide in water. *The Journal of Chemical Physics*, *110*(4), 2099–2108. Available from https://doi.org/10.1063/1.477819.

Asher, S. A., Mikhonin, A. V., & Bykov, S. (2004). UV Raman Demonstrates that α-Helical Polyalanine Peptides Melt to Polyproline II Conformations. *Journal of the American Chemical Society*, *126*(27), 8433–8440. Available from https://doi.org/10.1021/ja049518j.

Avbelj, F., & Baldwin, R. L. (2002). Role of backbone solvation in determining thermodynamic β propensities of the amino acids. *Proceedings of the National Academy of Sciences of the United States of America, 99*(3), 1309–1313. Available from https://doi.org/10.1073/pnas.032665499.

Avbelj, F., & Moult, J. (1995). Role of electrostatic screening in determining protein main chain conformational preferences. *Biochemistry, 34*(3), 755–764. Available from https://doi.org/10.1021/bi00003a008.

Balaram, P. (1992). The design and construction of synthetic protein mimics. *Pure and Applied Chemistry, 64*(8), 1061–1066. Availalable from https://doi.org/10.1351/pac199264081061.

Balaram, P. (1999). *De novo* design: backbone conformational constraints in nucleating helices and ß-hairpins. *The Journal of Peptide Research, 54*(3), 195–199. Availalable from https://doi.org/10.1034/j.1399-3011.1999.00119.x.

Baltzer, L., Nilsson, H., & Nilsson, J. (2001). De Novo Design of Proteins What Are the Rules? *Chemical Reviews, 101*(10), 3153–3164. Available from https://doi.org/10.1021/cr0000473.

Baumeister, B., Sakai, N., & Matile, S. (2001). *p*-Octiphenyl β-Barrels with Ion Channel and Esterase Activity. *Organic Letters, 3*(26), 4229–4232. Available from https://doi.org/10.1021/ol016914n.

Bhattacharyya, R., & Chakrabarti, P. (2003). Stereospecific interactions of proline residues in protein structures and complexes. *Journal of Molecular Biology, 331*(4), 925–940. Available from https://doi.org/10.1016/S0022-2836(03)00759-9.

Blow, D. M., Birkoft, J. J., & Hartley, B. S. (1969). Role of a buried acid group in the mechanism of action of chymotrypsin. *Nature, 221*, 337–340. Available from https://doi.org/10.1038/221337a0.

Bobde, V., Beri, S., & Durani, S. (1993). Stable type II'reverse turn-310 helix conformation of Boc-D-Glu-Ala-Gly-Lys-Ala-Leu-OMe in apolar solvents. *Tetrahedron, 49*(24), 5397–5406. Available from https://doi.org/10.1016/S0040-4020(01)82388-7.

Branden, C. I., & Tooze, J. (1998). *Introduction to Protein Structure* (2nd ed.). Garland Science. Available from https://doi.org/10.1201/9781136969898.

Breslow, R. (1995). Biomimetic chemistry and artificial enzymes: catalysis by design. *Accounts of Chemical Research, 28*(3), 146–153. Available from https://doi.org/10.1021/ar00051a008.

Breslow, R. (2006). *Artificial Enzymes. Artificial Enzymes* (pp. 1–35). Wiley. Available from https://doi.org/10.1002/3527606645.ch1.

Broo, K. S., Nilsson, H., Flodberg, A., & Baltzer, L. (1998). Cooperative Nucleophilic and General-Acid Catalysis by the $HisH^+$ – His Pair and Arginine Transition State Binding in Catalysis of Ester Hydrolysis Reactions by Designed Helix – Loop – Helix Motifs. *Journal of the American Chemical Society, 120*(17), 4063–4068. Available from https://doi.org/10.1021/ja9737580.

Burkhard, P., Meier, M., & Lustig, A. (2001). Design of a minimal protein oligomerization domain by a structural approach. *Protein Science, 9*(12), 2294–2301. Available from https://doi.org/10.1110/ps.9.12.2294.

Burley, S. K., & Petsko, G. A. (1985). Aromatic-aromatic interaction: a mechanism of protein structure stabilization. *Science, 229*(4708), 23–28. Available from https://doi.org/10.1126/science.3892686.

Butcher, D. J., & Moe, G. R. (1996). Role of hydrophobic interactions and desolvation in determining the structural properties of a model alpha beta peptide. *Proceedings of the National Academy of Sciences of the United States of America, 93*(3), 1135–1140. Available from https://doi.org/10.1073/pnas.93.3.1135.

Butcher, D. J., Bruch, M. D., & Moe, G. R. (1995). Design and characterization of a model αβ peptide. *Biopolymers*, *36*, 109–120. Available from https://doi.org/10.1002/bip.360360202.

Chan, H. S., & Dill, K. A. (1991). Solvation: how to obtain microscopic energies from partitioning and solvation experiments. *Annual Review of Biophysics and Biomolecular*, *26*(1), 425–459. Available from https://doi.org/10.1146/annurev.biophys.26.1.425.

Chothia, C. (1973). Conformation of twisted β-pleated sheets in proteins. *Journal of Molecular Biology*, *75*(2), 295–302. Available from https://doi.org/10.1016/0022-2836(73)90022-3.

Chothia, C. (1984). Principles that determine the structure of proteins. *Annual Review of Biochemistry*, *53*, 537–572. Available from https://doi.org/10.1146/annurev.bi.53.070184.002541.

Chothia, C. (1992). One thousand families for the molecular biologist. *Nature*, *357*, 543–544. Available from https://doi.org/10.1038/357543a0.

Chothia, C., Levitt, M., & Richardson, D. (1977). Structure of proteins: packing of alpha-helices and pleated sheets. *Proceedings of the National Academy of Sciences of the United States of America*, *74*(10), 4130–4134. Available from https://doi.org/10.1073/pnas.74.10.4130.

Chou, K. C., Ne'methy, G., & Scheraga, H. A. (1983). Role of interchain interactions in the stabilization of the right-handed twist of β-sheets. *Journal of Molecular Biology*, *168*(2), 389–407. Available from https://doi.org/10.1016/S0022-2836(83)80025-4.

Chou, K.-C., Ne'methy, G., Rumsey, S., Tuttle, R. W., & Scheraga, H. A. (1986). Interactions between two β-sheets energetics of β/β packing in proteins. *Journal of Molecular Biology*, *188*(4), 641–649. Available from https://doi.org/10.1016/S0022-2836(86)80012-2.

Cochran, A. G., Skelton, N. J., & Starovasnik, M. A. (2001). Tryptophan zippers: Stable, monomeric β-hairpins. *Proceedings of the National Academy of Sciences of the United States of America*, *98*(10), 5578–5583. Available from https://doi.org/10.1073/pnas.091100898.

Corey, D. R., & Phillips, M. A. (1994). Cyclic peptides as proteases: a reevaluation. *Proceedings of the National Academy of Sciences of the United States of America*, *91*(10), 4106–4109. Available from https://doi.org/10.1073/pnas.91.10.4106.

Cram, D, J (1988a). The design of molecular hosts, guests, and their complexes (Nobel lecture). *Angew. Chem. Int. Ed. Engl*, *27*, 1009–1020. Available from https://doi.org/10.1002/anie.198810093.

Cram, D, J (1988b). The design of molecular hosts, guests, and their complexes. *Science*, *240*(4853), 760–767. Available from https://doi.org/10.1126/science.3283937.

Creighton, T. E. (1993). *Proteins: structure and molecular properties* (pp. 1–507). W. H. Freeman and Company.

Dahiyat, B., & Mayo, S. (1997a). Probing the role of packing specificity in protein design. *Proceedings of the National Academy of Sciences of the United States of America*, *94*(19), 10172–10177. Available from https://doi.org/10.1073/pnas.94.19.10172.

Dahiyat, B, & Mayo, S (1997b). De novo protein design: fully automated sequence selection. *Science*, *278*(5335), 82–87. Available from https://doi.org/10.1126/science.278.5335.82.

Darbre, T., & Reymond, J.-L. (2006). Peptide dendrimers as artificial enzymes, receptors, and drug-delivery agents. *Accounts of Chemical Research*, *39*(12), 925–934. Available from https://doi.org/10.1021/ar050203y.

Daura, X., Gunsteren., & Mark, A. E. (1999). Folding–unfolding thermodynamics of a β-heptapeptide from equilibrium simulations. *Proteins*, *34*, 269–280. Available from https://doi.org/10.1002/(SICI)1097-0134(19990215)34:3%3C269::AID-PROT1%3E3.0.CO;2-3.

Daura, X., Gademann., Schafer, H., Jaun, B., Seebach, D., & Gunsteren. (2001). The β-peptide hairpin in solution: conformational study of a β-hexapeptide in methanol by NMR spectroscopy and MD simulation. *Journal of the American Chemical Society*, *123*(10), 2393–2404. Available from https://doi.org/10.1021/ja003689g.

Davies, D. R. (1990). The structure and function of the aspartic proteinases. *Annual Review of Biophysics and Biophysical Chemistry, 19*, 189–215. Available from https://doi.org/10.1146/annurev.bb.19.060190.001201.

DeAlba, E., Jimenez, M. A., & Rico, M. (1997). Turn residue sequence determines β-hairpin conformation in designed peptides. *Journal of the American Chemical Society, 119*(1), 175–183. Available from https://doi.org/10.1021/ja962325e.

DeGrado, W. F., Wasserman, Z. R., & Lear, J. D. (1989). Protein design, a minimalist approach. *Science (New York, N.Y.), 243*(4891), 622–628. Available from https://doi.org/10.1126/science.2464850.

Dill, K. A. (1990). Dominant forces in protein folding. *Biochemistry, 29*(31), 7133–7155. Available from https://doi.org/10.1021/bi00483a001.

Dill, K. A., Truskett, T. M., Vlachy, V., & Hribar-Lee, B. (2005). Modeling water, the hydrophobic effect, and ion solvation. *Annual Review of Biophysics and Biomolecular Structure, 34*, 173–199. Available from https://doi.org/10.1146/annurev.biophys.34.040204.144517.

Du, D., Zhu, Y., Huang, C.-H., & Gai, F. (2004). Understanding the key factors that control the rate of β-hairpin folding. *Proceedings of the National Academy of Sciences of the United States of America, 101*(45), 15915–15920. Available from https://doi.org/10.1073/pnas.0405904101.

Durani, S. (2008). Protein Design with l- and d-α-Amino Acid Structures as the Alphabet. *Accounts of Chemical Research, 41*(10), 1301–1308. Available from https://doi.org/10.1021/ar700265t.

Dwyer, M. A., & Hellinga, H. W. (2004). Periplasmic binding proteins: a versatile superfamily for protein engineering. *Current Opinion in Structural Biology, 14*(4), 495–504. Available from https://doi.org/10.1016/j.sbi.2004.07.004.

Edwards, C. M. B., Cohen, M. A., & Bloom, S. R. Q. (1999). Peptides as drugs. *Journal of Medicine, 92*(1), 1–4. Available from https://doi.org/10.1093/qjmed/92.1.1.

Eker, F., Griebenov, K., & Schweitzer-Stenner, R. (2003). Stable conformations of tripeptides in aqueous solution studied by UV circular dichroism spectroscopy. *Journal of the American Chemical Society, 125*(27), 8178–8185. Available from https://doi.org/10.1021/ja034625j.

Espinosa, J. F., & Gellman, S. H. (2000). A Designed β-Hairpin Containing a Natural Hydrophobic Cluster. *Angewandte Chemie International Ed., 39*(13), 2330–2333. Available from https://doi.org/10.1002/1521-3773(20000703)39:13%3C2330::AID-ANIE2330%3E3.0.CO;2-C.

Fabiola, F., Pattabhi, V., Rawale, S., Raju, E. B., & Durani, S. (1997) Configurationally guided peptide conformational motifs: crystal structure of a ldßlßd dßl type hexapeptide fold. Chemical Communications 15, 1379–1380. Availalable from https://doi.org/10.1039/A702562I.

Farrera, S. J., Giralt, E., Royo, M., & Albericio, F. (2007). Cell-penetrating proline-rich peptidomimetics. *Methods in Molecular Biology, 386*, 241–267. Available from https://doi.org/10.1007/978-1-59745-430-8_9.

Felcy, F. G., Pattabhi, V., Raju, E. B., & Durani, S. (1997). Reengineering a type II β-turn as a potential helix nucleator. Part I. Crystal structure of Boc-Val-Pro-(D) Asp-Asp-Val-OMe monohydrate. *The Journal of Peptide Research: Official Journal of the American Peptide Society, 50*(5), 352–356. Available from https://doi.org/10.1111/j.1399-3011.1997.tb01194.x.

Fersht, A. R. (1972). Conformational equilibria in α-and δ-chymotrypsin: the energetics and importance of the salt bridge. *Journal of Molecular Biology, 64*(2), 497–509. Available from https://doi.org/10.1016/0022-2836(72)90513-X.

Fersht, A. R. (1998). *Structure and mechanism in protein science: A guide to enzyme catalysis and protein folding*. W. H. Freeman and Company.

Fezoui, Y., Weaver, D. L., & Osterhout. (1994). De novo design and structural characterization of an alpha-helical hairpin peptide: a model system for the study of protein folding intermediates. *Proceedings of the National Academy of Sciences of the United States of America, 91*(9), 3675–3679. Available from https://doi.org/10.1073/pnas.91.9.3675.

Forood, B., Feliciano, E. J., & Nambiar, K. P. (1993). Stabilization of alpha-helical structures in short peptides via end capping. *Proceedings of the National Academy of Sciences of the United States of America, 90*(3), 838–842. Available from https://doi.org/10.1073/pnas.90.3.838.

Gellman, S. H. (1998). Foldamers: a manifesto. *Accounts of Chemical Research, 31*(4), 174–180. Available from https://doi.org/10.1021/ar960298r.

Gibney, B. R., & Dutton, P. L. (1999). Histidine placement in de novo–designed heme proteins. *Protein Science, 8*(9), 1888–1898. Available from https://doi.org/10.1110/ps.8.9.1888.

Graf, J., Nguyen, P. H., Stock, G., & Schwalbe, H. (2007). Structure and Dynamics of the Homologous Series of Alanine Peptides: A Joint Molecular Dynamics/NMR Study. *Journal of the American Chemical Society, 129*(5), 1179–1189. Available from https://doi.org/10.1021/ja0660406.

Griffiths-Jones, S. R., Maynard, A. J., & Searle, M. S. (1999). Dissecting the stability of a β-hairpin peptide that folds in water: NMR and molecular dynamics analysis of the β-turn and β-strand contributions to folding. *Journal of Molecular Biology, 292*(5), 1051–1069. Available from https://doi.org/10.1006/jmbi.1999.3119.

Gunasekaran, K., Ramakrishnan, C., & Balaram, P. (1997). Beta-hairpins in proteins revisited: lessons for de novo design. *Prot, & Eng, 10*(10), 1131–1141. Available from https://doi.org/10.1093/protein/10.10.1131.

Guruprasad, K., & Rajkumar, S. J. (2000). gβ-and gγ-turns in proteins revisited: A new set of amino acid turn-type dependent positional preferences and potentials. *Biosciences, 25*(2), 143–156. Available from https://doi.org/10.1007/BF03404909.

Han, W. G., Jalkanen, K. J., Elstner, M., & Suhai, S. (1998). Theoretical Study of Aqueous *N*-Acetyl-l-alanine *N'*-Methylamide: Structures and Raman, VCD, and ROA Spectra. *The Journal of Physical Chemistry B, 108*(14), 2587–2602. Available from https://doi.org/10.1021/jp972299m.

Handel, T., & DeGrado, W. F. (1990). De novo design of a Zn2 + -binding protein. *Journal of the American Chemical Society, 112*(18), 6710–6711. Available from https://doi.org/10.1021/ja00174a039.

Haque, T. S., Little, J. C., & Gellman, S. H. (1994). "Mirror image" reverse turns promote. beta.-hairpin formation. *Journal of the American Chemical Society, 116*(9), 4105–4106. Available from https://doi.org/10.1021/ja00088a067.

Head-Gordon, T., Stillinger, F. H., Wright, M. H., & Gay, D. M. (1992). Poly (L-alanine) as a universal reference material for understanding protein energies and structures. *Proceedings of the National Academy of Sciences of the United States of America, 89*(23), 11513–11517. Available from https://doi.org/10.1073/pnas.89.23.11513.

Henzler, W. K. A., Wilson, E. E., Lee, D. –, & Ramamoorthy, A. (2003). Determination of the conformation and stability of simple homopolypeptides using solid-state NMR. *Solid State Nuclear Magnetic Resonance, 24*(2–3), 94–109. Available from https://doi.org/10.1016/S0926-2040(03)00048-1.

Hill, C. P., Anderson, D. H., Wesson, L., DeGrado, W. F., & Eisenberg, D. (1990). Crystal Structure of α_1: Implications for Protein Design. *Science (New York, N.Y.), 249*(4968), 543–546. Available from https://doi.org/10.1126/science.2382133.

Horne, W. S., Stout, C. D., & Ghadiri, M. R. (2003). A heterocyclic peptide nanotube. *Journal of the American Chemical Society, 125*(31), 9372–9376. Available from https://doi.org/10.1021/ja034358h.

Hutchison, E. G., & Thornton, J. M. (1994). A revised set of potentials for β-turn formation in proteins. *Protein Science*, *3*(12), 2207–2216. Available from https://doi.org/10.1002/pro.5560031206.

Imamura, Y., Watanabe, N., Umezawa, N., Iwatsubo, T., Kato, N., Tomita, T., & Higuchi, T. (2009). Inhibition of γ-Secretase Activity by Helical β-Peptide Foldamers. *Journal of the American Chemical Society*, *131*(21), 7353–7359. Available from https://doi.org/10.1021/ja9001458.

Imperiali, B., & Ottensen, J. J. (1999). Uniquely folded mini-protein motifs. *The Journal of Peptide Research*, *54*(3), 177–184. Available from https://doi.org/10.1034/j.1399-3011.1999.00121.x.

Jain, R. M., Singh, M., & Chauhan, V. S. (1994). Conformationally restricted peptides: solution conformation of tetra and hepta peptides containing α, β-dehydrophenylalanine residues in alternate positions. *Tetrahedron*, *50*(3), 907–920. Available from https://doi.org/10.1016/S0040-4020(01)80805-X.

Jiang, L., & Lai, L. (2002). CH··· O Hydrogen Bonds at Protein-Protein Interfaces* 210. *The Journal of Biological Chemistry*, *277*(40), 37732–37740. Available from https://doi.org/10.1074/jbc.M204514200.

Kaplan, J., & DeGrado, W. F. (2004). De novo design of catalytic proteins. *Proceedings of the National Academy of Sciences of the United States of America*, *101*(32), 11566–11570. Available from https://doi.org/10.1073/pnas.0404387101.

Karle, I. L., Awasthi, S. K., & Balaram, P. (1996). A designed beta-hairpin peptide in crystals. *Proceedings of the National Academy of Sciences of the United States of America*, *93*(16), 8189–8193. Available from https://doi.org/10.1073/pnas.93.16.8189.

Karle, I. L., & Balaram, P. (1990). Structural characteristics of. alpha.-helical peptide molecules containing Aib residues. *Biochemistry*, *29*(29), 6747–6756. Available from https://doi.org/10.1021/bi00481a001.

Karle, I. L., Das, C., & Balaram, P. (2000). De novo protein design: crystallographic characterization of a synthetic peptide containing independent helical and hairpin domains. *Proceedings of the National Academy of Sciences of the United States of America*, *97*(7), 3034–3037. Available from https://doi.org/10.1073/pnas.97.7.3034.

Karle, I. L., Gopi, H. N., & Balaram. (2001). Peptide hybrids containing α-and β-amino acids: structure of a decapeptide β-hairpin with two facing β-phenylalanine residues. *Proceedings of the National Academy of Sciences of the United States of America*, *98*(7), 3716–3719. Available from https://doi.org/10.1073/pnas.071050198.

Kauzmann, W. (1954). *The Mechanism of Enzyme Action*. Johns Hopkins Press.

Kauzmann, W. (1959). Some factors in the interpretation of protein denaturation. *Advances in Protein Chemistry*, *14*, 1–63. Available from https://doi.org/10.1016/S0065-3233(08)60608-7.

Kemp, D. S., Boyd, J. G., & Muendel, C. C. (1991). The helical s constant for alanine in water derived from template-nucleated helices. *Nature*, *352*, 451–454. Available from https://doi.org/10.1038/352451a0.

Kemp, D. S., Curran, T. P., Davis, W. M., Boyd, J. G., & Muendel, C. (1991). Studies of N-terminal templates for .alpha.-helix formation. Synthesis and conformational analysis of (2S,5S,8S,11S)-1-acetyl-1,4-diaza-3-keto-5-carboxy-10-thiatricyclo[2.8.1.04,8]tridecane (Ac-Hel1-OH). *The Journal of Organic Chemistry*, *56*(23), 6672–6682. Available from https://doi.org/10.1021/jo00023a037.

Klimov, D. K., & Thirumalai, D. (2000). Mechanisms and kinetics of β-hairpin formation. *Proceedings of the National Academy of Sciences of the United States of America*, *94*(6), 2544–2549. Available from https://doi.org/10.1073/pnas.97.6.2544.

Koder, R. L., & Dutton, P. L. (2006). Intelligent design: the de novo engineering of proteins with specified functions. *Dalton Transactions*, *26*, 3045–3051. Available from https://doi.org/10.1039/B514972J.

Kortemme, T., Ramirez-Alvarado, M., & Serrano, L. (1998). Design of a 20-amino acid, three-stranded β-sheet protein. *Science, 281*(5374), 253–256. Available from https://doi.org/10.1126/science.281.5374.253.

Kumar, A., Ramakrishnan, V., Ranbhor, R., Patel, K., & Durani, S. (2009). Homochiral stereochemistry: the missing link of structure to energetics in protein folding. *The Journal of Physical Chemistry B, 113*(51), 16435–16442. Available from https://doi.org/10.1021/jp906811k.

Lahr, S. J., Engel, D. E., Stayrook, S. E., Maglio, O., North, B., Geremia, S., ... DeGrado, W. F. (2005). Analysis and design of turns in α-helical hairpins. *Journal of Molecular Biology, 346*(5), 1441–1454. Available from https://doi.org/10.1016/j.jmb.2004.12.016.

Lau, S. Y. M., Taneja, A. K., & Hodges, R. S. (1984). Synthesis of a model protein of defined secondary and quaternary structure. Effect of chain length on the stabilization and formation of two-stranded alpha-helical coiled-coils. *The Journal of Biological Chemistry, 259*(21), 13253–13261. Available from https://doi.org/10.1016/S0021-9258(18)90686-1.

Lazar, G. A., Marshall, S. A., Plecsy, J. J., Mayo, S. L., & Desjarlais, J. R. (2003). Designing proteins for therapeutic applications. *Current Opinion in Structural Biology, 13*(4), 513–518. Available from https://doi.org/10.1016/S0959-440X(03)00104-0.

Lee, D. G., Kim, H. N., Park, Y., Kim, H. K., Choi, B. H., Choi, C. H., & Hahm, K. S. (2002). Design of novel analogue peptides with potent antibiotic activity based on the antimicrobial peptide, HP (2–20), derived from N-terminus of Helicobacter pylori ribosomal protein L1. *Biochimica et Biophysica Acta, 1598*(1–2), 185–194. Available from https://doi.org/10.1016/S0167-4838(02)00373-4.

Lerner, R. A., Benjovic, S. J., & Schultz, P. G. (1991). At the crossroads of chemistry and immunology: catalytic antibodies. *Science, 252*(5006), 659–667. Available from https://doi.org/10.1126/science.2024118.

Lewis, C. A., Chiu, A., Kubryk, M., Balsells, J., Pollard, D., Esser, C. K., ... Miller, S. J. (2006). Remote desymmetrization at near-nanometer group separation catalyzed by a miniaturized enzyme mimic. *The Journal of the American Chemical Society, 128*(51), 16454–16455. Available from https://doi.org/10.1021/ja067840j.

Lombardi, A., Summa, C. M., Geremia, S., Randaccio, L., Pavone, V., & DeGrado, W. F. (2000). Retrostructural analysis of metalloproteins: application to the design of a minimal model for diiron proteins. *Proceedings of the National Academy of Sciences of the United States of America, 97*(12), 6298–6305. Available from https://doi.org/10.1073/pnas.97.12.6298.

Lu, S. M., & Hodges, R. S. (2004). Defining the minimum size of a hydrophobic cluster in two-stranded α-helical coiled-coils: Effects on protein stability. *Protein Science, 13*(3), 714–726. Available from https://doi.org/10.1110/ps.03443204.

Ma, J. C., & Dougherty, D. A. (1997). The cation − π interaction. *Chemical Reviews, 97*(5), 1303–1324. Available from https://doi.org/10.1021/cr9603744.

Marino, S. F., & Regan, L. (1999). Secondary ligands enhance affinity at a designed metal-binding site. *Chemistry and Biology, 6*(9), 649–655. Available from https://doi.org/10.1016/S1074-5521(99)80116-1.

Marshall, G. R., Hodgkin, E. E., Langs, D. A., Smith, G. D., Zabrocki, J., & Leplawy, M. T. (1990). Factors governing helical preference of peptides containing multiple alpha, alpha-dialkyl amino acids. *Proceedings of the National Academy of Sciences of the United States of America, 87*(1), 487–491. Available from https://doi.org/10.1073/pnas.87.1.487.

Masunov, A., & Lazaridis, T. (2003). Potentials of mean force between ionizable amino acid side chains in water. *Journal of the American Chemical Society, 125*(7), 1722–1730. Available from https://doi.org/10.1021/ja025521w.

Maynard, A. J., Sharman, G. J., & Searle, M. S. (1998). Origin of β-hairpin stability in solution: structural and thermodynamic analysis of the folding of a model peptide supports hydrophobic stabilization in water. *Journal of the American Chemical Society*, *120*(9), 1996–2007. Available from https://doi.org/10.1021/ja9726769.

McGaughey, G. B., Gagne, M., & Rappe, A. K. (1998). π-stacking interactions: alive and well in proteins. *The Journal of Biological Chemistry*, *273*(25), 15458–15463. Available from https://doi.org/10.1074/jbc.273.25.15458.

Meyer, E. A., Castellano, R. K., & Diederich, F. (2003). Interactions with aromatic rings in chemical and biological recognition. *Angewandte Chemie International (Ed.)*, *42*(11), 1210–1250. Available from https://doi.org/10.1002/anie.200390319.

Micklatcher, C., & Chmielewski, J. (1993). Helical peptide and protein design. *Current Opinion in Structural Biology*, *3*(1), 724–729. Available from https://doi.org/10.1016/S1367-5931(99)00031-9.

Mikhonin, A. V., & Asher, S. A. (2006). Direct UV Raman Monitoring of 3_{10}-Helix and π-Bulge Premelting during α-Helix Unfolding. *Journal of the American Chemical Society*, *128*(42), 13789–13795. Available from https://doi.org/10.1021/ja062269 + .

Milner -White, E. J., & Poet, R. (1986). Four classes of β-hairpins in proteins. *Biochemical Journal*, *240*(1), 289–292. Available from https://doi.org/10.1042/bj2400289.

Moodie, S. L., Mitchell, J. B. O., & Thornton, J. M. (1996). Protein recognition of adenylate: an example of a fuzzy recognition template. *Journal of Molecular Biology*, *263*(3), 486–500. Available from https://doi.org/10.1006/jmbi.1996.0591.

Mu, Y., & Stock, G. (2002). Conformational dynamics of trialanine in water: a molecular dynamics study. *The Journal of Physical Chemistry B*, *106*(20), 5294–5301. Available from https://doi.org/10.1021/jp013977e.

Munoz, G. S., Puiggali, J., Rodriguez, A., & Subirana, J. A. (1983). Crystals of polyglycine in the β form. *Journal of Molecular Biology*, *167*(1), 223–225. Available from https://doi.org/10.1016/S0022-2836(83)80045-X.

Musgrave, K. B., Laplaza, C. E., Holm, R. H., Hedman, B., & Hodgson, K. O. (2002). Structural Characterization of Metallopeptides Designed as Scaffolds for the Stabilization of Nickel (II)-Fe_4S_4 Bridged Assemblies by X-ray Absorption Spectroscopy. *Journal of the American Chemical Society*, *124*(12), 3083–3092. Available from https://doi.org/10.1021/ja011861q.

Nagaraj, R., Shamala, N., & Balaram, P. (1979). Stereochemically constrained linear peptides. Conformations of peptides containing. alpha.-aminoisobutyric acid. *Journal of the American Chemical Society*, *101*(1), 16–20. Available from https://doi.org/10.1021/ja00495a003.

Nanda, V., & DeGrado, W. F. (2006). Computational design of heterochiral peptides against a helical target. *Journal of the American Chemical Society*, *128*(3), 809–816. Available from https://doi.org/10.1021/ja054452t.

Nemethy, G., & Scheraga, H. A. (1981). Strong interaction between disulfide derivatives and aromatic groups in peptides and proteins. *Biochemical and Biophysical Research Communications*, *98*(2), 482–487. Available from https://doi.org/10.1016/0006-291X(81)90865-2.

Neurberger, A., & Brocklehurst, K. (1987). *Hydrolytic Enzymes*. Amsterdam: Elsevier Science.

Nguyen, H. D., Marchut, A. J., & Hall, C. K. (2004). Solvent effects on the conformational transition of a model polyalanine peptide. *Protien Science*, *13*(11), 2909–2924. Available from https://doi.org/10.1110/ps.04701304.

Ohnishi, S., Kamikubo, H., Onitsuka, M., & Shortle, D. (2006). Conformational preference of polyglycine in solution to elongated structure. *Journal of the American Chemical Society*, *128*(50), 16338–16344. Available from https://doi.org/10.1021/ja066008b.

Olsen, K. A., Fesinmeyer, M. R., Stewart, J. M., & Andersen, N. H. (2005). Hairpin folding rates reflect mutations within and remote from the turn region. *Proceedings of the National Academy of Sciences of the United States of America, 102*(43), 15483–15487. Available from https://doi.org/10.1073/pnas.0504392102.

Ottesen, J. J., & Imperiali, B. (2001). Design of a discretely folded mini-protein motif with predominantly β-structure. *Nature Structural Biology, 8*, 535–539. Available from https://doi.org/10.1038/88604.

Ozturk, G., & Akkaya, E. U. (2004). Differential and substrate-selective reactivity of calix [4] arene derivatives with cyclenyl-Zn (II) modifications at the upper rim. *Organic Letters, 6*(2), 241–243. Available from https://doi.org/10.1021/ol0360694.

Park, S., Yang, X., Saven, J. G., & Curr. (2004). Advances in computational protein design. *Current Opinion in Structural Biology, 14*(4), 487–494. Available from https://doi.org/10.1016/j.sbi.2004.06.002.

Pauling, L., & Corey, R. B. (1951a). Configuration of polypeptide chains. *Nature, 168*, 550–551. Available from https://doi.org/10.1038/168550a0.

Pauling, L., & Corey, R. B. (1951b). Configurations of Polypeptide Chains With Favored Orientations Around Single Bonds: Two new pleated sheets. *Proceedings of the National Academy of Sciences of the United States of America, 37*(11), 729–740. Available from https://doi.org/10.1073/pnas.37.11.729.

Perutz, M. F., & Raidt, H. (1975). Stereochemical basis of heat stability in bacterial ferredoxins and in haemoglobin A2. *Nature, 255*, 256–259. Available from https://doi.org/10.1038/255256a0.

Phillips, S, T, Peirsanti, G, & Bartlett, Paul, A (2005). Quantifying amino acid conformational preferences and side chain-side chain interactions in β-hairpins. *Proceedings of National Academy of Sciences USA, 102*(173), 13737–13742. Available from https://doi.org/10.1073/pnas.0506646102. https://www.pnas.org/doi/full/10.1073/pnas.0506646102.

Pollard, T. D., & Earnshaw, W. C. (2017). *Cell Biology*. Elsevier.

Poon, C. D., Samulski, E. T., Weise, C. F., & Weisshaar, J. C. (2000). Do bridging water molecules dictate the structure of a model dipeptide in aqueous solution? *Journal of the American Chemical Society, 122*(23), 5642–5643. Available from https://doi.org/10.1021/ja993953 + .

Raghothama, S., Awasthi, S. K., & Balaram, P. J. (1998). β-Hairpin nucleation by Pro-Gly β-turns. Comparison of D-Pro-Gly and L-Pro-Gly sequences in an apolar octapeptide. *Journal of the Chemical Society Perkin, 2*, 137–144. Available from https://doi.org/10.1039/A703331A.

Ragsdale, S. W. (2006). Metals and their scaffolds to promote difficult enzymatic reactions. *Chemical Reviews, 106*(8), 3317–3337. Available from https://doi.org/10.1021/cr0503153.

Rajashankar, K. R., Ramakumar, S., Jain, R. M., & Chauhan, V. S. (1995). Observation of Water-Mediated Helix-Terminating Conformation in a Dehydrophenylalanine Peptide: Crystal and Solution Structure of the Octapeptide Ac-.DELTA.Phe-Val-.DELTA.Phe-Phe-Ala-Val-.DELTA.Phe-Gly-OMe. *Journal of the American Chemical Society, 117*(47), 11773–11779. Available from https://doi.org/10.1021/ja00152a020.

Rajashankar, K. R., Ramakumar, S., Mal, T. K., Jain, R. M., & Chauhan, V. S. (1995). Synthesis, and crystal and molecular structure of the 3_{10}-helical α,β-dehydro pentapeptide Boc-Leu-Phe-Ala-ΔPhe-Leu-Ome. *Biopolymers, 35*(2), 141–147. Available from https://doi.org/10.1002/bip.360350202.

Ramachandran, G. N., Ramakrishnan, C., & Sasisekaran, V. (1963). Stereochemistry of polypeptide chain configurations. *Journal of Molecular Biology, 7*(1), 95–99. Available from https://doi.org/10.1016/S0022-2836(63)80023-6.

Ramachandran, G. N., & Sasisekaran, V. (1968). Conformation of polypeptides and proteins. *Advances in Protein Chemistry*, *23*, 283–437. Available from https://doi.org/10.1016/S0065-3233(08)60402-7.

Ramagopal, U. A., Ramakumar, S., Joshi, R. M., & Chauhan, V. S. (1998). Crystal structure of Boc-LAla-ΔPhe-ΔPhe-ΔPhe-ΔPhe-NHMe: a left-handed helical peptide. *The Journal of Peptide Research*, *52*(3), 208–215. Available from https://doi.org/10.1111/j.1399-3011.1998.tb01478.x.

Ramagopal, U. A., Ramakumar, S., Sahal, D., & Chauhan, V. S. (2001). De novo design and characterization of an apolar helical hairpin peptide at atomic resolution: Compaction mediated by weak interactions. *Proceedings of the National Academy of Sciences of the United States of America*, *98*(3), 870–874. Available from https://doi.org/10.1073/pnas.98.3.870.

Ramakrishnan, V., Ranbhor, R., & Durani, S. (2004). Existence of specific "folds" in polyproline II ensembles of an "unfolded" alanine peptide detected by molecular dynamics. *Journal of the American Chemical Society*, *126*(50), 16332–16333. Available from https://doi.org/10.1021/ja045787y.

Ramakrishnan, V., Ranbhor, R., Kumar, A., & Durani, S. (2006). The interplay of sequence and stereochemistry in defining conformation in proteins and polypeptides. *Biopolymers*, *83*(5), 537–545. Available from https://doi.org/10.1002/bip.20584.

Ramakrishnan, V., Ranbhor, R., Kumar, A., & Durani, S. (2006). The link between sequence and conformation in protein structures appears to be stereochemically established. *The Journal of Physical Chemistry B*, *110*(18), 9314–9323. Available from https://doi.org/10.1021/jp056417e.

Rana, S., Kundu, B., & Durani, S. (2004). Stereospecific peptide folds. A rationally designed molecular bracelet. *Chemical Communications*, 2462–2463. Available from https://doi.org/10.1039/B410532J.

Rana, S., Kundu, B., & Durani, S. (2005). A small peptide stereochemically customized as a globular fold with a molecular cleft. *Chemical Communications*, 207–209. Available from https://doi.org/10.1039/B413802C.

Rana, S., Kundu, B., & Durani, S. (2007a). A mixed-α, β miniprotein stereochemically reprogrammed to high-binding affinity for acetylcholine. *Biopolymers*, *87*(4), 231–243. Available from https://doi.org/10.1002/bip.20829.

Rana, S., Kundu, B., & Durani, S. (2007b). A double catgrip mixed l and d mini protein only 20 residues long. *Bioorganic and Medicinal Chemistry*, *15*(11), 3874–3882. Available from https://doi.org/10.1016/j.bmc.2007.03.030.

Ravi, A., Prasad, B. V. V., & Balaram, P. (1983). Cyclic peptide disulfides. Solution and solid-state conformation of Boc-Cys-Pro-Aib-Cys-NHMe with a disulfide bridge from Cys to Cys, a disulfide-bridged peptide helix. *Journal of the American Chemical Society*, *105*(1), 105–109. Available from https://doi.org/10.1021/ja00339a019.

Regan, L., & Clarke, N. D. (1990). A tetrahedral zinc (II)-binding site introduced into a designed protein. *Biochemistry*, *29*(49), 10878–10883. Available from https://doi.org/10.1021/bi00501a003.

Revell, J. D., & Wennemers, H. (2007). Peptidic catalysts developed by combinatorial screening methods. *Current Opinion in Chemical Biology*, *11*(3), 269–278. Available from https://doi.org/10.1016/j.cbpa.2007.05.015.

Richardson, J. S. (1981). The anatomy and taxonomy of protein structure. *Advances in Protein Chemistry*, *34*, 167–339. Available from https://doi.org/10.1016/S0065-3233(08)60520-3.

Roberts, V. A., Iverson, B. L., Iverson, S. A., Benkovic, S. J., Lerner, R. A., Getzoff, E. D., & Tainer, J. A. (1990). Antibody remodeling: a general solution to the design of a metal-coordination site in an antibody binding pocket. *Proceedings of the National Academy of Sciences of the United States of America*, *87*(17), 6654–6658. Available from https://doi.org/10.1073/pnas.87.17.6654.

Rose, G. D., Gierasch, L. M., & Smith, J. A. (1985). Turns in peptides and proteins. *Advances in Protein Chemistry*, *37*, 1–109. Available from https://doi.org/10.1016/S0065-3233(08)60063-7.

Ruan, F., Chen, Y., & Hopkins, P. B. (1990). Metal ion-enhanced helicity in synthetic peptides containing unnatural, metal-ligating residues. *Journal of the American Chemical Society*, *112*(25), 9403–9404. Available from https://doi.org/10.1021/ja00181a058.

Santiveri, C. M., Santoro, J., Rico, M., & Jimenez, M. A. (2004). Factors involved in the stability of isolated β-sheets: Turn sequence, β-sheet twisting, and hydrophobic surface burial. *Protein Science: A Publication of the Protein Society*, *13*(4), 1134–1147. Available from https://doi.org/10.1110/ps.03520704.

Scheiner, S., Redfern, P., & Hillenbrand, E. A. (1986). Factors influencing proton positions in biomolecules. *International Journal of Quantum Chemistry*, *29*(4), 817–827. Available from https://doi.org/10.1002/qua.560290420.

Schmuck, C., & Dudaczek, J. (2007). Screening of a combinatorial library reveals peptide-based catalysts for phosphorester cleavage in water. *Organic Letters*, *9*(26), 5389–5392. Available from https://doi.org/10.1021/ol702191q.

Schultz, P. G., & Lerner, R. A. (1993). Antibody catalysis of difficult chemical transformations. *Accounts of Chemical Research*, *26*(8), 391–395. Available from https://doi.org/10.1021/ar00032a001.

Schultz, P. G., & Lerner, R. A. (1995). From molecular diversity to catalysis: lessons from the immune system. *Science*, *269*(5232), 1835–1842. Available from https://doi.org/10.1126/science.7569920.

Schweitzer-Stenner, R. (2002). Dihedral angles of tripeptides in solution directly determined by polarized Raman and FTIR spectroscopy. *Biophysical Journal*, *83*(1), 523–532. Available from https://doi.org/10.1016/S0006-3495(02)75188-6.

Seebach, D., Beck, A. K., & Bierbaum, D. J. (2004). The World of β- and γ-Peptides Comprised of Homologated Proteinogenic Amino Acids and Other Components. *Chemistry & Biodiversity*, *1*(8), 1111–1239. Available from https://doi.org/10.1002/cbdv.200490087.

Shamala, N., Nagaraj, R., & Balaram, P. (1977). The crystal and molecular structure of the amino terminal tetrapeptide of alamethicin. A novel 3_{10} helical conformation. *Biochemical and Biophysical Research Communications*, *79*(1), 292–298. Available from https://doi.org/10.1016/0006-291X(77)90094-8.

Shamala, N., Nagaraj, R., & Balaram, P. (1978). The 3_{10} helical conformation of a pentapeptide containing α-aminoisobutyric acid (Aib): X-ray crystal structure of Tos–(Aib) 5–OMe. *Chemical Communications*, *22*, 996–997. Available from https://doi.org/10.1039/C39780000996.

Sharman, G. J. (1997). Dissecting the effects of cooperativity on the stabilisation of a de novo designed three stranded anti-parallel β-sheet. *Chemical Communications*, 1955–1956. Available from https://doi.org/10.1039/A705376B.

Sharman, G. J., & Searle, M. S. (1998). Cooperative Interaction between the Three Strands of a Designed Antiparallel β-Sheet. *Journal of the American Chemical Society*, *120*(21), 5291–5300. Available from https://doi.org/10.1021/ja9705405.

Sheinerman, F. B., Norel, R., & Honig, B. (2000). Electrostatic aspects of protein–protein interactions. *Current Opinion in Structural Biology*, *10*(2), 153–159. Available from https://doi.org/10.1016/S0959-440X(00)00065-8.

Shi, Z., Olson, C. A., Rose, G. D., Baldwin, R. L., & Kallenbach, N. R. (2002). Polyproline II structure in a sequence of seven alanine residues. *Proceedings of the National Academy of Sciences of the United States of America*, *99*(14), 9190–9195. Available from https://doi.org/10.1073/pnas.112193999.

Shi, Z., Woody, R. W., & Kallenbach, N. R. (2002). Is polyproline II a major backbone conformation in unfolded proteins? *Advances in Protein Chemistry*, *62*, 163–240. Available from https://doi.org/10.1016/S0065-3233(02)62008-X.

Shifman, J. M., Gibney, B. R., Sharp, R. E., & Dutton, P. L. (2000). Heme redox potential control in de novo designed four-α-helix bundle proteins. *Biochemistry*, *39*(48), 14813–14821. Available from https://doi.org/10.1021/bi000927b.

Sibanda, B. L., Blundell, T. L., & Thornton, J. M. (1989). Conformation of β-hairpins in protein structures: A systematic classification with applications to modelling by homology, electron density fitting and protein engineering. *Journal of Molecular Biology*, *206*(4), 759–777. Available from https://doi.org/10.1016/0022-2836(89)90583-4.

Sibanda, B. L., & Thornton, J. M. (1985). β-Hairpin families in globular proteins. *Nature*, *316*, 170–174. Available from https://doi.org/10.1038/316170a0.

Soto, P., Baumketner, A., & Shea, J. –E. (2006). Aggregation of polyalanine in a hydrophobic environment. *The Journal of Chemical Physics*, *124*(13), 134904–134906. Available from https://doi.org/10.1063/1.2179803.

Stanger, H. E., & Gellman, S. H. (1998). Rules for Antiparallel β-Sheet Design: d-Pro-Gly Is Superior to l-Asn-Gly for β-Hairpin Nucleation. *Journal of the American Chemical Society*, *120*(17), 4236–4237. Available from https://doi.org/10.1021/ja973704q.

Stavrakoudis, A., Demetropoulos, I. N., Sakarellos, C., Sakarellos-Daitsiotis, M., & Tsikaris, V. (1997). Design, synthesis and catalytic activity of a serine protease synthetic model. *Letters in Peptide Science*, *4*, 481–487. Available from https://doi.org/10.1007/BF02442921.

Struthers, M. D., Cheng, R. P., & Imperiali, B. (1996). Design of a monomeric 23-residue polypeptide with defined tertiary structure. *Science*, *271*(5247), 342–345. Available from https://doi.org/10.1126/science.271.5247.342.

Suh, J. (2001). Synthesis of polymeric enzyme-like catalysts. *Synthetic Letters*, *9*, 1343–1363. Available from https://doi.org/10.1055/s-2001-16770.

Suh, J. (2003). Synthetic artificial peptidases and nucleases using macromolecular catalytic systems. *Accounts of Chemical Research*, *36*(7), 562–570. Available from https://doi.org/10.1021/ar020037j.

Sujatha, M. S., Sasidhar, Y. U., & Balaji, P. V. (2007). MP2/6-311++ G (d, p) study on galactose–aromatic residue analog complexes in different position-orientations of the saccharide relative to aromatic residue. *Journal of Molecular Structure (Theochem)*, *814*(1–3), 11–24. Available from https://doi.org/10.1016/j.theochem.2007.02.032.

Syud, F. A., Stanger, H. E., & Gellman, S. H. (2001). Interstrand side chain – side chain interactions in a designed β-hairpin: significance of both lateral and diagonal pairings. *Journal of the American Chemical Society*, *123*(36), 8667–8677. Available from https://doi.org/10.1021/ja0109803.

Tanuichi, H. (1970). Formation of randomly paired disulfide bonds in des-(121–124)-ribonuclease after reduction and reoxidation. *The Journal of Biological Chemistry*, *245*(20), 5459–5468. Available from https://doi.org/10.1016/S0021-9258(18)62776-0.

Tashiro, S., Kobayashi, M., & Fujita, M. (2006). Folding of an Ala-Ala-Ala tripeptide into a β-turn via hydrophobic encapsulation. *Journal of the American Chemical Society*, *128*(29), 9280–9281. Available from https://doi.org/10.1021/ja061722e.

Tatko, C. D., & Waters, M. L. (2003). The geometry and efficacy of cation−π interactions in a diagonal position of a designed β-hairpin. *Protein Science*, *12*(11), 2443−2452. Available from https://doi.org/10.1110/ps.03284003.

Thorpe, I. F., Zhou, J., & Voth, G. A. (2008). Peptide folding using multiscale coarse-grained models. *The Journal of Physical Chemistry B*, *112*(41), 13079−13090. Available from https://doi.org/10.1021/jp8015968.

Tobias, D. J., & Brooks, C. L. (1992). Conformational equilibrium in the alanine dipeptide in the gas phase and aqueous solution: A comparison of theoretical results. *The Journal of Physical Chemistry*, *96*(9), 3864−3870. Available from https://doi.org/10.1021/j100188a054.

Toniolo, C., & Benedetti, E. (1991). The polypeptide 3_{10}-helix. *Trends in Biochemical Sciences*, *16*, 350−353. Available from https://doi.org/10.1016/0968-0004(91)90142-I.

Venkatachalapathi, Y. V., & Balaram, P. (1979). An incipient 310 helix in Piv-Pro-Pro-Ala-NHMe as a model for peptide folding. *Nature*, *281*, 83−84. Available from https://doi.org/10.1038/281083a0.

Venkatraman, J., Shankaramma, S. C., & Balaram, P. (2001). Design of folded peptides. *Chemical Reviews*, *101*(10), 3131−3152. Available from https://doi.org/10.1021/cr000053z.

Warshel, A., & Russell, S. T. (1984). Calculations of electrostatic interactions in biological systems and in solutions. *Quarterly Reviews of Biophysics*, *17*(3), 283−422. Available from https://doi.org/10.1017/S0033583500005333.

Weiss, M. S., Brandl, M., Suhnel, J., Pal, D., & Hilgenfeld, R. (2001). More hydrogen bonds for the (structural) biologist. *Trends in Biochemical Sciences*, *26*(9), 521−523. Available from https://doi.org/10.1016/S0968-0004(01)01935-1.

Wells, J. A., Fairbrother, W. J., Otlewski, J., Jr., & Burnier, J. (1994). A reinvestigation of a synthetic peptide (TrPepz) designed to mimic trypsin. *Proceedings of the National Academy of Sciences of the United States of America*, *91*(10), 4110−4114. Available from https://doi.org/10.1073/pnas.91.10.4110.

Wentworth, P., & Janda, K. D. (2001). Catalytic antibodies. *Cell Biochemistry and Biophysics*, *35*, 63−87. Available from https://doi.org/10.1385/CBB:35:1:63.

Wetlaufer, D. B. (1990). Nucleation in protein folding-confusion of structure and process. *Trends in Biochemical Sciences*, *15*(11), 414−415. Available from https://doi.org/10.1016/0968-0004(90)90275-G.

Wieczorek, R., & Dannenberg, J. J. (2003). H-Bonding Cooperativity and Energetics of α-Helix Formation of Five 17-Amino Acid Peptides. *Journal of the American Chemical Society*, *125*(27), 8124−8129. Available from https://doi.org/10.1021/ja035302q.

Wieczorek, R., & Dannenberg, J. J. (2004). Comparison of Fully Optimized α- and 3_{10}-Helices with Extended β-Strands. An ONIOM Density Functional Theory Study. *Journal of the American Chemical Society*, *126*(43), 14198−14205. Available from https://doi.org/10.1021/ja048831i.

Wieczorek, R., & Dannenberg, J. J. (2005). Enthalpies of hydrogen-bonds in α-helical peptides. An ONIOM DFT/AM1 study. *Journal of the American Chemical Society*, *127*(42), 14534−14535. Available from https://doi.org/10.1021/ja053839t.

Yamada, T., Nakao, M., Miyazawa, T., Kuwata, S., Sugiura, M., In, Y., & Ishida, T. (1993). Conformational difference between diastereomers of Dnp-Val-Aib-Gly-Leu-pNA studied by x-ray crystal analyses. *Biopolymers*, *33*(5), 813−822. Available from https://doi.org/10.1002/bip.360330509.

Yan, S., Gawlak, G., Makabe, K., Tereshko, V., Koide, A., & Koide, S. (2007). Hydrophobic surface burial is the major stability determinant of a flat, single-layer β-sheet. *Journal of Molecular Biology*, *368*(1), 230−243. Available from https://doi.org/10.1016/j.jmb.2007.02.003.

Zagrovic, B., Lipfert, J., Sorin, E. J., Millett, I. S., Gunsteren., Doniach, S., & Pande, V. S. (2005). Unusual compactness of a polyproline type II structure. *Proceedings of the National Academy of Sciences of the United States of America*, *102*(33), 11698–11703. Available from https://doi.org/10.1073/pnas.0409693102.

Zhou, H.-X. (2018). Electrostatic interactions in protein structure, folding, binding, and condensation. *Xiaodong Pang Chemical Reviews*, *118*(4), 1691–1741. Available from https://doi.org/10.1021/acs.chemrev.7b00305.

Chapter 2

Modeling and simulation of peptides

Amay Redkar and Vibin Ramakrishnan
Department of Biosciences and Bioengineering, Indian Institute of Technology Guwahati, Guwahati, Assam, India

2.1 Introduction

An increasing number of researchers are utilizing the dynamic and hierarchical self-assembling properties of proteins and peptides to design novel biofunctional nanostructures with therapeutic or catalytic applications (Craik & Kan, 2021; Katyal, Meleties, & Montclare, 2019; Otvos, 2008; Ulijn & Smith, 2008). Peptides represent an important class of biological molecules composed of short chains of amino acids with a wide range of functions in signaling and regulation. Biological systems already incorporate them to resolve many physiological and pharmacological quandaries. Apart from peptide-based natural hormone analogs, peptides drug candidates have been employed to disrupt protein – protein interactions and target or inhibit intracellular molecules (Birk et al., 2013; Chang et al., 2013; Wang et al., 2022). These avenues have made peptide therapeutics a promising industry, with nearly 20 new peptide drugs entering clinical trials annually. Over 400 peptide drugs are currently under clinical development worldwide, and more than 60 are already approved for clinical use in the United States, Europe, and Japan. Since 2017, more than 10 peptide drugs have entered the market. Peptide drugs like liraglutide (Victoza) and glucagon-like peptide I (GLP-1) used to treat metabolic diseases have a gross revenue crossing 2 billion (USD). Peptide drugs with hormone analogs properties like leuprolide (Lupron), gosarelin (Zoladex), somatostatin analogs, octreotide, and lanreotide added up to over 4 billion USD in the market. Sprifermin (FGF18) is one of the few available strategies to repair osteoarthritis damage, which has cleared phase 3 clinical trial (Hochberg et al., 2019) (Fig. 2.1).

Despite these successful scientific and economic ventures, the interest in peptide drug therapeutics has waned due to their poor bioavailability (Drucker, 2020), relatively high manufacturing cost (Zompra, Galanis,

De Novo Peptide Design. DOI: https://doi.org/10.1016/B978-0-323-99917-5.00009-3

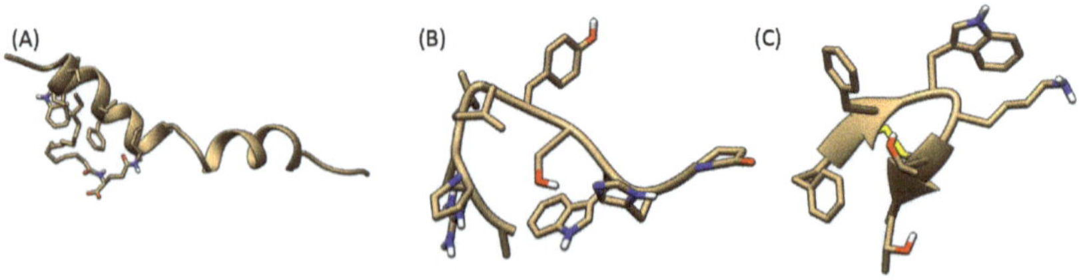

FIGURE 2.1 Structures of peptide-based therapeutics. (A) Liraglutide (PDB ID: 4APD). (B) Leuprolide (PDB ID: 1YY2). (C) Octreotide (PDB ID: 1SOC).

Werbitzky, & Albericio, 2009), short biological half-lives (Pollaro & Heinis, 2010), and limited tissue uptake. Multiple approaches like incorporating disulfide rich regions, introducing the use of constrained or cyclic peptides, and stapled peptides have been tried to amend these limitations. Applications of in silico tools have become an unavoidable approach for optimizing and designing new peptide drugs.

This chapter will summarize some contemporary approaches to peptide design, emphasizing peptide modeling, structure prediction and simulations, and their unique potential for the design of identifying novel, safe, and effective therapeutic solutions.

2.2 Peptide design

Peptide design represents a massive multioptimization search. For example, ideally, a therapeutic peptide should be deliverable to a biological target with enough concentration to induce the expected therapeutic effect with the capacity to navigate various biochemical and biophysical barriers. It should also be ethically, legally, and economically sensible. In this situation, one searches through the vastness of theoretical amino acids and their modifications space to recognize the relatively few peptides that compromise between the multiple and often uncorrelated or even anticorrelated properties essential for a viable candidate (Nicolaou & Brown, 2013).

The enormity of chemical space available to peptides is due to two factors, its sequence space and its conformational space. Just like in silico methods, cell-free methods exist to generate and screen millions of sequences, yet the activity will not represent a fraction of the available sequence configurations without even accounting for the multitude of modifications (Bashiruddin & Suga, 2015; Lian, Upadhyaya, Rhodes, Liu, & Pei, 2013; Sohrabi, Foster, & Tavassoli, 2020). Conformational diversity in a molecule is due to rotatable (chi) torsions. The number of rotatable torsions in standard amino acids is between 1 and 4. Amino acid links add additional conformational flexibility because of the rotatable phi/psi backbone dihedral angles. Peptides will not usually appear as a conformation or even a small ensemble of conformations under physiological conditions. They typically prefer to present an ensemble of various conformations with different peptide activity and functions such as target binding, crossing biological membranes, or forming a site for catalytic action. NMR and crystallography can be used to determine peptide structure, but it is safe to say that a multitude of functionally relevant peptide conformations remain masked from direct experimental elucidation. Hence it is vital to incorporate rational limits based on potency and other functionally relevant properties and model them in silico.

One approach for peptide design and discovery is rational design (Fig. 2.2). By rational design of peptides, we mean the conscious use of scientific principles to drastically and efficiently decrease the theoretical amino acid search space to a more manageable number of effective molecules. Although principles and phenomena typically drive rational peptide design projects observed *in vitro* by chemical and biological means, integrating *in silico* methods presents an important cost and time reduction option, dramatically improving the peptide development process. The successful application of in silico tools is critically dependent on a feedback loop between meticulous validation of the tools and wet lab experiments. Their use without appreciation for their limits should be therefore avoided.

Drug designers have followed this approach for decades and recently by antibody/protein designers (Choong, Lee, Soong, Law, & Lim, 2017; Kuroda, Shirai, Jacobson, & Nakamura, 2012). Computer-aided drug design guides for developing peptide modeling and design's general philosophy and approach. Nevertheless, there is evidence that direct application of small molecule and protein design strategies are not viable for peptide drug design (Audie & Swanson, 2012; Das, 2011; Pike & Nanda, 2015; Rentzsch & Renard, 2015). Among the myriad of complications, two important reasons for this pertain to the building blocks of peptides. The conformational space

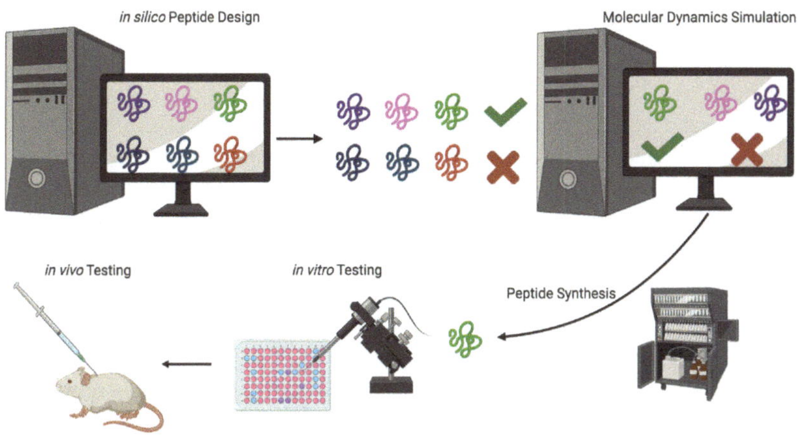

FIGURE 2.2 Rational peptide design framework. Figure created using BioRender (https://biorender.com/).

accessible to peptides limits the application of small-molecule computational methods at one end. On the other hand, the need of including nonstandard amino acids and diverse cyclization chemistries in peptide drug design imposes a constraint on established protein modeling software. Hence, computational approaches based on peptides are needed.

By efficiently generating testable predictions, suggesting, refining, and evaluating design hypotheses while gaining structural understanding and insight, research is increasingly focused on expanding and implementing peptide-based computational methods to curtail conceptual peptide chemical space to be attainable and associated with particular moieties. These approaches can be divided into two categories: ligand-based peptide design and target-based peptide design. Ligand-based peptide design methods, to design new peptides, exclusively exploit the ligands' information, especially its Structure – Activity Relationship (SAR). Meanwhile, target-based drug design methods wholly utilize the target's properties, typically proteins. Hybrid techniques are frequently used in practice.

2.2.1 Ligand-based peptide design

The design of ligand-based peptides can be classified into sequence-based, properties-based, and conformation-based strategies.

2.2.1.1 Sequence-based methods

Sequence-based methods base their strategies on assuming that the conserved sites and motifs are essential for function and structure (Diller, Swanson, Bayden, Jarosinski, & Audie, 2015). These methods rely on multiple sequence alignments and conserved regions and are routinely used in peptide drug design. Rivera et al. reported a peptide-specific quantitative structure-activity relationship (QSAR) model to determine their correlation with activity and sequence using a Point Accepted Mutation (PAM) 250 matrix (Rivera et al., 2011). They demonstrated it by analyzing a group of collagen peptides and their antiangiogenic activity and presenting regions of sequences of interest. SARvision is a computational program that can support SAR analysis and design of peptide based on their sequence (Hansen, Villar, & Feyfant, 2013). In terms of the general orientation for peptide SAR analysis, SARvision distinguishes out as it integrates advanced visualization potentialities with extensive sequence/activity analysis functionality.

2.2.1.2 Property-based methods

Principles for property-based design have not yet matured, especially when compared to sequence-based methods. Du et al. modeled peptides capable of binding to protein targets using multiple amino acid physical – chemical descriptors (volume, hydrophobicity, and hydrophilicity) and ΔG

decomposition for each residue (Du, Huang, & Chou, 2008; Du, Ma, Xie, & Huang, 2014; Du, Wei, Pang, Chou, & Huang, 2007). Property-based techniques have been popular for analyzing antimicrobial peptides (Bhonsle, Clark, Bartolotti, & Hicks, 2013). Giguere et al. and Wang et al. designed new antimicrobial peptides using a machine-learning approach (Giguère et al., 2015; Wang, Garlick, & Zloh, 2021).

2.2.1.3 Conformation-based methods

Conformational ensembles or predicted peptide structures can be analyzed, and SARs established based on them. Inverse Design and Automation Software (IDeAS) for peptide-design retrofits side chains on a backbone with residues, containing both L and D stereochemistry (Ranbhor et al., 2018). Using IDeAS, Prakash et al. designed a 21-residues heterochiral peptide capable of blue fluorescent protein behavior (Prakash, Ranbhor, & Ramakrishnan, 2020).

PEP-FOLD is a computational framework established by Thevenet et al. for determining the tertiary structure of 9–36 amino acids with linear and disulfide-bonded peptides under the hypothesis that they assume well-defined structures (Thévenet et al., 2012). It could generate near-native structures for 24 linear peptides shorter than 25 amino acids on a benchmark, with the best rigid core of the predicted structures deviating by 1.5 Å compared to NMR structures. Benchmarking 13 longer linear peptides predicted average rigid core structural deviations of 2.8 Å from their native forms. The results were rather less impressive for cyclised peptides, which is surprising considering that the constrained peptides are intuitively easier to model. Judging from the data, it is noteworthy that the tool can resolve problems associated with conformational sampling better than the scoring problem. The latest version, PEP-FOLD3 developed by Lamiable et al., can predict the 3D structure of linear peptides of about 5–50 amino acids but cannot predict the same for disulfide bonded peptides (Lamiable et al., 2016). The benchmark could generate poses deviating by 3.4 Å on average for 61 peptides from their native state.

Long-time-scale molecular dynamics simulations have also been used to investigate peptide conformational space (Klepeis, Lindorff-Larsen, Dror, & Shaw, 2009; Lindorff-Larsen, Piana, Dror, & Shaw, 2011). Quantum mechanical calculations, specifically for tackling the scoring deficiency-related problems in peptide conformational analysis, also have potential application in in silico peptide design (Denisiuk et al., 2013; Han & Wang, 2012; Improta, Vitagliano, & Esposito, 2015). It is fair to say that much more theoretical and technical innovation is required to effectively apply these computationally expensive and complicated methods to impact peptide drug design projects positively.

Avenues for 3D pharmacophore hypothesis generation are also enabled by peptide conformational modeling and prediction. The approach broadly screens peptides with desired potency or ADME/Tox activities for a given pharmacophore hypothesis. In order to analyze agonists of melanocortin-4 receptor isoforms, Haslach et al. used nonbiased peptide library scanning (Haslach et al., 2014). They also generated a SAR platform and peptide-based ligand template for melanocortin-4 receptor isoforms.

2.2.2 Target-based peptide design

Target-based peptide design is perhaps better developed and well-studied due to the transferability of protein complex than the ligand-based approach to peptide design. The first step of target-based design is analyzing the unbound or ligand-bound target using computer-enabled analysis. This analysis reveals properties such as potential binding surfaces' potential specificity sites, among other pharmacological activity determinants. A design hypothesis step follows this step. With the help of computational tools, peptide design ideas are refined and tested. The scope of the peptide design efforts depends on the quantity and quality of target information alongside the design philosophy and can range from template-based to de novo.

Databases of protein – peptide complex structures such as PepX (Vanhee et al., 2010), PepBind (A. A. Das, Sharma, Kumar, Krishna, & Mathur, 2013), and peptidDB (London, Movshovitz-Attias, & Schueler-Furman, 2010) databases have been curated to assist efforts for target-based drug design. The PepX and PepBind database contain unique protein – peptide interface clusters and complex structures from the PDB, respectively. PeptidDB is a curated database of 103 protein – peptide complexes. Apart from structural data, the binding affinity of high throughput screening experiments and kinetics of protein – peptide interaction can also provide insights related to the structural basis of protein – peptide recognition.

Binding interaction between proteins and ligands can be modeled and predicted by docking algorithms. However, unlike small-molecule algorithms, peptide docking algorithms tend to be computationally expensive for virtual screening experiments like this. It has also been shown that docking programs do not provide a general solution for peptides (Audie & Swanson, 2012). Conventional small molecule docking programs like AutoDock, Vina, and MOE-Dock have a poor success rate when the ligand is more than three amino acids long (Kellenberger, Rodrigo, Muller, & Rognan, 2004). Therefore it is imperative to follow peptide-focused docking protocols.

Rosetta FlexPepDock *ab-initio* is a peptide docking program that samples coarse-grained representations of the peptide with the coarse-grained representation of the receptor alongside rigid body orientations around the target's peptide-binding surface. (London, Raveh, Cohen, Fathi, & Schueler-Furman, 2011; Raveh, London, Zimmerman, & Schueler-Furman, 2011).

Subsequently, the coarse-grained model of protein and peptide undergoes all-atom refinement, including the side chains. Another iteration of the basic FlexPepDock algorithm is PaFlexPepDock (Li et al., 2014). In the PaFlexPepDock algorithm, three steps of the protein – peptide docking, the *ab initio* peptide folding and peptide docking with its receptor, and refinement of some flexible areas of the protein are performed parallelly.

In the AnchorDock algorithm, developed by Ben-Shimon and Niv, the free peptide structure and anchoring points on the protein surface will be identified first computationally (Ben-Shimon & Niv, 2015). The free peptide undergoes anchor-constrained simulated annealing molecular dynamics at the predicted anchoring spot. This approach eliminates any need for information about the peptide-binding site. In their blind docking test, it was able to model 10 from 13 cases with a mean square deviation less than 2.24 Å.

Using PatchDock (Schneidman-Duhovny, Inbar, Nussinov, & Wolfson, 2005) and combinatorial peptide library, Mehra et al. designed IB peptides capable of disrupting the nonspecific binding of lymphocytes to endothelial cells (Mehra, Jerath, Ramakrishnan, & Trivedi, 2015). Using similar methods, they also investigated the potential antimalarial activity by targeting PFI1625c using peptide drugs (Lhouvum, Ramakrishnan, & Trivedi, 2013).

2.2.3 *De novo* peptide design

De novo peptide design has a long pedigree dating as early as the 1980s to generate diverse candidates for screening using multifarious small and big toolkits. Unal et al. described the VitAL algorithm, which designs peptides for protein target binding sites using Autodock and a Viterbi algorithm (Unal, Gursoy, & Erman, 2010) by docking its residue pair by pair. Another target-based de novo design Monte Carlo-based procedure was developed by Bhattacherjee and Wallin, which explores peptide sequence and conformational space against target protein (Bhattacherjee & Wallin, 2013). To assess the method, they attempted to determine the peptide binding specificity profiles of three different peptide – protein domains. Bhardwaj et al. designed 12 conformationally restrained peptides with defined folds by assembling the backbone using fragment assembly method or fragment-independent kinematic closure-driven method and sequence design (Bhardwaj et al., 2016). They validated their designed peptides, including the heterochiral peptide, using X-ray crystallography and NMR.

2.3 Prediction of peptide structure

Predicting protein structures from their primary sequences represents one of the most challenging problems in computational biology today. Predicted peptide structures or conformational ensembles can be analyzed, and SAR established based on them (Ramakrishnan, Ranbhor, & Durani, 2004; Ramakrishnan, Ranbhor, & Durani, 2005; Ramakrishnan, Ranbhor, Kumar,

& Durani, 2006). Hence it is imperative to know the structural information of a novel peptide to coopt toward the desired therapeutic properties. Many attempts have been made to solve the protein structure prediction problem, with many software applications being developed for this purpose, including I-TASSER (Roy, Kucukural, & Zhang, 2010), Rosetta (Simons, Kooperberg, Huang, & Baker, 1997), HHpred (Soding, Biegert, & Lupas, 2005), NovaFold, and most notably AlphaFold 2, which recently performed excellently in the CASP 14 experiment. Ko and Lee recently modeled the structures of 203 protein – peptide complexes from the PepBDB DB and 183 from the PepSet using AlphaFold 2 and concluded that AlphaFold 2 learned how to extract interaction information between receptors and peptides from evolutionary information (Ko & Lee, 2021).

2.4 *In silico* validation of design by molecular dynamic simulations

Alder and Wainwright originally proposed the concept of molecular dynamics in 1959. It was developed to simulate a system of colliding particles with a rigid core (Alder & Wainwright, 1959). Rahman later expanded it to include continuous potentials and uniform time steps (Rahman, 1964). According to classical mechanics, the system's state can be specified by the positions and velocities (momenta) of all particles, each of which can assume any value. The probability that a given state with energy E is occupied in equilibrium at constant particle number N, volume V, and temperature T (the "canonical" ensemble) is proportional to the Boltzmann factor. The equilibrium value can thus be obtained by averaging over all states accessible to the system, weighing each state by this factor. The averaging over all states in the classical molecular dynamics simulation is performed by *integrating* these *continuous* variables.

Molecular dynamics simulations incorporate Newton's equations of motion to replicate the time-based evolution of N associated atoms with masses m_i of a system.

$$m_i \frac{d^2 r_i}{dt^2} = F_i$$

for $i = 1 \ldots N$ where $r_i(t)$ represents ith particle position, F_i is the momentary force acting upon each atom calculated from the system's interactions. The forces are the negative derivatives (slopes) of the potential energy function $V\ (r_1, r_2, \ldots. r_N)$, which depend on all atoms' positions.

$$F_i = -\left(\frac{\partial V}{\partial r_i}\right)$$

Calculating the potential energy function V is crucial in molecular dynamics simulations. The equation shows that the force will try to drive the particle to a state of low potential energy. The resulting acceleration can establish the approximate positions of the atoms in a short time interval. This is termed the integration step of the equation of motion. The method is repeated for a large number of small successive steps. The coordinates thus generated describing the position, velocity, and forces on all atoms as a function of time represent a trajectory of the system. If the potential energy function is a good estimation of the actual interacting particles, it can produce a detailed illustration of the equilibrium and dynamics properties in the investigated system.

Generation of new configuration at $(t + \Delta t)$th time is followed by calculating the forces acting on a specific configuration of atoms at the current time t according to equation 1.

Several numerical algorithms like Verlet (Verlet, 1967) and Leapfrog are available, with good performance for long time steps requiring only a single force evaluation per step. Groningen MAchine for Chemical Simulation (GROMACS) (Van Der Spoel et al., 2005) molecular dynamics program utilizes Leapfrog scheme, a slightly modified but theoretically equivalent Verlet algorithm for integration. The Leapfrog algorithm uses positions r at time t and velocities v at time $t - \Delta t/2$. It updates positions and velocities using the forces $F(t)$ determined by the positions at time t:

$$r_i(t + \Delta t) \approx r_i(t) + \Delta t v_i\left(t + \frac{\Delta t}{2}\right)$$

$$v_i\left(t + \frac{\Delta t}{2}\right) \approx v_i\left(t - \frac{\Delta t}{2}\right) \approx v_i\left(t - \frac{\Delta t}{2}\right) + \frac{\Delta t}{m_i}F_i$$

The name comes from the generation of positions and velocities at the whole and half-time steps, respectively; they are leaping like frogs over each other's back. The equations of motion are modified for temperature coupling and pressure coupling and extended to include the conservation of constraints.

2.4.1 Temperature and pressure coupling

The system's temperature should be maintained constant but can drift from the set value due to force truncation, integration errors, and heating due to external or frictional forces.

Berendsen's weak coupling scheme is used in these studies to maintain temperature, in which atom velocities are rescaled at every step with a calculated factor λ.

$$\lambda = \sqrt{l + \frac{\Delta t}{\tau_T}\left(\frac{T_o}{T} - 1\right)}$$

T_0 represents the reference temperature, and τ_T is the temperature coupling time constant. This weak coupling scheme will produce exponential relaxation to the reference temperature.

The system's pressure is similarly controlled through a coupling method by rescaling the coordinates of the box vectors at every time step with a factor μ.

$$\mu = \sqrt[3]{1 + \frac{\Delta t}{\tau_P}\beta(P - P_0)}$$

where β refers to the isothermal compressibility of the system, this will produce exponential pressure relaxation, similar to weak temperature coupling.

2.4.2 Energy minimization

The energy minimization process minimizes the strain energy of the molecule by altering the atomic position to optimal geometry, that is to its local minima. As these procedures are "downhill" methods, they cannot cross the energy barriers; they will end up in local minima close to the point from where the minimization process started. It is infrequent for a direct minimization method to find the global minimum of the structure. So the minimization process reduces the potential energy of a given conformation by relieving local strains in the structure; hence this can also be called an optimization process. In molecular dynamics, the structures are allowed to simulate by thermal motion as a function of time. This is done by applying a force field on the atoms to drive the motion. The acceleration and velocities are then used to calculate new positions for the atoms over a short time step (femtoseconds). This process is iterated thousands of times to generate a series of conformations of the structures forming as a trajectory. Simulation at optimum temperature (300K) provides information on structural fluctuations around the starting conformation and the conformational transitions pathway. High-temperature molecular dynamics can search the conformational space as this provides more energy to cross the energetic barriers and reach the otherwise unexplored regions in the energy landscape.

The most widely used methods can be categorized, based on the derivatives of strain energy they employ, as first derivative-based like steepest descent and conjugate gradient and first and second derivative based like Newton – Raphson procedures. Generally, the first 100 – 500 steps are carried out using the Steepest gradient method and convergence is achieved using the conjugate gradient method or Newton – Raphson procedures.

2.4.2.1 Steepest descent method

In the steepest descent method, the first derivative of strain energy is calculated or estimated for each coordinate of each atom and then moving it while retaining the resultant change in strain energy. The same step is done for the next atom after returning it to its original placement until all atoms have been tested. The atoms are moved from their position proportionally to the derivative calculated. Then the entire process is repeated until a threshold of energy is reached. The biggest challenge with this method relates to step size for atom movement. This step size dictates the quality of the minimization and the efficiency of the process.

2.4.2.2 Conjugate gradient method

Here the previous search direction and current gradient are used to achieve minimization. This method converges faster as it incorporates the history of minimization to calculate search direction and contains a scaling factor for optimal step size. However, the required computing cycles scale proportionally with the number of atoms.

2.4.2.3 Newton – Raphson methods

Newton – Raphson procedures employ the strain energy of the curvature to locate local minima. The computations involved are considerably more complex, but will completely apply all information presented and converge quickly.

2.4.3 Force field

The force field is used to describe the time evolution of bond length, bond angles and torsions, the nonbonding van der Waals, and the electrostatic interaction between atoms. Force field generally takes the following form for computing the potential energy.

Total energy:

$$E_{(potential\ energy)} = E_{bonded} + E_{non-bonded}$$

$$E_i = E_{bond} + E_{angle} + E_{dihedral} + E_{electrostatic} + E_{van\ der\ waals}$$

Bond length:

$$E_{bond-stretch} = \sum_{1,2\ pairs} K_b(b-b_0)^2$$

Bond angle:

$$E_{bond-bend} = \sum_{angles} K_\theta(\theta-\theta_0)^2$$

Dihedral angle:

$$E_{rotate-along-bond} = \sum_{1,4\ pairs} K_\phi\left(1 - \cos(n\phi)\right)$$

Kr, *Kθ*, and *Kφ* are force constants for bond, angle, and dihedral angle, respectively. *R0*, *θ0*, and *φ0* define the given type's equilibrium distance, equilibrium angle, and dihedral angle. *n* is the periodicity of the Fourier term. These parameter values are derived from model molecules and vary among force fields (Table 2.1).

2.4.3.1 Lennard-Jones potential

Lennard-Jones. 6–12 potential is most commonly used for van der Waals interactions (*EvdW*) in *Enb*.

$$E_{van-der-Waals} = \sum_{non-bonded\ pairs}\left(\frac{A_{ik}}{r_{ik}^{12}} - \frac{C_{ik}}{r_{ik}^{6}}\right)$$

The interactions between nonbound but interacting atoms *i* and *k* with the distance *rik*. *Aik* and *Cik* are van der Waal parameters.

2.4.3.2 Coulomb potential

Biochemical molecules are often charged, so an electrostatic energy (*Eelec*) term is added to E_{nb}.

$$E_{electrostatic} = \sum_{non-bonded\ pairs}\frac{q_i q_k}{Dr_{ik}}$$

D is a molecular dielectric constant that accounts for the environmental attenuation of electrostatic interaction between the two atoms with the point charge q_i and q_k .

2.4.4 Conformational analysis

The technique of changing dihedral angles to find molecular systems with low-energy conformations is known as conformational search. Modifying dihedral angles and then minimizing energy for each angle were used to develop unique structures. Duplicate structures with high energy are rejected, but low-energy unique conformations are retained. Only dihedral angles are evaluated in the conformational search because the flexibility of molecules is often generated by the rotation of unhindered bond dihedrals with no change in bond angles or bond lengths. The goal is to locate the global minimum of the potential energy surface of a molecular system. The subject of identifying low-energy molecular conformations has been explored in several different

TABLE 2.1 Force fields and their sources.

Force field	Source	References
AMBER	University of California, San Francisco/HyperChem	Pérez et al. (2007)
CHARMM	Harvard University/Accelrys, Inc.	Vanommeslaeghe et al. (2010)
ECEPP	Cornell University	Arnautova, Jagielska, and Scheraga (2006)
GROMOS	University of Groningen/Biomos	Schmid et al. (2011)
OPLS	Yale University/Hyperchem	Robertson, Tirado-Rives, and Jorgensen (2015)

methods (Howard & Kollman, 1988). The following stages are included in many of these approaches, with minor variations:

1. Primary structure selection: The initial structure (e.g., energy minimized structure) is the most recently approved conformation and remains unchanged during the search. In Monte Carlo searches, this is known as a random walk scheme. Because low-energy conformations tend to be similar, beginning with an approved conformation keeps the search on the low-energy part of the potential surface. On the other hand, the usage-directed strategy aims to sample a low-energy region equally by going through all previously accepted conformations when selecting each beginning structure (Chang, Guida, & Still, 1989). The usage-directed strategy has been demonstrated to be preferable for swiftly finding low-energy conformations in comparative investigations.
2. Modifying the initial structure by varying geometric parameters: Systematic or arbitrary variations are possible. Systematic variations can investigate low-energy conformational space comprehensively. Except for the simplest systems, the amount of possible modifications becomes prohibitive. One approach of reducing systematic variation's dimensionality is to deplete fluctuations with a limited degree of resolution initially, then reduce the additional variations available by gradually raising the resolution. Random variations select a new value from a considerable span or combinations of distinct values for one or more geometric parameters. Before completing the energy reduction of the new structure, multiple random variants are compared to a collection of preceding structures that limit the number of recurrent conformations.
3. Geometry optimization for energy-minimized conformations of the modified structure: Dihedral angle variations result in structures that are

energy lowered in order to discover a possible surface local minimum. Regardless of the optimiser's preference (minimizer), choosing an optimiser quickly reaches a local minimum while avoiding potential surface obstacles.

4. Comparison of the conformation to previous reports: If the conformation is unique and the energy meets a set of criteria, it is accepted. The acceptability of the conformation is determined using two types of criteria. To eliminate duplication, preliminary geometrical comparisons to earlier acceptable conformations are performed. Internal coordinates, interatomic lengths, and least-squares superposition of conformers are commonly used to compare conformations using maximal torsions or Root Mean Square deviation (RMSD). Geometric optimization can invert chiral centers; hence the chiral centers of the new structures should be confirmed after they have been reduced in size. Second, the energetic test for accepting a new conformer can be performed using either a suitable cut-off in terms of the best energy identified so far or a Metropolis criterion. Structures with higher energy are more likely to be authorized if they have a temperature and energy difference (Fig. 2.3).

2.4.5 Case studies of molecular dynamics simulations and trajectory analysis

2.4.5.1 Antimicrobial activity

Hazam et al. designed six peptides of 12 amino acid length around the gramicidin helical conformation incorporating L and D peptides for added stability and present distinct polar and hydrophobic zones (Hazam, Jerath, Kumar, Chaudhary, & Ramakrishnan, 2017). To determine the rigidity of their sequences, they performed a 10-ns simulation of the peptides in water and

FIGURE 2.3 Clusters of peptide conformations assumed by a peptide(IHIHIYI) after 10 ns molecular dynamics simulation. Each strand represents the central conformation of each cluster.

determined the number of microstates at equilibrium. In another investigation, Hazam, Singh, et al. determined the stability of another set of four peptides based on cluster analysis and RMSD for a 10-ns simulation in water and determined the structural stability of the peptide (Hazam, Singh, Chaudhary, & Ramakrishnan, 2019).

Molecular simulation can also investigate the potential mechanism and factors driving the potency of various antimicrobial peptides. As an extension of their earlier investigation, Hazam, Akhil, et al. studied the interaction of single (seven amino acid) and double (12 amino acid) turn versions of their peptides with Phosphatidylcholine: Palmitoyloleoylphosphatidylglycerol membranes (3:1 ratio), under identical conditions using 160-ns molecular dynamics simulation in order to elucidate upon the differential effect of the peptides (Hazam, Akhil, Jerath, Saikia, & Ramakrishnan, 2019). Based on the multimeric complex formation, the interaction of positively charged residues, and the adsorption of peptides into the membrane and the relative stability of each peptide, indicated by the cluster analysis, the reduction in potency between the peptide subsets becomes more evident (Fig. 2.4).

2.4.5.2 *Drug delivery*

To determine the cell specificity of Stereo-chemical Amphipathic chaRged Tumor Homing and Internalizing peptides, Jerath et al. used molecular dynamics simulations (Jerath, Goyal, Trivedi, Santhoshkumar, & Ramakrishnan, 2019). Using RMSD and radius of gyration, they inferred that the syndiotactic peptides', peptides with alternating L and D amino acids, backbone are semirigid, with the ability to maintain helical structure in different solvents but unwind on interaction with membrane. In another study, Jerath et al. used molecular dynamics simulations to assess the effect of incorporation D-Pro, replacing Gly, into their design based on Arginine Glycine Aspartic Acid (RGD) and Asparagine Glycine Arginine (NGR) for

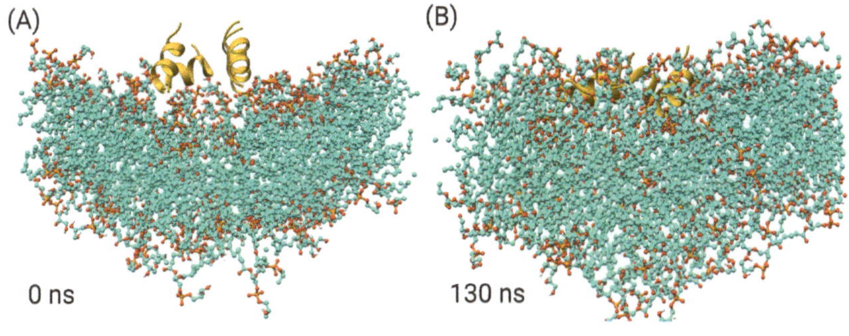

FIGURE 2.4 The peptides here are shown at the beginning, 0 ns (A), and end, 130 ns (B), of the simulation interacting with membrane. Note the position of peptide shift from above the membrane to embedded into it.

drug delivery (Jerath, Goyal, Trivedi, Santhoshkumar, & Ramakrishnan, 2020). Based on the backbone dihedral distributions of the D-Pro incorporated into the peptide, they were able to show that it assists in locking the form of the RGD and NGR motifs.

In order to design a set of tumor homing peptides with conformational locks, Goyal et al. employed a set pruning procedure based on electrostatic profiling, docking and finally, molecular dynamics simulations (Goyal et al., 2021). Their molecular dynamics simulation studies investigated the receptor peptide complex interaction determined by docking. They ran a position restrained molecular dynamics of 1 ns followed by a 10 ns production run. The simulations were analyzed using RMSD, Root Mean Square Fluctuations (RMSF), the radius of gyration, and each system's hydrogen bond distribution. They also determined the receptor's active site occupancy by using an index of the active site residues and the peptide.

2.5 Conclusion

Design and modeling peptides using rational methods allow researchers to bolster an already promising industry by incorporating various in silico methods discussed here to explore the vast chemical space intelligently. Despite the developments mentioned in this chapter, several avenues for investigation remain fertile. Scoring method problems and simultaneous prediction of backbone and side-chain for peptide conformations are the most noteworthy examples of such challenges. Molecular dynamics simulations of peptides also allow for validation of designs and exploration of the designed peptides' stability and mechanism of action.

References

Alder, B. J., & Wainwright, T. E. (1959). Studies in molecular dynamics. I. General method. *The Journal of Chemical Physics*, *31*(2), 459–466. Available from https://doi.org/10.1063/1.1730376.

Arnautova, Y. A., Jagielska, A., & Scheraga, H. A. (2006). A new force field (ECEPP-05) for peptides, proteins, and organic molecules. *The Journal of Physical Chemistry B*, *110*(10), 5025–5044. Available from https://doi.org/10.1021/jp054994x.

Audie, J., & Swanson, J. (2012). Recent work in the development and application of protein–peptide docking. *Future Medicinal Chemistry*, *4*(12), 1619–1644. Available from https://doi.org/10.4155/fmc.12.99.

Bashiruddin, N. K., & Suga, H. (2015). Construction and screening of vast libraries of natural product-like macrocyclic peptides using in vitro display technologies. *Current Opinion in Chemical Biology*, *24*, 131–138. Available from https://doi.org/10.1016/j.cbpa.2014.11.011.

Ben-Shimon, A., & Niv, M. Y. (2015). AnchorDock: Blind and flexible anchor-driven peptide docking. *Structure*, *23*(5), 929–940. Available from https://doi.org/10.1016/j.str.2015.03.010.

Bhardwaj, G., Mulligan, V. K., Bahl, C. D., Gilmore, J. M., Harvey, P. J., Cheneval, O., ... Baker, D. (2016). Accurate de novo design of hyperstable constrained peptides. *Nature*, *538* (7625), 329–335. Available from https://doi.org/10.1038/nature19791, PMID:27626386.

Bhattacherjee, A., & Wallin, S. (2013). Exploring protein-peptide binding specificity through computational peptide screening. *PLoS Computational Biology*, *9*(10), e1003277. Available from https://doi.org/10.1371/journal.pcbi.1003277.

Bhonsle, J., Clark, T., Bartolotti, L., & Hicks, R. (2013). A brief overview of antimicrobial peptides containing unnatural amino acids and ligand-based approaches for peptide ligands. *Current Topics in Medicinal Chemistry*, *13*(24), 3205–3224. Available from https://doi.org/10.2174/15680266113136660226.

Birk, A. V., Liu, S., Soong, Y., Mills, W., Singh, P., Warren, J. D., ... Szeto, H. H. (2013). The mitochondrial-targeted compound SS-31 re-energizes ischemic mitochondria by interacting with cardiolipin. *Journal of the American Society of Nephrology*, *24*(8), 1250–1261. Available from https://doi.org/10.1681/ASN.2012121216.

Chang, G., Guida, W. C., & Still, W. C. (1989). An internal-coordinate Monte Carlo method for searching conformational space. *Journal of the American Chemical Society*, *111*(12), 4379–4386. Available from https://doi.org/10.1021/ja00194a035.

Chang, Y. S., Graves, B., Guerlavais, V., Tovar, C., Packman, K., To, K.-H., ... Sawyer, T. K. (2013). Stapled α-helical peptide drug development: A potent dual inhibitor of MDM2 and MDMX for p53-dependent cancer therapy. *Proceedings of the National Academy of Sciences*, *110*(36), E3445–E3454. Available from https://doi.org/10.1073/pnas.1303002110.

Choong, Y. S., Lee, Y. V., Soong, J. X., Law, C. T., & Lim, Y. Y. (2017). Computer-aided antibody design: An overview. In T. S. Lim (Ed.), *Recombinant Antibodies for Infectious Diseases* (pp. 221–243). Springer International Publishing. Available from https://doi.org/10.1007/978-3-319-72077-7_11.

Craik, D. J., & Kan, M.-W. (2021). How can we improve peptide drug discovery? Learning from the past. *Expert Opinion on Drug Discovery*, *16*(12), 1399–1402. Available from https://doi.org/10.1080/17460441.2021.1961740.

Das, A. A., Sharma, O. P., Kumar, M. S., Krishna, R., & Mathur, P. P. (2013). PepBind: A comprehensive database and computational tool for analysis of protein–peptide interactions. *Genomics, Proteomics & Bioinformatics*, *11*(4), 241–246. Available from https://doi.org/10.1016/j.gpb.2013.03.002.

Das, R. (2011). Four small puzzles that Rosetta doesn't solve. *PLoS One*, *6*(5), e20044. Available from https://doi.org/10.1371/journal.pone.0020044.

Denisiuk, A., Schubert, V., Wolter, F. E., Irran, E., Trouillas, P., & Süssmuth, R. D. (2013). Conformational investigation of antibiotic proximicin by X-ray structure analysis and quantum studies suggest a stretched conformation of this type of γ-peptide. *Bioorganic & Medicinal Chemistry*, *21*(12), 3582–3589. Available from https://doi.org/10.1016/j.bmc.2013.02.051.

Diller, D. J., Swanson, J., Bayden, A. S., Jarosinski, M., & Audie, J. (2015). Rational, computer-enabled peptide drug design: Principles, methods, applications and future directions. *Future Medicinal Chemistry*, *7*(16), 2173–2193. Available from https://doi.org/10.4155/fmc.15.142.

Drucker, D. J. (2020). Advances in oral peptide therapeutics. *Nature Reviews Drug Discovery*, *19*(4), 277–289. Available from https://doi.org/10.1038/s41573-019-0053-0.

Du, Q.-S., Huang, R.-B., & Chou, K.-C. (2008). Recent advances in QSAR and their applications in predicting the activities of chemical molecules, peptides and proteins for drug design. *Current Protein & Peptide Science*, *9*(3), 248–259. Available from https://doi.org/10.2174/138920308784534005.

Du, Q. S., Ma, Y., Xie, N. Z., & Huang, R. B. (2014). Two-level QSAR network (2L-QSAR) for peptide inhibitor design based on amino acid properties and sequence positions. *SAR and QSAR in Environmental Research*, *25*(10), 837–851. Available from https://doi.org/10.1080/1062936X.2014.959049.

Du, Q.-S., Wei, Y.-T., Pang, Z.-W., Chou, K.-C., & Huang, R.-B. (2007). Predicting the affinity of epitope-peptides with class I MHC molecule HLA-A*0201: An application of amino acid-based peptide prediction. *Protein Engineering Design and Selection*, *20*(9), 417–423. Available from https://doi.org/10.1093/protein/gzm036.

Giguère, S., Laviolette, F., Marchand, M., Tremblay, D., Moineau, S., Liang, X., … Corbeil, J. (2015). Machine learning assisted design of highly active peptides for drug discovery. *PLOS Computational Biology*, *11*(4), e1004074. Available from https://doi.org/10.1371/journal.pcbi.1004074.

Goyal, R., Jerath, G., Akhil, R., Chandrasekharan, A., Puppala, E. R., Ponneganti, S., … Ramakrishnan, V. (2021). Geometry encoded functional programming of tumor homing peptides for targeted drug delivery. *Journal of Controlled Release*, *333*, 16–27. Available from https://doi.org/10.1016/j.jconrel.2021.03.010.

Han, C., & Wang, J. (2012). Influence of an unnatural amino acid side chain on the conformational dynamics of peptides. *ChemPhysChem*, *13*(6), 1522–1534. Available from https://doi.org/10.1002/cphc.201100995.

Hansen, M. R., Villar, H. O., & Feyfant, E. (2013). Development of an informatics platform for therapeutic protein and peptide analytics. *Journal of Chemical Information and Modeling*, *53*(10), 2774–2779. Available from https://doi.org/10.1021/ci400333x.

Haslach, E. M., Huang, H., Dirain, M., Debevec, G., Geer, P., Santos, R. G., … Haskell-Luevano, C. (2014). Identification of tetrapeptides from a mixture based positional scanning library that can restore nM full agonist function of the L106P, I69T, I102S, A219V, C271Y, and C271R human melanocortin-4 polymorphic receptors (hMC4Rs). *Journal of Medicinal Chemistry*, *57*(11), 4615–4628. Available from https://doi.org/10.1021/jm500064t.

Hazam, P. K., Akhil, R., Jerath, G., Saikia, J., & Ramakrishnan, V. (2019). Topological effects on the designability and bactericidal potency of antimicrobial peptides. *Biophysical Chemistry*, *248*, 1–8. Available from https://doi.org/10.1016/j.bpc.2019.02.005.

Hazam, P. K., Jerath, G., Kumar, A., Chaudhary, N., & Ramakrishnan, V. (2017). Effect of tacticity-derived topological constraints in bactericidal peptides. *Biochimica et Biophysica Acta (BBA)—Biomembranes*, *1859*(8), 1388–1395. Available from https://doi.org/10.1016/j.bbamem.2017.05.002.

Hazam, P. K., Singh, A., Chaudhary, N., & Ramakrishnan, V. (2019). Bactericidal potency and extended serum life of stereo-chemically engineered peptides against Mycobacterium. *International Journal of Peptide Research and Therapeutics*, *25*(2), 465–472. Available from https://doi.org/10.1007/s10989-018-9690-0.

Hochberg, M. C., Guermazi, A., Guehring, H., Aydemir, A., Wax, S., Fleuranceau-Morel, P., … Eckstein, F. (2019). Effect of intra-articular sprifermin vs placebo on femorotibial joint cartilage thickness in patients with osteoarthritis: The FORWARD randomized clinical trial. *JAMA*, *322*(14), 1360. Available from https://doi.org/10.1001/jama.2019.14735.

Howard, A. E., & Kollman, P. A. (1988). An analysis of current methodologies for conformational searching of complex molecules. *Journal of Medicinal Chemistry*, *31*(9), 1669–1675. Available from https://doi.org/10.1021/jm00117a001.

Improta, R., Vitagliano, L., & Esposito, L. (2015). Bond distances in polypeptide backbones depend on the local conformation. *Acta Crystallographica Section D Biological*

Crystallography, *71*(6), 1272–1283. Available from https://doi.org/10.1107/S1399004715005507.

Jerath, G., Goyal, R., Trivedi, V., Santhoshkumar, T. R., & Ramakrishnan, V. (2019). Syndiotactic peptides for targeted delivery. *Acta Biomaterialia*, *87*, 130–139. Available from https://doi.org/10.1016/j.actbio.2019.01.036.

Jerath, G., Goyal, R., Trivedi, V., Santhoshkumar, T. R., & Ramakrishnan, V. (2020). Conformationally constrained peptides for drug delivery. *Journal of Peptide Science*, *26* (4–5). Available from https://doi.org/10.1002/psc.3244.

Katyal, P., Meleties, M., & Montclare, J. K. (2019). Self-assembled protein- and peptide-based nanomaterials. *ACS Biomaterials Science & Engineering*, *5*(9), 4132–4147. Available from https://doi.org/10.1021/acsbiomaterials.9b00408.

Kellenberger, E., Rodrigo, J., Muller, P., & Rognan, D. (2004). Comparative evaluation of eight docking tools for docking and virtual screening accuracy. *Proteins: Structure, Function, and Bioinformatics*, *57*(2), 225–242. Available from https://doi.org/10.1002/prot.20149.

Klepeis, J. L, Lindorff-Larsen, K., Dror, R. O, & Shaw, D. E (2009). Long-timescale molecular dynamics simulations of protein structure and function. *Current Opinion in Structural Biology*, *19*(2), 120–127. Available from https://doi.org/10.1016/j.sbi.2009.03.004, PMID: 19361980.

Ko, J., & Lee, J. (2021). Can AlphaFold2 predict protein – peptide complex structures accurately? [Preprint]. *Bioinformatics*. Available from https://doi.org/10.1101/2021.07.27.453972.

Kuroda, D., Shirai, H., Jacobson, M. P., & Nakamura, H. (2012). Computer-aided antibody design. *Protein Engineering Design and Selection*, *25*(10), 507–522. Available from https://doi.org/10.1093/protein/gzs024.

Lamiable, A., Thévenet, P., Rey, J., Vavrusa, M., Derreumaux, P., & Tufféry, P. (2016). PEP-FOLD3: Faster *de novo* structure prediction for linear peptides in solution and in complex. *Nucleic Acids Research*, *44*(W1), W449–W454. Available from https://doi.org/10.1093/nar/gkw329.

Lhouvum, K., Ramakrishnan, V., & Trivedi, V. (2013). Insight into structural and biochemical determinants of substrate specificity of PFI1625c: Correlation analysis of protein-peptide molecular models. *Journal of Molecular Graphics and Modelling*, *43*, 21–30. Available from https://doi.org/10.1016/j.jmgm.2013.03.008.

Li, H., Lu, L., Chen, R., Quan, L., Xia, X., & Lü, Q. (2014). PaFlexPepDock: Parallel ab-initio docking of peptides onto their receptors with full flexibility based on Rosetta. *PLoS One*, *9* (5), e94769. Available from https://doi.org/10.1371/journal.pone.0094769.

Lian, W., Upadhyaya, P., Rhodes, C. A., Liu, Y., & Pei, D. (2013). Screening bicyclic peptide libraries for protein–protein interaction inhibitors: Discovery of a tumor necrosis factor-α antagonist. *Journal of the American Chemical Society*, *135*(32), 11990–11995. Available from https://doi.org/10.1021/ja405106u.

Lindorff-Larsen, K., Piana, S., Dror, R. O, & Shaw, D. E (2011). How fast-folding proteins fold. *Science*, *334*(6055), 517–520. Available from https://doi.org/10.1126/science.1208351, PMID: 22034434.

London, N., Movshovitz-Attias, D., & Schueler-Furman, O. (2010). The Structural basis of peptide-protein binding strategies. *Structure*, *18*(2), 188–199. Available from https://doi.org/10.1016/j.str.2009.11.012.

London, N., Raveh, B., Cohen, E., Fathi, G., & Schueler-Furman, O. (2011). Rosetta FlexPepDock web server—High resolution modeling of peptide–protein interactions. *Nucleic Acids Research*, *39*(Suppl. 2), W249–W253. Available from https://doi.org/10.1093/nar/gkr431.

Mehra, A., Jerath, G., Ramakrishnan, V., & Trivedi, V. (2015). Characterization of ICAM-1 biophore to design cytoadherence blocking peptides. *Journal of Molecular Graphics & Modelling*, *57*, 27–35. Available from https://doi.org/10.1016/j.jmgm.2015.01.004.

Nicolaou, C. A., & Brown, N. (2013). Multi-objective optimization methods in drug design. *Drug Discovery Today: Technologies*, *10*(3), e427–e435. Available from https://doi.org/10.1016/j.ddtec.2013.02.001.

Otvos, L. (Ed.), (2008). *Peptide-based drug design* (Vol. 494). Humana Press. Available from https://doi.org/10.1007/978-1-59745-419-3.

Pérez, A., Marchán, I., Svozil, D., Sponer, J., Cheatham, T. E., Laughton, C. A., & Orozco, M. (2007). Refinement of the AMBER force field for nucleic aAcids: Improving the description of α/γ conformers. *Biophysical Journal*, *92*(11), 3817–3829. Available from https://doi.org/10.1529/biophysj.106.097782.

Pike, D. H., & Nanda, V. (2015). Empirical estimation of local dielectric constants: Toward atomistic design of collagen mimetic peptides: Empirical estimation of local dielectric constants. *Biopolymers*, *104*(4), 360–370. Available from https://doi.org/10.1002/bip.22644.

Pollaro, L., & Heinis, C. (2010). Strategies to prolong the plasma residence time of peptide drugs. *MedChemComm*, *1*(5), 319–324. Available from https://doi.org/10.1039/C0MD00111B.

Prakash, V., Ranbhor, R., & Ramakrishnan, V. (2020). De novo designed heterochiral blue fluorescent protein. *ACS Omega*, *5*(41), 26382–26388. Available from https://doi.org/10.1021/acsomega.0c02574.

Rahman. (1964). Correlations in the Motion of Atoms in Liquid Argon. *Physical Review*.

Ramakrishnan, V., Ranbhor, R., & Durani, S. (2004). Existence of specific "folds" in polyproline II ensembles of an "unfolded" alanine peptide detected by molecular dynamics. *Journal of the American Chemical Society*, *126*(50), 16332–16333. Available from https://doi.org/10.1021/ja045787y.

Ramakrishnan, V., Ranbhor, R., & Durani, S. (2005). Simulated folding in polypeptides of diversified molecular tacticity: Implications for protein folding and de novo design. *Biopolymers*, *78*(2), 96–105. Available from https://doi.org/10.1002/bip.20241.

Ramakrishnan, V., Ranbhor, R., Kumar, A., & Durani, S. (2006). The link between sequence and conformation in protein structures appears to be stereochemically established. *The Journal of Physical Chemistry. B*, *110*(18), 9314–9323. Available from https://doi.org/10.1021/jp056417e.

Ranbhor, R., Kumar, A., Tendulkar, A., Patel, K., Ramakrishnan, V., & Durani, S. (2018). IDeAS: Automated design tool for hetero-chiral protein folds. *Physical Biology*, *15*(6), 066005. Available from https://doi.org/10.1088/1478-3975/aacdc3.

Raveh, B., London, N., Zimmerman, L., & Schueler-Furman, O. (2011). Rosetta FlexPepDock ab-initio: Simultaneous folding, docking and refinement of peptides onto their receptors. *PLoS ONE*, *6*(4), e18934. Available from https://doi.org/10.1371/journal.pone.0018934.

Rentzsch, R., & Renard, B. Y. (2015). Docking small peptides remains a great challenge: An assessment using AutoDock Vina. *Briefings in Bioinformatics*, *16*(6), 1045–1056. Available from https://doi.org/10.1093/bib/bbv008.

Rivera, C. G., Rosca, E. V., Pandey, N. B., Koskimaki, J. E., Bader, J. S., & Popel, A. S. (2011). Novel peptide-specific quantitative structure–activity relationship (QSAR) analysis applied to collagen IV peptides with antiangiogenic activity. *Journal of Medicinal Chemistry*, *54*(19), 6492–6500. Available from https://doi.org/10.1021/jm200114f.

Robertson, M. J., Tirado-Rives, J., & Jorgensen, W. L. (2015). Improved peptide and protein torsional energetics with the OPLS-AA force field. *Journal of Chemical Theory and Computation, 11*(7), 3499–3509. Available from https://doi.org/10.1021/acs.jctc.5b00356.

Roy, A., Kucukural, A., & Zhang, Y. (2010). I-TASSER: A unified platform for automated protein structure and function prediction. *Nature Protocols, 5*(4), 725–738. Available from https://doi.org/10.1038/nprot.2010.5.

Schmid, N., Eichenberger, A. P., Choutko, A., Riniker, S., Winger, M., Mark, A. E., & van Gunsteren, W. F. (2011). Definition and testing of the GROMOS force-field versions 54A7 and 54B7. *European Biophysics Journal, 40*(7), 843–856. Available from https://doi.org/10.1007/s00249-011-0700-9.

Schneidman-Duhovny, D., Inbar, Y., Nussinov, R., & Wolfson, H. J. (2005). PatchDock and SymmDock: Servers for rigid and symmetric docking. *Nucleic Acids Research, 33*(Suppl. 2), W363–W367. Available from https://doi.org/10.1093/nar/gki481.

Simons, K. T., Kooperberg, C., Huang, E., & Baker, D. (1997). Assembly of protein tertiary structures from fragments with similar local sequences using simulated annealing and bayesian scoring functions. *Journal of Molecular Biology, 268*(1), 209–225. Available from https://doi.org/10.1006/jmbi.1997.0959.

Soding, J., Biegert, A., & Lupas, A. N. (2005). The HHpred interactive server for protein homology detection and structure prediction. *Nucleic Acids Research, 33*, W244–W248. Available from https://doi.org/10.1093/nar/gki408. (Web Server).

Sohrabi, C., Foster, A., & Tavassoli, A. (2020). Methods for generating and screening libraries of genetically encoded cyclic peptides in drug discovery. *Nature Reviews Chemistry, 4*(2), 90–101. Available from https://doi.org/10.1038/s41570-019-0159-2.

Thévenet, P., Shen, Y., Maupetit, J., Guyon, F., Derreumaux, P., & Tufféry, P. (2012). PEP-FOLD: an updated de novo structure prediction server for both linear and disulfide bonded cyclic peptides. *Nucleic Acids Research, 40* (Web Server issue), W288–W293. Available from https://doi.org/10.1093/nar/gks419, PMID: 22581768.

Ulijn, R. V., & Smith, A. M. (2008). Designing peptide based nanomaterials. *Chemical Society Reviews, 37*(4), 664. Available from https://doi.org/10.1039/b609047h.

Unal, E. B., Gursoy, A., & Erman, B. (2010). VitAL: Viterbi algorithm for de novo peptide design. *PLoS One, 5*(6), e10926. Available from https://doi.org/10.1371/journal.pone.0010926.

Van Der Spoel, D., Lindahl, E., Hess, B., Groenhof, G., Mark, A. E, & Berendsen, H. J. C. (2005). GROMACS: fast, flexible, and free. *Journal of Computational Chemistry, 26*(16), 1701–1718. Available from https://doi.org/10.1002/jcc.20291, PMID: 16211538.

Vanhee, P., Reumers, J., Stricher, F., Baeten, L., Serrano, L., Schymkowitz, J., & Rousseau, F. (2010). PepX: A structural database of non-redundant protein–peptide complexes. *Nucleic Acids Research, 38*(Suppl. 1), D545–D551. Available from https://doi.org/10.1093/nar/gkp893.

Vanommeslaeghe, K., Hatcher, E., Acharya, C., Kundu, S., Zhong, S., Shim, J., … Mackerell, A. D., Jr. (2010). CHARMM general force field: A force field for drug-like molecules compatible with the CHARMM all-atom additive biological force fields. *Journal of Computational Chemistry, 31*(4), 671–690. Available from https://doi.org/10.1002/jcc.21367.

Verlet, L. (1967). Computer "Experiments" on classical fluids. I. thermodynamical properties of lennard-Jones molecules. *Physical Review*.

Wang, C., Garlick, S., & Zloh, M. (2021). Deep learning for novel antimicrobial peptide design. *Biomolecules, 11*(3), 471. Available from https://doi.org/10.3390/biom11030471.

Wang, L., Wang, N., Zhang, W., Cheng, X., Yan, Z., Shao, G., … Fu, C. (2022). Therapeutic peptides: Current applications and future directions. *Signal Transduction and Targeted Therapy*, *7*(1), 1–27. Available from https://doi.org/10.1038/s41392-022-00904-4.

Zompra, A. A., Galanis, A. S., Werbitzky, O., & Albericio, F. (2009). Manufacturing peptides as active pharmaceutical ingredients. *Future Medicinal Chemistry*, *1*(2), 361–377. Available from https://doi.org/10.4155/fmc.09.23.

Chapter 3

Solid phase peptide synthesis

Deepti Goyal[1] **and Bhupesh Goyal**[2]
[1]*Department of Chemistry, DAV College, Chandigarh, India,* [2]*School of Chemistry & Biochemistry, Thapar Institute of Engineering & Technology (Deemed to be University), Patiala, Punjab, India*

3.1 Introduction

The term "peptide synthesis" comprises various techniques and procedures that produce materials ranging from small peptides to large proteins. In 1963, R.B. Merrifield reported a novel strategy of solid phase peptide synthesis (SPPS) (Merrifield, 1963). The advent of SPPS has brought a revolution in the peptide synthesis. Merrifield was awarded Nobel Prize in Chemistry in 1984 for his work in the area of peptide synthesis.

The SPPS was based on the concept of successive steps of coupling (peptide bond formation) and deprotection of reactive functional groups in the amino acid residues to synthesize peptides. The peptide chain was assembled on a solid support in SPPS. The application of solid support facilitates easy work-up procedure that includes washing and filtration in a single reaction vessel. The SPPS has simplified the purification procedure associated with solution phase peptide synthesis. Furthermore, with the advancements in technology fully automated peptide synthesizers are now available. The peptide synthesizers can synthesize multiple peptide samples once the synthetic strategy and the peptide sequence has been defined. The SPPS has gained much popularity for the peptide synthesis, although solution phase can still be used for the synthesis of short peptides on industrial scale.

With the advent of solid-phase methodology, the large-scale synthesis of peptide-based "active pharmaceutical ingredients" become possible (Albericio & Kruger, 2012; Bruckdorfer, Marder, & Albericio, 2004; Zompra, Galanis, Werbitzky, & Albericio, 2009). With the help of SPPS, peptides of 30–50 amino acids can be synthesized on a multikilogram scale within few weeks, which, otherwise need up to ~100 synthetic steps. The ease of the synthetic methodology along with the development of methods to increase the stability of peptides lead to the commercialization of medium-sized peptides (25–50 residues), and number of peptides have been

De Novo Peptide Design. DOI: https://doi.org/10.1016/B978-0-323-99917-5.00007-X

advanced to the clinical trials (Fosgerau & Hoffmann, 2015; Ghosh, 2016; Kaspar & Reichert, 2013). The identical methodology has been implemented in the oligonucleotide field (Eritja, 2007; Wilk, Grajkowski, Srinivasachar, & Beaucage, 1999). In addition, solid-phase technology has led to the significant advances in the field of combinatorial sciences, which play a critical role in the drug design and discovery (Gil & Bräse, 2009; Nicolaou & Pfefferkorn, 2005).

3.2 Principles of solid phase peptide synthesis

In the SPPS, the *C*-terminal amino acid residue of the desired peptide is linked to an insoluble solid support by its carboxyl functionality. The target peptide is synthesized linearly from the *C*-terminal to the *N*-terminal ($C \rightarrow N$ strategy) by repetition of successive deprotection/coupling steps after incorporating the first amino acid. The various functional groups in amino acid side-chains must be protected with permanent protecting groups (P_n), which are not affected by the reaction conditions employed during peptide chain assembly. The α-amino group is protected by a temporary protecting group (T, usually a urethane derivative) and mild conditions are used to cleave the temporary protecting group. The mild conditions preserve the peptide integrity and decrease epimerization rate that can proceed through the formation of 5(4H)-oxazolone of the activated amino acid during the coupling process (Fig. 3.1) (Bergmann & Zervas, 1928; Goodman & Levine, 1964). The protective role of urethanes against epimerization highlights the prevalence of the $C \rightarrow N$ strategy.

The carboxy group of the second amino acid (added in large amount) is activated in the form of activated ester by carrying out the reaction in the presence of a coupling agent. After the completion of the coupling reaction, washing with solvents is done to remove the excess reagents. Before the addition of third amino acid, the protecting group is removed from the *N*-terminal of the dipeptide. The deprotection of the *N*-terminal protecting group and the coupling reaction are repeated till the formation of the target peptide sequence. The side-chain protecting groups are removed concomitantly during the final step of cleavage of desired peptide from the resin. The

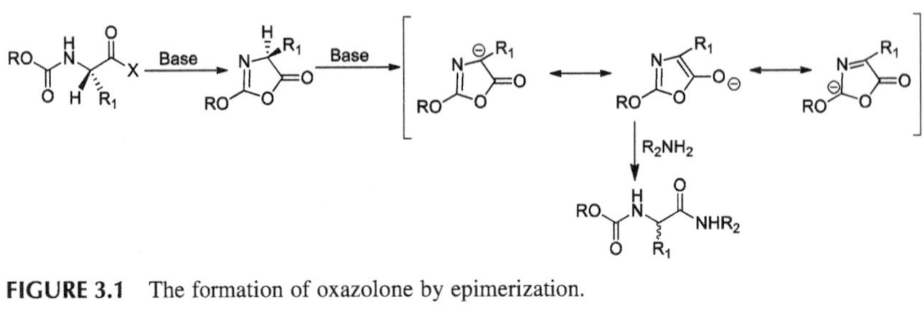

FIGURE 3.1 The formation of oxazolone by epimerization.

protecting groups of the reactive functionalities of amino acid residues and resin linkage are selected on the basis that both the removal of the protecting groups as well as cleavage of the assembled peptide can take place in the identical conditions. In SPPS, two main strategies are based on the concourse of tert-butyloxycarbonyl (Boc) and fluorenylmethoxycarbonyl (Fmoc) as temporary protecting groups.

3.2.1 Merrifield solid phase peptide synthesis or Boc/Bzl

During the 1960s and 1970s, the Boc strategy was developed. In Boc strategy, both Boc and permanent side-chain protecting groups during the chain elongation step are removed by acidolysis. The Merrifield technique (Barany, Kneib-Cordnier, & Mullen, 1987; Kent, 1988; Stewart & Young, 1984) for the peptide synthesis is shown in Fig. 3.2. The *C*-terminal amino acid is affixed to the support through formation of a benzyl ester with hydroxymethylphenylacetamidomethyl polystyrene (PAM resin). The temporary protection of the α-amino group was done by Boc group and it can be easily removed with neat trifluoroacetic (TFA) or TFA in dichloromethane (DCM). The trifluoroacetate is neutralized with diisopropylethylamine (DIPEA) in DCM before the coupling step, or neutralized in situ during the coupling step.

The use of preformed amino acid symmetrical anhydrides or benzotriazolyl esters in *N*,*N*-dimethylformamide (DMF) or *N*-methylpyrrolidone (NMP) is favored currently as compared to the activation of the incoming amino acid with dicyclohexylcarbodiimide in DCM earlier. Different benzyl-based protecting groups have been developed to protect the trifunctional amino acids side-chains. The side-chain protecting groups are chemically tweaked for particular functional groups by substitution of the benzyl ring with electron donating or withdrawing groups. The anhydrous hydrogen fluoride at

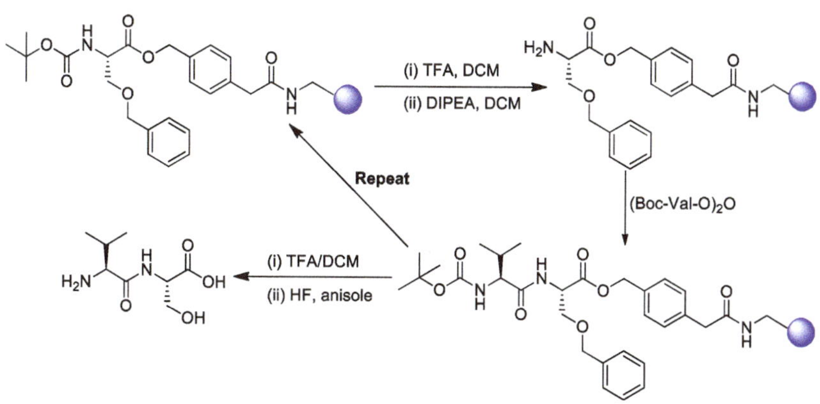

FIGURE 3.2 The various steps involved in the Merrifield SPPS are shown.

low temperature (−20°C) in the presence of scavengers (Tam, Heath, & Merrifield, 1983) is used to cleave the peptide from the resin as well as for removing side-chain protecting groups.

The Merrifield method has become a powerful tool due to the improved quality of base-resin material, Boc-protected amino acids, as well as better hydrofluoric acid (HF) cleavage protocols, which, in turn, led to the efficient synthesis of a number of large peptides and small proteins (Kent, 1988). However, Merrifield technique has the following limitations that limit its utility: (1) the use of highly toxic HF for the removal of the peptide − resin linkages; (2) the requirement for special polytetrafluoroethylene-lined apparatus; (3) significant changes can take place in the structural integrity of peptides possessing fragile sequences in the strongly acidic conditions.

3.2.2 Fmoc/tBu solid phase peptide synthesis

The Fmoc/*t*Bu method (Atherton, Fox, Harkiss, & Sheppard, 1978) is established on an orthogonal protecting group strategy (Carpino & Han, 1972) as compared to Merrifield SPPS (Fig. 3.2), which employs acidolysis to achieve selectivity in the elimination of temporary and permanent protecting groups. In the 1980s, with the advent of the Fmoc group, both Boc and Fmoc strategies were commonly used for SPPS. The α-amino functionality has been protected by base-labile *N*-Fmoc group. The acid-labile side-chain protecting groups and acid-labile linkers are used in the orthogonal protecting group strategy. The above strategy has the advantage that the removal of temporary and permanent protection is effected by different mechanisms, facilitating the use of milder acidic conditions for the final deprotection and cleaving the peptide − resin linkage as compared to the conditions used in the Merrifield method. The Fmoc/*t*Bu method for the peptide synthesis is depicted in Fig. 3.3. The *C*-terminal residue is attached to a TFA-labile linkage agent and the side-chain functional groups are protected with TFA-labile protecting groups. A solution of 20% piperidine in DMF is used for

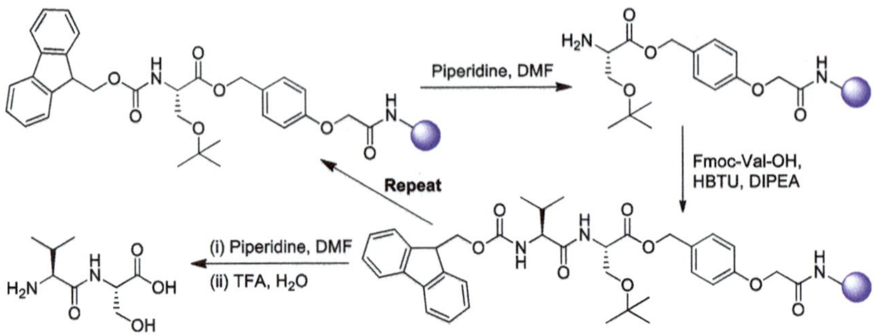

FIGURE 3.3 The steps that lead to the formation of a dipeptide using Fmoc SPPS are shown.

removing the temporary N^{α}-Fmoc protecting group. The coupling reaction is performed in DMF or NMP with preformed active esters or using activation reagents, which forms in situ benzotriazolyl esters. The side-chain deprotection and the cleavage of the peptide from resin were done using 95% TFA. The development of the Fmoc approach to SPPS has been listed in a number of excellent reviews (Fields & Noble, 1990; Grant, 1992).

Usually, *t*-butyl and trityl-based side-chain protection and alkoxybenzyl-based linkers are utilized due to their ease of removal with TFA. TFA is a good solvent for peptides as it can be used in standard laboratory glassware and is easily evaporated due to its volatile nature. The ease of the cleavage reaction and the simplicity with which the method can be used to multiple peptide synthesis makes Fmoc/*t*Bu method an attractive approach for the peptide synthesis. The Fmoc SPPS does not involve the repetitive use of TFA in each cycle and also avoids the use of anhydrous HF. The above mentioned advantages associated with Fmoc SPPS accounts for the popularity of this method for the routine synthesis of peptides (Behrendt, White, & Offer, 2016). However, a significant drawback of this method is the formation of incomplete sequences, which affects the purity of the final desired peptide synthesized. In some cases, the separation of the desired peptide from the incomplete sequences is quite cumbersome due to very small difference in their polarity (Paradis-Bas, Tulla-Puche, & Albericio, 2016). The above conditions arise due to the removal of temporary protecting group at any stage during peptide synthesis because of steric effects or peptide aggregation on the resin. The steric effects are generally observed in α,α-disubstituted amino acids, *N*-methylated amino acids, and bulky side-chain protecting groups. The steric effects restrict the accessibility of the reagents into the reaction site on the peptidyl resin (Bedford et al., 1992). The hydrophobic peptides and long sequences generally encounter peptide aggregation during the peptide synthesis (Paradis-Bas et al., 2016).

3.3 Resins

The first step of SPPS is the loading of the N-protected *C*-terminal amino acid to the solid support *via* an ester or an amide bond, which, in turn, depend on the *C*-terminal functional group of desired peptide. The use of a solid support during peptide synthesis has its own advantages as a large excess of reagents at high concentration can drive the peptide coupling reactions to completion. The extra reagents and side products are easily removed from the growing peptide chain by filtration and washings. Moreover, all steps involved in the peptide synthesis can be performed in the same reaction vessel.

Various commercially available linkers are anchored on the different solid supports [polystyrene (PS), PA, polyethylene glycol (PEG)-PS]. Although SPPS can be carried out on various solid supports, however, currently microporous gel-type solid supports are used (Garcia-Martin & Albericio, 2008; Jensen, Tofteng Shelton, & Pedersen, 2013). In the SPOT technique, cellulose membranes have been efficiently used for the parallel synthesis of a large number of peptides (Frank, 2007). The syntheses are performed concurrently on different areas of the cellulose membrane and can be automatized.

The cross-linked PS beads have been commonly used for peptide synthesis after the initial work of Merrifield. The PEG being less hydrophobic has also been utilized as solid support for SPPS. It can be used alone or grafted to PS or polyamides. PS is physically stable and does not solubilize in commonly used solvents. These have high solvation power, and PS swells in DCM, which was traditionally used during initial years of SPPS development. However, DMF and NMP are used nowadays. However, PS does not swell in hexane, methanol, or water.

PEG–PS resins, namely PEG–PS and TentaGel, were introduced independently by Barany et al. (1992) and Rapp, Zhang, Haebich, and Bayer (1989). The carboxylic acid containing PEG is integrated into aminomethyl-PS (AM-PS) through an amide bond in PEG–PS, whereas, polymerization of ethylene oxide onto a styrene and divynylbenzene copolymer is performed in TentaGel. TentaGel is well-solvated in both polar and nonpolar solvents due to high conformation flexibility of PEG chains. Various studies highlighted the superiority of PEG over PS-based solid support (Kates, Solé, Beyermann, Barany, & Albericio, 1996). In addition, both PEG–PS and TentaGel are stable to moderate pressure and can be employed in columns for continuous-flow peptide synthesis. The solvation in aqueous media has permitted the use of these supports for the synthesis of libraries of mixture of peptides and further screening of peptides in solid phase when these peptides were anchored to the solid support (Buchardt et al., 2000). The PEG–PS solid supports have lower loading capacity and are more expensive as compared to PS.

The resins preloaded with the first *C*-terminal N-protected amino acid are preferred for SPPS because the anchoring of an amino acid to the solid support by esterification is difficult. Moreover, it can result in the epimerization, dipeptide formation, low substitution, and can be hazardous for some residues. The various resins employed in Fmoc/*t*Bu strategy for the synthesis of *C*-terminal peptide acid are shown in Fig. 3.4 (Barlos et al., 1989; Flörsheimer & Riniker, 1991; Mergler, Nyfeler, Tanner, Gosteli, & Grogg, 1988; Wang, 1973). It should be noted that the anchoring reactions must be carried out in an anhydrous condition.

For hydroxymethyl-based resins, formation of the ester linkage is easier with unhindered resins such as Wang resin as compared to Sasrin and 4-(4-Hydroxymethyl-3-methoxyphenoxy)butyric acid resins possessing methoxy groups. The symmetrical anhydride method is employed for the esterification process. The Fmoc release measurement can be used to check the loading.

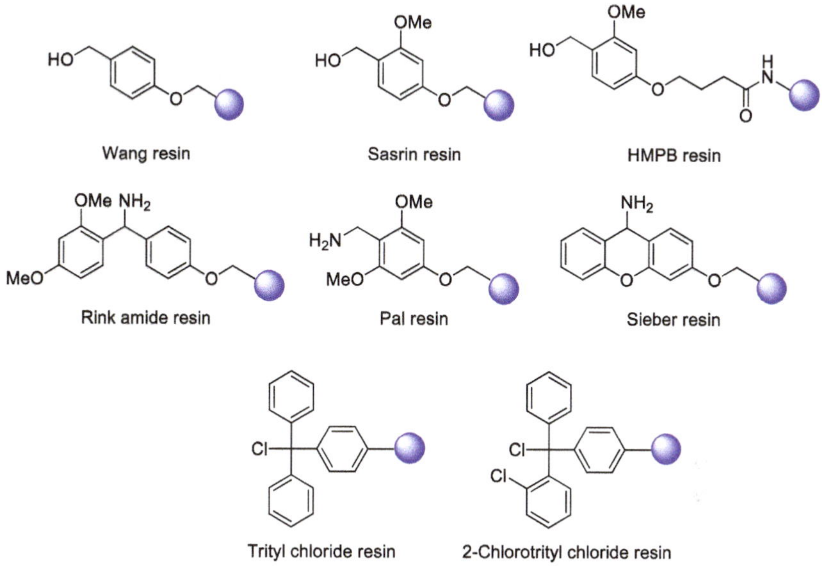

FIGURE 3.4 The various resins that are employed for Fmoc SPPS.

The esterification needs to be performed with fresh reactants in case of difficult anchoring. Arginine derivatives require three esterification steps to yield correct loading. The unreacted resin-bound hydroxyl groups should be capped by benzoic anhydride or acetic anhydride after anchoring.

For the synthesis of *C*-terminal peptide amides, the commonly used resins are Rink Amide (Rink, 1987), Pal (Bernatowicz, Daniels, & Köster, 1989), and Sieber (Sieber, 1987) resins (Fig. 3.4). These resins are well-suited with Fmoc chemistry as well as TFA cleavage. The standard peptide coupling methods can be employed for the attachment of the first urethane N-protected residue. These resins are Fmoc-protected and need to be deprotected before adding first amino acid residue. A double coupling reaction needs to be performed with bulky *C*-terminal amino acids.

Trityl-based resins are highly acid-labile and can be employed for Pro, Gly *C*-terminal peptides (Fig. 3.4). The bulky groups present in the linker hinder the formation of diketopiperazine (DKP). The cleavage of protected peptide segments from the resin can be done in extremely mild acidolysis conditions. The chloride or alcohol precursors of the trityl-based resins are available. The glassware and reagents need to be dried in order to prevent the hydrolysis into the alcohol form, as trityl chloride resin is extremely moisture-sensitive. The trityl alcohol precursor should be activated and it is highly recommended to reactivate the chloride just before use. After activation, the first residue is attached by reaction with the Fmoc amino acid derivative in the presence of a base. The reaction is free from epimerization as activated species are not involved in the reaction.

3.4 Linkers

The linker links the growing peptide chain with the solid support and protects the *C*-terminal α-carboxyl group during the peptide synthesis. The final peptide is released as an acid or amide upon treatment with TFA, which depends upon the choice of linker (Fig. 3.5). Further, the concentration of TFA used decides whether the fully deprotected peptide is released or the acid sensitive side-chain protecting groups are still retained. The *C*-terminal modified peptides such as esters and secondary amides are achieved when peptide – resin linkage is cleaved by nucleophiles.

3.5 Side-chain protecting groups

The protecting groups play a key role during SPPS as a large number of amino acids generally used in peptide synthesis have reactive functional groups in their side-chains (Lloyd-Williams, Giralt, & Albericio, 1997). These reactive functional groups can be affected by the rather harsh conditions employed during SPPS. Hence, to achieve high purity and efficiency during the peptide synthesis, it is necessary to mask these reactive functional groups by suitable protecting groups.

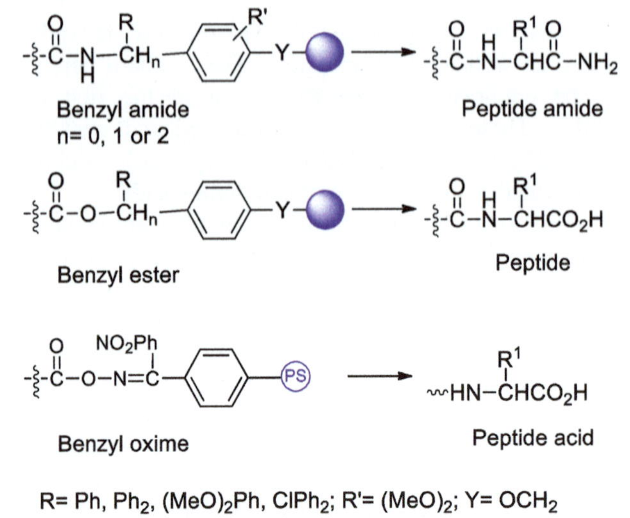

FIGURE 3.5 The final peptide formed depends upon the type of linker. The formation of peptide amide from benzyl amide and peptide acid from benzyl ester or oxime is shown.

Most commonly, the TFA labile protecting groups are used, which facilitate the formation of completely deprotected peptide as it is cleaved from the support. However, the protecting groups are available that can be removed selectively, which allows the side-chain modification of each amino acid residue within the peptide. The above strategy is useful for the synthesis of side-chain cyclized peptides, biotinylated peptides, and phosphor peptides (Albericio, 2000). The commonly employed side-chain protecting groups during the peptide synthesis are listed in Table 3.1. In addition, the cleavage conditions of various protecting groups are also listed in Table 3.1.

3.6 Coupling reaction

The SPPS is the simplest method for the synthesis of peptides by stepwise coupling of *N*-protected amino acids. The SPPS involves the activation of carboxylic group *in situ*. Generally, the activated amino acid is taken in excess (2–10 times) as compared to the resin. This facilitates the effective diffusion of reactants and ensures good product yield. The peptide sequence that is already attached to the resin, nature of the activated species, and concentration of reagents determine the total time needed for a complete acylation reaction.

The addition of additives such as HOBt (König & Geiger, 1970), 60 HOAt (Carpino, 1993), Oxyma-Pure (Subirós-Funosas, Prohens, Barbas, El-Faham, & Albericio, 2009), and Oxyma-B (Jad et al., 2014) results in the formation of active esters. The esters formed in the presence of additives are relatively less reactive than the *O*-acyl isourea (active species formed in the presence of carbodiimides). The lower reactivity of esters helps to reduce the racemization and hence enhances coupling efficiency, prevents the formation of inactive *N*-acyl urea, and prevents other side reactions (El-Faham & Albericio, 2011). The benzotriazole-based additives are labeled as class 1 explosive materials due to their thermal instability (Wehrstedt, Wandrey, & Heitkamp, 2005). A great care needs to be taken during their transportation and storage, and thus, restrict their applications. The OxymaPure [Ethyl 2-cyano-2-(hydroxyimino)acetate] has been reported to be highly efficient and suppresses racemization remarkably in the carbodiimide approach (Subirós-Funosas et al., 2009). Besides these, iminium/uronium salts have been reported as another class of coupling agents. The iminium/uronium salts were obtained by reaction of HOXt with the chloroformamidinium salt (Dourtoglou, Gross, Lambropoulou, & Zioudrou, 1984; Dourtoglou, Ziegler, & Gross, 1978).

The favored coupling reagents are benzotriazol-1-yl-oxytripyrrolidino phosphonium hexafluorophosphate (PyBOP) (Coste, Le Nguyen, & Castro, 1990) for phosphonium-based activation and *O*-(benzotriazol-1-yl)-1,1,3,3-tetramethyluronium tetrafluoroborate (TBTU) (Knorr, Trzeciak, Bannwarth, & Gillessen, 1989) or *N*-[(1H-benzotriazol-1-yl) (dimeth-ylamino)

TABLE 3.1 The side-chain protecting groups that are currently used in the Fmoc-based SPPS.

Amino acid and side-chain functionality	Protecting groups	Cleavage conditions	Remarks and side reactions
Arg	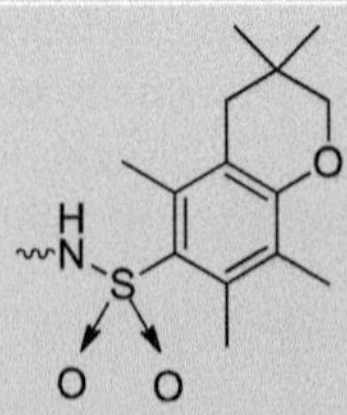	50% v/v TFA (Ramage & Green, 1987) TFA-anisole-EDT-EMS (95:3:1:1) (Ramage, Green, & Blake, 1991)	The cleavage is accelerated in the presence of thiols.
	Pmc		
	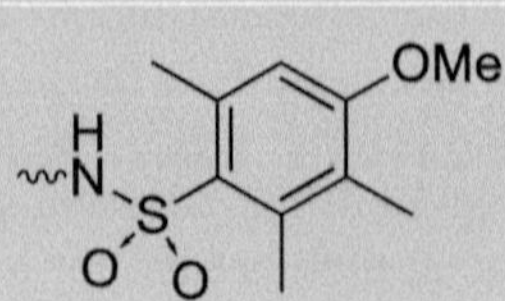	90%–95% v/v TFA (Atherton, Sheppard, & Wade, 1983), TFA-anisole (9:1) (Fujino, Wakimsu, & Kitadu, 1981)	
	Mtr		
	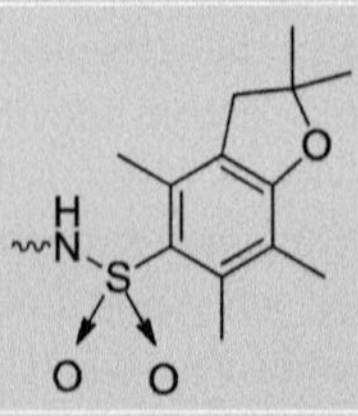	95% v/v TFA (Carpino et al., 1993)	
	Pbf		
Asp/Glu	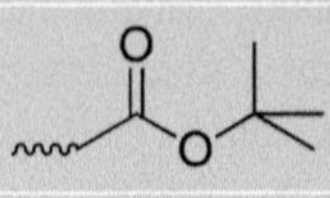	90% v/v TFA (Schwyzer & Dietrich, 1961)	The formation of Aspartimide may take place.
	O^{t}Bu		
	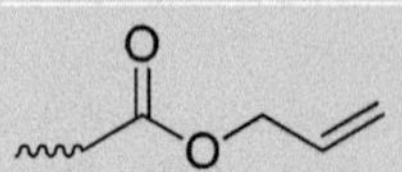	$Pd(Ph_3P)_4$/ $PhSiH_3$ (Dessolin, Guillerez, Thieriet, Guibe, & Loffet, 1995; Kates et al., 1993; Trzeciak & Bannwarth, 1992)	

(*Continued*)

TABLE 3.1 (Continued)

Amino acid and side-chain functionality	Protecting groups	Cleavage conditions	Remarks and side reactions
	OAll		
Asn/Gln	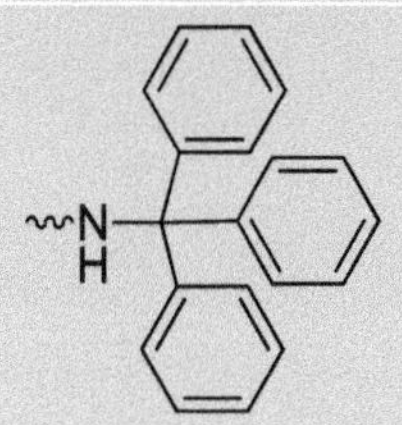	90% v/v TFA (Sieber & Riniker, 1991)	Protections avoid dehydration of the carboxamide side-chain during activation and help to solubilize Fmoc-Asn-OH and Fmoc-Gln-OH. pyrGlu formation for *N*-terminal glutamine peptides.
	Trt		
Cys	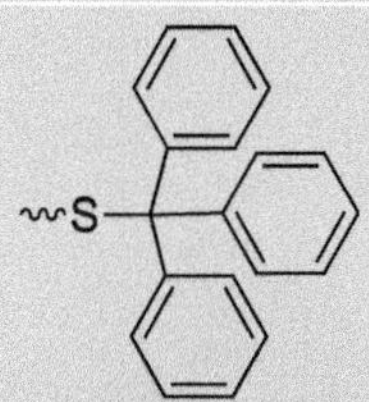	90% v/v TFA (Sieber & Riniker, 1991)	During activation, the epimerization can occur to a greater extent. Possibility of peptide folding due to the formation of disulfide linkage.
	Trt		
His	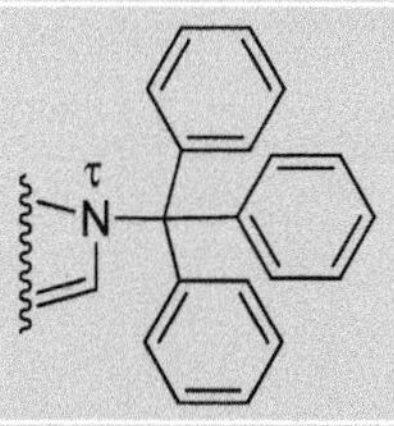	50% TFA in DCM (Sieber & Riniker, 1987)	Epimerization is observed during activation.
	Trt (NHτ)		

(*Continued*)

TABLE 3.1 (Continued)

Amino acid and side-chain functionality	Protecting groups	Cleavage conditions	Remarks and side reactions
	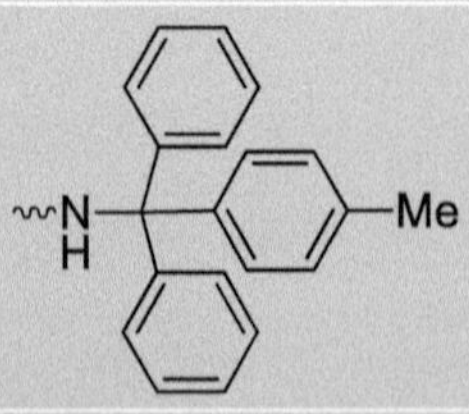	1% TFA in DCM (Aletras, Barlos, Gatos, Koutsogianni, & Mamos, 1995)	
	Mtt		
Lys	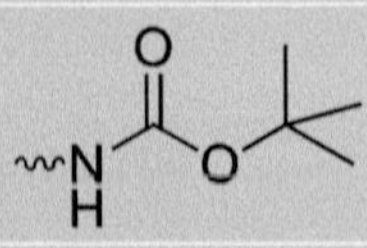	90% v/v TFA (Chang et al., 1980)	
	Boc		
	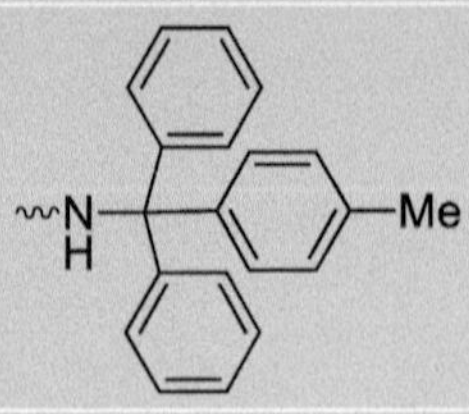	1% TFA in DCM (Aletras et al., 1995)	
	Mtt		
	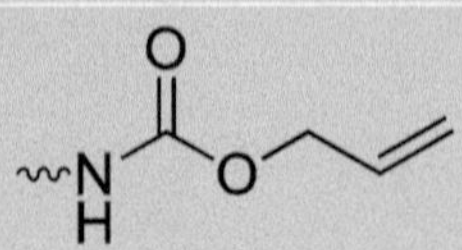	$Pd(Ph_3P)_4$/ $PhSiH_3$ (Thieriet, Alsina, Giralt, Guibe, & Albericio, 1997; Trzeciak & Bannwarth, 1992)	
	Alloc		
Ser/Thr/Tyr	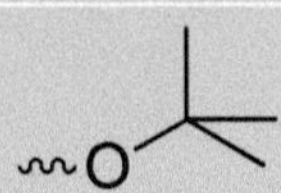	90% v/v TFA (Beyerman & Bontekoe, 1961; Callahan, Anderson, Paul, & Zimmermann, 1963)	

(*Continued*)

TABLE 3.1 (Continued)

Amino acid and side-chain functionality	Protecting groups	Cleavage conditions	Remarks and side reactions
	*t*Bu		
	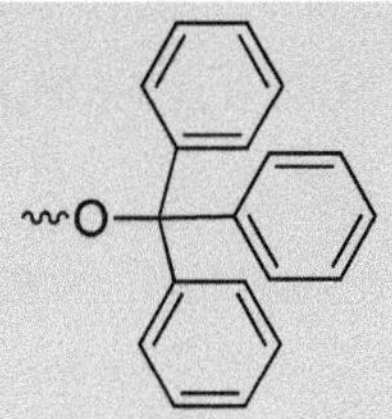	1%–5% TFA in DCM (Barlos et al., 1991; Barlos, Gatos, & Koutsogianni, 1998)	
	Trt		
Trp	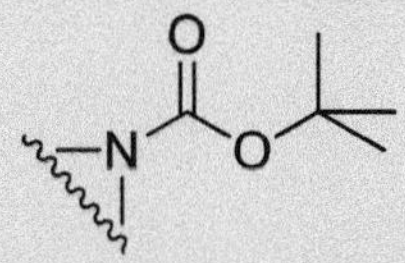	90% v/v TFA (White, 1992)	Although Trp can be used unprotected, however, side reactions may take place in case Arg(Pmc) or Arg(Pbf) is in the peptide sequence.
	Boc		

methylene]-*N*-methyl-methanaminium hexafluorophosphate N-oxide (HBTU) (Dourtouglou, Ziegler, & Gross, 1978) for aminium/uronium-based activation (Fig. 3.6). These coupling reagents are used to make the OBt ester of the corresponding N-protected amino acids. Generally, the carboxylate (reacts with coupling reagents) of N-protected amino acids was synthesized by reacting with DIPEA.

In recent years, *N*-[1H-benzotriazol-1-yl(dimethylamino)methylene]-*N*-methylmethanaminium hexafluorophosphate (HATU) (Carpino, 1993) and 7-azabenzotriazol-1-yl-oxytris(pyrrolidino)phosphonium hexafluorophosphate (PyAOP) (Carpino, El-Faham, Minor, & Albericio, 1994) have been reported to be more efficient. These coupling agents generate OAt esters and their use reduces epimerization.

The uronium/aminium-based reagents (HBTU, HATU, TBTU) are believed to show similar mechanistical function as their phosphonium

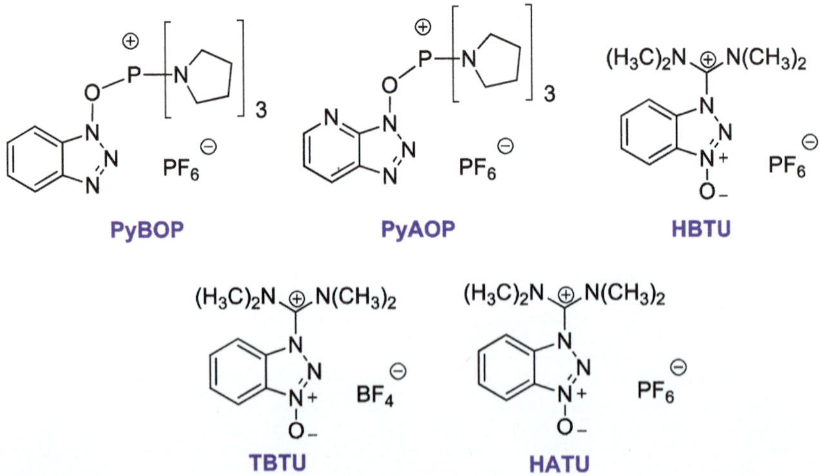

FIGURE 3.6 The phosphonium and uronium/aminium coupling reagents employed in the SPPS are shown.

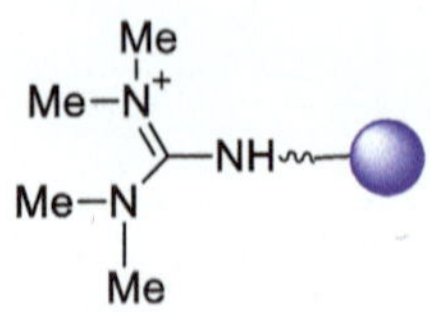

FIGURE 3.7 *N*-terminal tetramethylguanidinated peptides.

analogs. However, they can irreversibly block the free *N*-terminal amino function of the peptide resin by forming tetramethylguanidinium derivatives (Fig. 3.7) (Story & Aldrich, 1994). In order to prevent the formation of tetramethylguanidinium derivatives, it is suggested to generate the carboxylate of the amino acid before adding the coupling reagents at a slightly less equivalent compared to the amino acid.

A double coupling step with fresh reagents needs to be performed in case a colorimetric test indicates the presence of free amino groups that highlight incomplete coupling. A capping procedure by employing acetic anhydride can be done to prevent the elongation of these less-reactive amino groups, when acylation is not complete even after second coupling. The aggregation of peptide might be the reason if the colorimetric test still indicates the presence of free amino groups after capping.

3.7 Fmoc deprotection

A solution of 20% piperidine in DMF is used to remove Fmoc group from the *N*-terminal of the peptidyl resin. The mechanism of the Fmoc

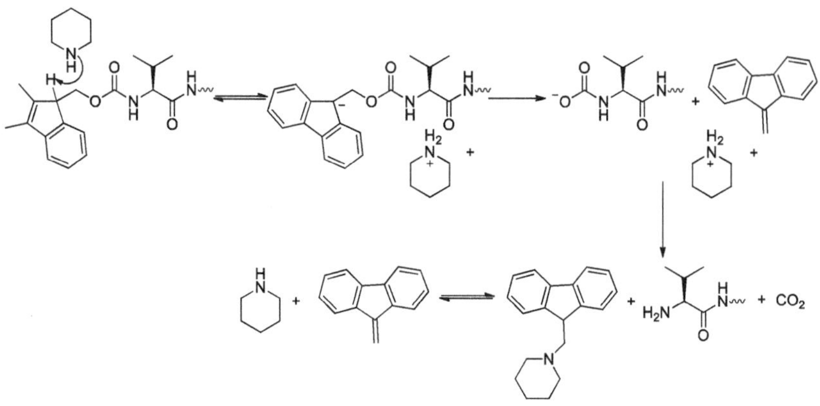

FIGURE 3.8 *N*-Fmoc removal reaction.

deprotection reaction is shown in Fig. 3.8. The reaction is generally complete within 10 minutes and may take more time in some cases. A deprotection time of 20 minutes is normally recommended for the safe removal of Fmoc group. The generation of cyclopentadienetype intermediate by the initial deprotonation of fluorene ring is the key step. The cyclopentadienetype intermediate rapidly eliminates to yield dibenzofulvene that is scavenged by piperidine. The dibenzofulvene-piperidine adduct shows strong absorbtion in the UV range, which facilitates the monitoring of the Fmoc group by UV spectroscopy in automatic synthesizers.

The partial deprotection of Fmoc group is carried out by treating with a mixture of 20% piperidine and 1%–5% 1,8-diazabicyclo[5.4.0]undec-7-ene (DBU) in DMF. Although in most of the cases the deprotection is effective with piperidine, complete Fmoc deprotection does not take place even under high concentrations of piperidine for larger peptides. In these cases, stronger tertiary base such as DBU can be employed (Wade, Bedford, Sheppard, & Tregear, 1991). DBU causes fast deprotection, less enantiomerization of resin-bound *C*-terminal Cys(Trt), and decreases the extent of broadening of UV Fmoc-deprotection peaks. In batch synthesis, addition of 2% piperidine to the deprotection mixture containing DBU is recommended in order to scavenge the dibenzofulvene formed during Fmoc removal and prevention of the alkylation of resin amino groups (Kates et al., 1996). DBU can lead to the formation of aspartimide, thus it is not suitable for Asp or Asn-containing sequences.

3.8 Final cleavage

The concentrated TFA is commonly employed for cleaving the peptide – resin linkage and the removal of the side-chain protecting groups. Under these conditions, the side-chain protecting groups yield stabilized carbocations that react rapidly with the electron-rich side-chain of amino acids (Cys, Met, Tyr, Thr, Ser, Trp), which, in turn, yield by-products. In order to trap the carbocations, scavengers are added to the cleavage mixture. A number of cleavage cocktails has been described to standardize the cleavage conditions of particular sequences. As compared to thiol-based cocktails, silane-based cocktails yield good results as they are nonodorous and less toxic (Pearson, Blanchette, Baker, & Guindon, 1989). Special attention should be given for the preparation and usage of the cleavage cocktails. All the steps need to be performed in the fume hood using gloves and safety glasses.

3.9 Side reactions

During SPPS, side reactions are normally encountered. Most of them are well-known in the literature and can be easily prevented by using appropriate protecting groups and linker. The possible side reactions encountered during peptide synthesis are discussed below.

3.9.1 Diketopiperazine formation

The intramolecular cleavage of the resin-bound ester linkage by free amino group of the preceding amino acid residue under basic conditions at the *C*-terminal deprotected dipeptide stage leads to the formation of DKP (Fig. 3.9). The formation of DKP is commonly encountered in Fmoc SPPS strategy as the deprotection of the Fmoc group by piperidine forms a free amino group. The nature of the *C*-terminal amino acid (particularly proline or glycine) and the type of peptide linker ester anchor determine the amount of DKP formation. To block the formation of DKP, hindered trityl-based resins need to be used.

3.9.2 Aspartimide formation

The common side reaction that is encountered in routine SPPS is the cyclization of aspartic acid residues to form aspartimide (Fig. 3.10). The aspartimide formation involves attack of the nitrogen attached to the α-carboxy group of aspartic acid or asparagine residues on the side-chain ester or amide group, respectively, which, in turn, leads to the formation of an imide ring.

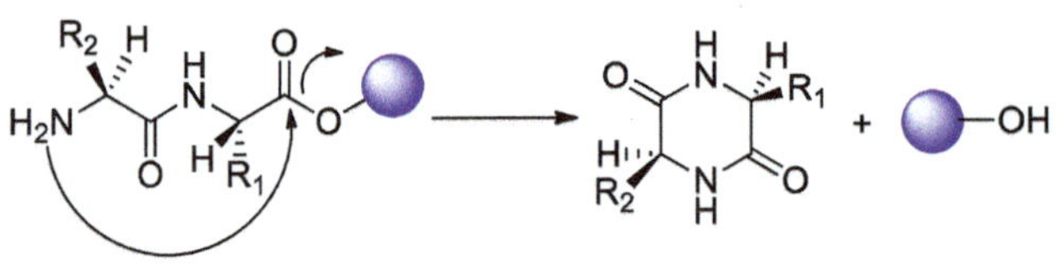

FIGURE 3.9 DKP formation.

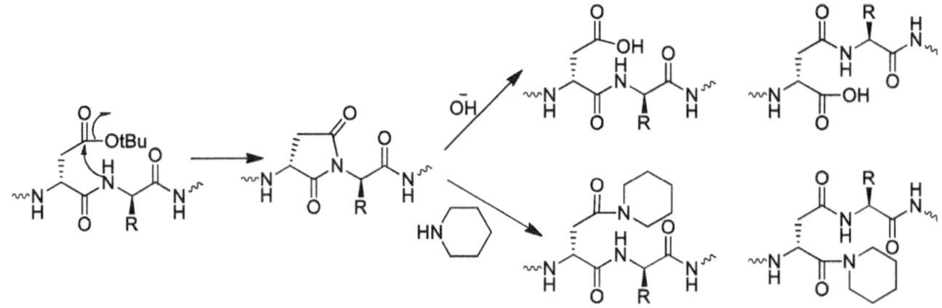

FIGURE 3.10 The formation of aspartimide and succinimide ring reopening in basic medium is illustrated.

The formation of aspartimide is a sequence-dependent side reaction, which takes place either during the peptide chain elongation or during final TFA cleavage in the peptides containing Asp(O*t*Bu)-AA motif [AA = Gly, Ala, Asn(Trt), Ser] (Lauer, Fields, & Fields, 1995). A mixture of both α and β-peptides are formed by the hydrolysis of the aspartimide. Also it leads to the formation of α and β-piperidides on reaction with piperidine employed in the removal of Fmoc. Although Asp (O*t*Bu) provides sufficient protection in Fmoc-based SPPS, sequences containing Asp(O*t*Bu)-Asn(Trt) are highly prone to the formation of aspartimide. The aspartimide formation can be prevented by adding HOBt to the piperidine solution or masking of the aspartyl amide bond with 2-hydroxy-4-methoxybenzyl (Hmb) (Quibell, Owen, Packman, & Johnson, 1994).

As discussed in the chapter, the continuous development of synthetic methods has enabled the preparation of a wide range of peptides having vast applications in the biomedical field. Peptide chemists need to focus on the more sustainable methods, with a key attention on the use of green solvents, and to improve atom economy in protection/deprotection processes.

References

Albericio, F., & Kruger, H. G. (2012). Therapeutic peptides. *Future Medicinal Chemistry*, *4*, 1527–1531.

Albericio, F. (2000). Orthogonal protecting groups for N^{α}-amino and *C*-terminal carboxyl functions in solid-phase peptide synthesis. *Biopolymers Peptide Science*, *55*, 123–139.

Aletras, A., Barlos, K., Gatos, D., Koutsogianni, S., & Mamos, P. (1995). Preparation of the very acid-sensitive Fmoc-Lys(Mtt) –OH application in the synthesis of side-chain to side-chain cyclic peptides and oligolysine cores suitable for the solid-phase assembly of MAPs and TASPs. *International Journal of Peptide and Protein Research*, *45*, 488–496.

Atherton, E., Fox, H., Harkiss, D., & Sheppard, R. C. (1978). A mild procedure for solid phase peptide synthesis: use of fluorenylmethoxycarbonylamino-acids. *Journal of the Chemical Society. Chemical Communications*, 537–539.

Atherton, E., Sheppard, R. C., & Wade, J. D. (1983). Side chain protected Nα-fluorenylmethoxycarbonylamino-acids for solid phase peptide synthesis. Nα-fluorenylmethoxycarbonyl-NG-

4-methoxy-2,3,6-trimethylbenzenesulphonyl-L-arginine. *Journal of the Chemical Society, Chemical Communications*, 1060–1062.

Barany, G., Albericio, F., Biancalana, S., Bontems, S. L., Chang, J. L., Eritja, R., . . . Hudson, D. (1992). Biopolymer syntheses on novel polyethylene glycol–polystyrene (PeG–PS) graft supports. In J. A. Smith & J. Rivier (Eds.), *Peptides–chemistry and biology: Proceedings of the twelfth American peptide symposium* (pp. 603 – 604). Minneapolis, MN.

Barany, G., Kneib-Cordnier, N., & Mullen, D. G. (1987). Solid-phase peptide synthesis: A silver anniversary report. *International Journal of Peptide and Protein Research*, *30*, 705–739.

Barlos, K., Gatos, D., Kallitsis, J., Papaphotiu, G., Sotiriu, P., Wenqing, Y., & Schäfer, W. (1989). Darstellung geschützter peptid-fragmente unter einsatz substituierter triphenylmethyl-harze. *Tetrahedron Letters*, *30*, 3943–3946.

Barlos, K., Gatos, D., & Koutsogianni, S. (1998). Fmoc/Trt-amino acids: Comparison to Fmoc/tBu-amino acids in peptide synthesis. *Journal of Peptide Research*, *51*, 194–200.

Barlos, K., Gatos, D., Koutsogianni, S., Schafer, W., Stavropoulos, G., & Yao, W. Q. (1991). Darstellung und einsatz von N–Fmoc–O–Trt–hydroxyaminosäuren zur "solid phase" synthese von peptide. *Tetrahedron Letters*, *32*, 471–474.

Bedford, J. O. Y., Hyde, C., Johnson, T., Jun, W. E. N., Owen, D., Quibell, M., & Sheppard, R. C. (1992). Amino acid structure and "difficult sequences" in solid phase peptide synthesis. *International Journal of Peptide and Protein Research*, *40*, 300–307.

Behrendt, R., White, P., & Offer, J. (2016). Advances in Fmoc solid-phase peptide synthesis. *Journal of Peptide Science: An Official Publication of the European Peptide Society*, *22*, 4–27.

Bergmann, M., & Zervas, L. (1928). Über katalytische racemisation von aminosäuren und peptiden. *Biochemische Zeitschrift*, *203*, 280–292.

Bernatowicz, M. S., Daniels, S. B., & Köster, H. (1989). A comparison of acid labile linkage agents for the synthesis of peptide C-terminal amides. *Tetrahedron Letters*, *30*, 4645–4648.

Beyerman, H. C., & Bontekoe, J. S. (1961). The t-butoxy group, a novel hydroxyl-protecting group for use in peptide synthesis with hydroxy-amino acids. *Proceedings of the Chemical Society*, *249*.

Bruckdorfer, T., Marder, O., & Albericio, F. (2004). From production of peptides in milligram amounts for research to multi-tons quantities for drugs of the future. *Current Pharmaceutical Biotechnology*, *5*, 29–43.

Buchardt, J., Schiodt, C. B., Ferreras, M., Foged, N. T., Delaisse, J. M., & Meldal, M. (2000). Combinatorial library of phosphinic peptides for discovery of MMP inhibitors on solid-phase. In G. B. Fields, J. P. Tam, & G. Barany (Eds.), *Peptides for the new millennium, proceedings of the sixteenth American peptide symposium* (Vol. 6, pp. 443–444). Minneapolis, MN and Netherlands: Springer.

Callahan, F. M., Anderson, G. W., Paul, R., & Zimmermann, J. E. (1963). The tertiary butyl group as a blocking agent for hydroxyl, sulfhydryl and amido functions in peptide synthesis. *Journal of the American Chemical Society*, *85*, 201–207.

Carpino, L. A. (1993). 1-Hydroxy-7-azabenzotriazole. An efficient peptide coupling additive. *Journal of the American Chemical Society*, *115*, 4397–4398.

Carpino, L. A., El-Faham, A., Minor, C. A., & Albericio, F. (1994). Advantageous applications of azabenzotriazole (triazolopyridine)-based coupling reagents to solid-phase peptide synthesis. *Journal of the Chemical Society. Chemical Communications*, *2*, 201–204.

Carpino, L. A., & Han, G. Y. (1972). 9-Fluorenylmethoxycarbonyl amino-protecting group. *The Journal of Organic Chemistry*, *37*, 3404–3409.

Carpino, L. A., Shroff, H., Triolo, S. A., Mansour, M. E., Wenschuch, H., & Albericio, F. (1993). The 2,2,4,6,7-pentamethyldihydrobenzofuran-5-sulfonyl group (Pbf) as arginine side chain protectant. *Tetrahedron Letters*, *34*, 7829–7832.

Chang, C. D., Wakai, M., Ahmed, M., Meienhofer, J., Lundell, E. D., & Huang, J. D. (1980). Preparation and properties of Nα-9-fluorenylmethyloxycarbonylamino acids bearing tert-butyl side chain protection. *International Journal of Peptide and Protein Research*, *15*, 59–60.

Coste, J., Le Nguyen, D., & Castro, B. (1990). PyBOP®: A new peptide coupling reagent devoid of toxic by-product. *Tetrahedron Letters*, *31*, 205–208.

Dessolin, M., Guillerez, M.-G., Thieriet, N., Guibe, F., & Loffet, A. (1995). New allyl group acceptors for palladium catalyzed removal of allylic protections and transacylation of allyl carbamates. *Tetrahedron Letters*, *36*, 5741–5744.

Dourtoglou, V., Gross, B., Lambropoulou, V., & Zioudrou, C. (1984). O-Benzotriazolyl-N,N,N′, N′-tetramethyluronium hexafluorophosphate as coupling reagent for the synthesis of peptides of biological interest. *Synthesis*, *1984*, 572–574.

Dourtoglou, V., Ziegler, J. C., & Gross, B. (1978). L′hexafluorophosphate de o-Benzotriazolyl-N,N-tetramethyluronium: un réactif de couplage peptidique nou-veau et efficace. *Tetrahedron Letters*, *19*, 1269–1272.

Dourtouglou, V., Ziegler, J. C., & Gross, B. (1978). L′hexafluorophosphate de O-benzotriazolyl-N,N-tetramethyluronium: un réactif de couplage peptidique nou-veau et efficace. *Tetrahedron Letters*, *19*, 1269–1272.

El-Faham, A., & Albericio, F. (2011). Peptide coupling reagents, more than a letter soup. *Chemical Reviews*, *111*, 6557–6602.

Eritja, R. (2007). Solid-phase synthesis of modified oligonucleotides. *International Journal of Peptide Research and Therapeutics*, *13*, 53–68.

Fields, G. B., & Noble, R. L. (1990). Solid phase peptide synthesis utilizing 9-fluorenylmethoxycarbonyl amino acids. *International Journal of Peptide and Protein Research*, *35*, 161.

Flörsheimer, A., & Riniker, B. (1991). Solid-phase synthesis of peptides with the highly acid-sensitive HMPB linker. In E. Giralt & D. Andreu (Eds.), *Peptides 1990: Proceedings of the 21st European peptide symposium* (pp. 131–133). Lieden: ESCOM.

Fosgerau, K., & Hoffmann, T. (2015). Peptide therapeutics: Current status and future directions. *Drug Discovery Today*, *20*, 122–128.

Frank, R. (2007). Segmented solid supports: My personal addiction to Merrifield's solid phase synthesis. *International Journal of Peptide Research and Therapeutics*, *13*, 45–52.

Fujino, M., Wakimsu, M., & Kitadu, C. (1981). Further studies on the use of multi-–substituted benzenesulfonyl groups for protection of the guanidino function of arginine. *Chemical and Pharmaceutical Bulletin*, *29*, 2825–2831.

Garcia-Martin, F., & Albericio, F. (2008). Solid supports for the synthesis of peptides. *Chimica Oggi – Chemistry Today*, *26*, 29–34.

Ghosh, S. (2016). Peptide therapeutics market: Forecast and analysis 2015–2025. *Chimica Oggi – Chemistry Today*, *34*, V–VII.

Gil, C., & Bräse, S. (2009). *Solid-phase synthesis of biologically active benzoannelated nitrogen heterocycles: an update, . Journal of Combinatorial Chemistry* (11, pp. 175–197).

Goodman, M., & Levine, L. (1964). Peptide synthesis via active esters. IV. Racemization and ring-opening reactions of optically active oxazolones. *Journal of the American Chemical Society*, *86*, 2918–2922.

Grant, G. A. (Ed.), (1992). *Synthetic peptides*. New York: W. H. Freeman & Co.

Jad, Y. E., Khattab, S. N., de la Torre, B. G., Govender, T., Kruger, H. G., El-Faham, A., & Albericio, F. (2014). Oxyma-B, an excellent racemization suppressor for peptide synthesis. *Organic & Biomolecular Chemistry, 12*, 8379–8385.

Jensen, K. J., Tofteng Shelton, P., & Pedersen, S. L. (2013). *Peptide synthesis and applications* (1047, pp. 23–41). Humana Press.

Kaspar, A. A., & Reichert, J. M. (2013). Future directions for peptide therapeutics development. *Drug Discovery Today, 18*, 807–817.

Kates, S. A., Solé, N. A., Beyermann, M., Barany, G., & Albericio, F. (1996). Optimized preparation of deca(L-alanyl)-L-valinamide by 9-fluorenylmethyloxycarbonyl (Fmoc) solid-phase synthesis on polyethylene glycol–polystyrene (PEG–PS) graft supports, with 1,8-diazobicyclo[5.4.0]—undec-7-ene (DBU) deprotection. *Peptide Research, 9*, 106–113.

Kates, S. A., Sole, N. A., Johnson, C. R., Hudson, D., Barany, G., & Albericio, F. (1993). A novel, convenient, three-dimensional orthogonal strategy for solid-phase synthesis of cyclic peptides. *Tetrahedron Letters, 34*, 1549–1552.

Kent, S. B. H. (1988). Chemical synthesis of peptides and proteins. *Annual Review of Biochemistry, 57*, 957–989.

Knorr, R., Trzeciak, A., Bannwarth, W., & Gillessen, D. (1989). New coupling reagents in peptide chemistry. *Tetrahedron Letters, 30*, 1927–1930.

König, W., & Geiger, R. (1970). A new method for synthesis of peptides: activation of the carboxyl group with dicyclohexylcarbodiimide using 1-hydroxybenzotriazoles as additives. *Chemische Berichte, 103*, 788–798.

Lauer, J. L., Fields, C. G., & Fields, G. B. (1995). Sequence dependence of aspartimide formation during 9-fluorenylmethoxycarbonyl solid–phase peptide synthesis. *Letters in Peptide Science, 1*, 197–205.

Lloyd-Williams, P., Giralt, E., & Albericio, F. (1997). *Chemical approaches to the synthesis of peptides and proteins* (pp. 1–304). New York: CRC Press.

Mergler, M., Nyfeler, R., Tanner, R., Gosteli, J., & Grogg, P. (1988). Peptide synthesis by a combination of solid-phase and solution methods II synthesis of fully protected peptide fragments on 2-methoxy-4-alkoxy-benzyl alcohol resin. *Tetrahedron Letters, 29*, 4009–4012.

Merrifield, R. B. (1963). Solid phase peptide synthesis. I. The synthesis of a tetrapeptide. *Journal of the American Chemical Society, 85*, 2149–2154.

Nicolaou, K. C., & Pfefferkorn, J. A. (2005). Solid-phase synthesis of natural products and natural product-like libraries. In K. C. Nicolaou, R. Hanko, & W. Hartwig (Eds.), *Handbook of combinatorial chemistry. drugs, catalysts, materials* (2, pp. 611–642). Wiley.

Paradis-Bas, M., Tulla-Puche, J., & Albericio, F. (2016). The road to the synthesis of "difficult peptides.". *Chemical Society Reviews, 45*, 631–654.

Pearson, D. A., Blanchette, M., Baker, M. L., & Guindon, C. A. (1989). Trialkyl-silanes as scavengers for the trifluoroacetic acid deblocking of protecting groups in peptide synthesis. *Tetrahedron Letters, 30*, 2739–2742.

Quibell, M., Owen, D., Packman, L. C., & Johnson, T. (1994). Suppression of piperidine-mediated side product formation for Asp(OBut)-containing peptides by the use of N-(2-hydroxy-4-methoxybenzyl) (Hmb) backbone amide protection. *Journal of the Chemical Society. Chemical Communications, 20*, 2343–2344.

Ramage, R., Green, J., & Blake, A. J. (1991). An acid labile arginine derivative for peptide synthesis: N^G-2,2,5,7,8-pentamethylchroman-6-sulphonyl-L-arginine. *Tetrahedron, 47*, 6353–6370.

Ramage, R., & Green, J. (1987). N_G-2,2,5,7,8-pentamethylchroman-6-sulphonyl-L-arginine: A new acid labile derivative for peptide synthesis. *Tetrahedron Letters, 28*, 2287–2290.

Rapp, W., Zhang, L., Haebich, R., & Bayer, E. (1989) Polystyrene–polyoxyethylene graft copolymers for high speed peptide synthesis. In G. Jung & E. Bayer (Eds.), *Peptides 1988: Proceedings of the twentieth European peptide symposium* (pp. 199–201). German Federal Republic, Tuebingen.

Rink, H. (1987). Solid-phase synthesis of protected peptide fragments using a trialkoxy-diphenyl-methylester resin. *Tetrahedron Letters*, *28*, 3787–3790.

Schwyzer, R., & Dietrich, H. (1961). L–Asparaginsäure–β–t–butylester und Derivate. *Helvetica Chimica Acta*, *44*, 2003–2006.

Sieber, P. (1987). A new acid-labile anchor group for the solid-phase synthesis of C-terminal peptide amides by the Fmoc method. *Tetrahedron Letters*, *28*, 2107–2110.

Sieber, P., & Riniker, B. (1987). Protection of histidine in peptide synthesis: A reassessment of the trityl group. *Tetrahedron Letters*, *28*, 6031–6034.

Sieber, P., & Riniker, B. (1991). Protection of carboxamide functions by the trityl residue. Application to peptide synthesis. *Tetrahedron Letters*, *32*, 739–742.

Stewart, J. M., & Young, J. D. (Eds.), (1984). *Solid phase peptide synthesis* (2nd ed.). Rockford: Pierce Chemical Company.

Story, S. C., & Aldrich, J. V. (1994). Side-product formation during cyclization with HBTU on a solid support. *International Journal of Peptide and Protein Research*, *43*, 292–296.

Subirós-Funosas, R., Prohens, R., Barbas, R., El-Faham, A., & Albericio, F. (2009). Oxyma: An efficient additive for peptide synthesis to replace the benzotriazole-based HOBt and HOAt with a lower risk of explosion. *Chemistry – A European Journal*, *15*, 9394–9403.

Tam, J. P., Heath, W. F., & Merrifield, R. (1983). An SN2 deprotection of synthetic peptides with a low concentration of HF in dimethylsulfide: Evidence and application in peptide synthesis. *Journal of the American Chemical Society*, *105*, 6442–6455.

Thieriet, N., Alsina, J., Giralt, E., Guibe, F., & Albericio, F. (1997). Use of Alloc-amino acids in solid-phase peptide synthesis. Tandem deprotection-coupling reactions using neutral conditions. *Tetrahedron Letters*, *38*, 7275–7278.

Trzeciak, A., & Bannwarth, W. (1992). Synthesis of 'head-to-tail' cyclized peptides on solid support by Fmoc chemistry. *Tetrahedron Letters*, *33*, 4557–4560.

Wade, J. D., Bedford, J., Sheppard, R. C., & Tregear, G. W. (1991). DBU as an N alpha-deprotecting reagent for the fluorenylmethoxycarbonyl group in continuous flow solid-phase peptide synthesis. *Peptide Research*, *4*, 194–199.

Wang, S. S. (1973). p-Alkoxybenzyl alcohol resin and p-alkoxybenzyloxycarbonylhydrazide resin for solid phase synthesis of protected peptide fragments. *Journal of the American Chemical Society*, *95*, 1328–1333.

Wehrstedt, K. D., Wandrey, P. A., & Heitkamp, D. (2005). Explosive properties of 1-hydroxybenzotriazoles. *Journal of Hazardous Materials*, *126*, 1–7.

White, P. (1992). In J. A. Smith, & J. E. Rivier (Eds.), *Peptides: Chemistry and biology* (p. 537). Leiden: ESCOM.

Wilk, A., Grajkowski, A., Srinivasachar, K., & Beaucage, S. L. (1999). Improved chemistry for the production of synthetic oligodeoxyribonucleotides. *Antisense & Nucleic Acid Drug Development*, *9*, 361–366.

Zompra, A. A., Galanis, A. S., Werbitzky, O., & Albericio, F. (2009). Manufacturing peptides as active pharmaceutical ingredients. *Future Medicinal Chemistry*, *1*, 361–377.

Chapter 4

Peptide-based Antibiotics

Ruchika Goyal and Vibin Ramakrishnan
Department of Biosciences and Bioengineering, Indian Institute of Technology Guwahati, Guwahati, Assam, India

4.1 Introduction

Peptides are polymers of amino acids capable to function as signaling molecules, enzymes, hormones, and carriers in living beings. Their therapeutic potential was acknowledged after the introduction of insulin for diabetes. Later, peptides like octreotide, calcitonin, oxytocin, vasopressin, exenatide, and teriparatide were introduced with advancements in peptide chemistry. At present, considering the pharmaceutical landscape, there are more than 80 peptide drugs marketed worldwide (Muttenthaler, King, Adams, & Alewood, 2021). They have been used for therapeutic indications, involving metabolic diseases, osteoporosis, cancer, chronic pain, cardiovascular, and endocrinology.

Multiple medicinal approaches have been utilized to improve peptide's stability, selectivity, potency, and pharmacokinetics. For instance, strategies like the use of unnatural amino acids, N-terminal modifications, deamination, cyclization, and nanoconjugations have been employed to improve the pharmacokinetic properties of peptidic drugs. A change in amino acid stereochemistry from L to D can make a huge impact in the structure and electrostatics of the peptide (Kumar, Ramakrishnan, Ranbhor, Patel, & Durani, 2009; Kumar, Ranbhor, Patel, Ramakrishnan, & Durani, 2017; Ramakrishnan, Ranbhor, & Durani, 2004; Ramakrishnan, Ranbhor, & Durani, 2005; Ramakrishnan, Ranbhor, Kumar, & Durani, 2006; Ranbhor, Ramakrishnan, Kumar, & Durani, 2006). Some of the attempts were successful in emerging peptide-based solutions as theranostic agents, including radio conjugates, nanoformulations, and fluorophore-tagged peptides (Goyal & Ramakrishnan, 2019; Goyal, Jerath, Akhil, et al., 2021; Goyal, Jerath, Chandrasekharan, et al., 2021; Goyal, Jerath, Chandrasekharan, Kumar, & Ramakrishnan, 2020; Jerath, Goyal, Trivedi, Santhoshkumar, & Ramakrishnan, 2019; Jerath, Goyal, Trivedi, Santhoshkumar, & Ramakrishnan, 2020).

De Novo Peptide Design. DOI: https://doi.org/10.1016/B978-0-323-99917-5.00004-4

Peptides like cell-penetrating peptides (CPPs), tumor homing peptides, anticancer peptides, and antimicrobial peptides (AMPs) have been used in drug delivery due to less side effects, biocompatibility, selectivity, and predictable metabolism being natural components of living beings. In addition to the naturally occurring peptides, rationally designed peptides have advanced the peptide therapeutic regime toward selective and defined functions. For instance, CPPs penetrate and deliver a wide range of cargoes in mammalian cells (Copolovici, Langel, Eriste, & Langel, 2014). On the other hand, tumor homing peptides are capable of selective targeting to tumors. Similarly, based on the functional properties, peptides like tumor penetrating peptides, lymphatic homing peptides, and anticancer peptides have also been introduced (Goyal & Ramakrishnan, 2019).

4.2 Antimicrobial peptides

AMPs or host defense peptides are evolutionarily conserved components of the innate immune system (Hancock & Scott, 2000). They are initially isolated from the venoms of arthropods and show potent broad-spectrum activity against Gram-positive bacteria, Gram-negative bacteria, viruses, and fungi (Hancock & Diamond, 2000; Reddy, Yedery, & Aranha, 2004). In general, AMPs are short amino acid polymers having a chain length of 20–50 residues and $+1 - +9$ net charge (Mishra & Wang, 2012; Wang, Li, & Wang, 2015). At present, there are more than 5000 known AMPs, which led to the formation of several antimicrobial databases (Table 4.1) (Fan et al., 2016; Wang et al., 2015; Zhao, Wu, Lu, Li, & Huang, 2013). For example, Antimicrobial Peptide Database is a manually created database based on the set of data-collection criteria. It contains 3257 antimicrobial peptides from six kingdoms (bacteria, fungi, archaea, protists, plants, animals) having various properties like antibacterial, antifungal, antiviral, wound healing, spermicidal, insecticidal, antioxidant, anticancer, and antidiabetic properties (Wang et al., 2015).

They are classified based on their structure, source, composition, and activity (Fig. 4.1). They are ubiquitously found in all life forms like bacteria, fungi, protists, fish, amphibians, insects, and mammals. Generally, they are rich in residues like tryptophan, arginine, proline, histidine, and glycine. Their applications are not limited to medicine (gramicidin, daptomycin, colistin, and others) but to the food industry (nisin, polylisine), animal husbandry (caerin 1.1, dicentracin), and agriculture (mastoparan-S, PAF26, thanatin, etc.) (Huan, Kong, Mou, & Yi, 2020). Structurally, the known AMPs have secondary structures like α-helices (magainin 2, cathelicidins), β-sheets (gramicidin S, defensins, hepcidins), extended (histatins, indolicidin), and combined forms (protegrin-3) (Fig. 4.2).

TABLE 4.1 Antimicrobial databases.

S. no.	Database	References
1	APD	Wang, Li, and Wang (2009), Wang et al. (2015), Wang & Wang (2004)
2	DBAASP	Gogoladze et al. (2014)
3	DRAMP	Fan et al. (2016)
4	CAMP	Thomas, Karnik, Barai, Jayaraman, and Idicula-Thomas (2009)
5	LAMP	Zhao et al. (2013)
6	BaAMPs	Di Luca, Maccari, Maisetta, and Batoni (2015)
7	Cybase	Wang, Kaas, Chiche, and Craik (2008)
8	DAMPD	Seshadri Sundararajan et al. (2011)
9	BACTIBASE	Hammami, Zouhir, Ben Hamida, and Fliss (2007), Hammami, Zouhir, Le Lay, Ben Hamida, and Fliss (2010)
10	PhytAMP	Hammami, Ben Hamida, Vergoten, and Fliss (2008)
11	DADP	Novković, Simunić, Bojović, Tossi, and Juretić (2012)
12	Defensins	Seebah et al. (2006)
13	EnzyBase	Wu, Lu, Huang, Li, and Huang (2012)
14	InverPep	Gómez, Giraldo, and Orduz (2017)
15	MilkAMP	Théolier, Fliss, Jean, and Hammami (2014)
16	Peptaibols	Whitmore and Wallace (2004)
17	THIOBASE	Li et al. (2012)
18	YADAMP	Piotto, Sessa, Concilio, and Iannelli (2012)

4.3 De novo design of antimicrobial peptides

The essential design features for AMPs' functionality include net positive and amphipathic character (Wimley, 2010). Around 50% hydrophobicity is desirable for the interaction of AMPs with bacterial membranes (Mishra & Wang, 2012; Wang et al., 2015). Their secondary structure plays an important role in mediating interaction and penetration through membranes. They have been reported to form diverse secondary structures, including helices, beta-sheet, and random coil, for stability (Fig. 4.2). Rational design strategies have been adopted to improve their antimicrobial activity based on the principles of electrostatics, hydrogen bonding, tacticity, and conformational preferences of amino acids (Prakash, Ranbhor, & Ramakrishnan, 2020;

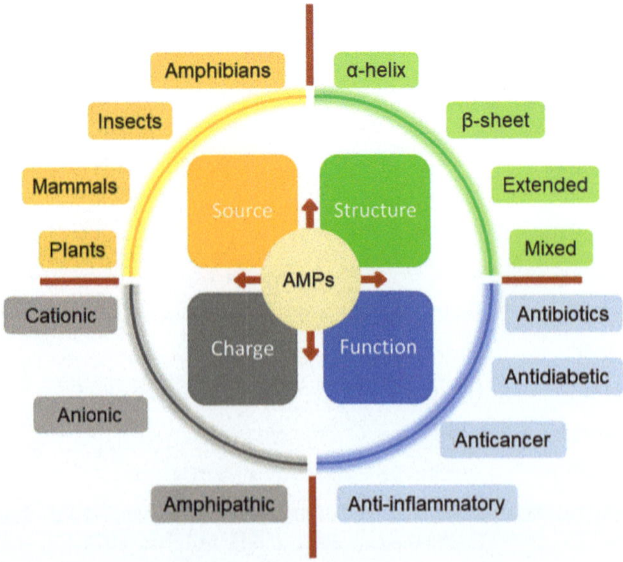

FIGURE 4.1 **AMPs classification.** AMPs, classified based on their origin, structure, functions, and charge. *AMPs*, Antimicrobial peptides.

Ranbhor, Kumar, Patel, Ramakrishnan, & Durani, 2018; Ranbhor, Kumar, Tendulkar, et al., 2018). In addition, it also includes the incorporation of unnatural amino acids, cyclization, stapling, N- or C-terminal modifications, and disulfide linkages.

4.3.1 Unnatural amino acids

Protease instability is one important limiting factor of peptide therapeutics. The incorporation of unnatural amino acids in peptides helps in increasing their stability by providing protection to proteolytic cleavage (Hazam, Akhil, Jerath, Saikia, & Ramakrishnan, 2019; Hazam, Goyal, & Ramakrishnan, 2019; Hazam, Jerath, Chaudhary, & Ramakrishnan, 2018; Hazam, Jerath, Kumar, Chaudhary, & Ramakrishnan, 2017; Hazam, Phukan, Akhil, Singh, & Ramakrishnan, 2021; Hazam, Singh, Chaudhary, & Ramakrishnan, 2019; Saikia, Sravani, Ramakrishnan, & Chaudhary, 2017). In fact, few AMPs like gramicidin are composed of unnatural amino acids naturally. Merrifield et al. replaced L-amino acids with D-amino acids in melittin, magainin 2, and cecropin A (Wade et al., 1990). Few other reports on the de novo design of AMPs demonstrate the use of unnatural amino acids for improved bioactivity. Hazam et al. employed unnatural amino acids in their designed AMPs using the knowledge of stereochemistry, tacticity, and principles of electrostatics (Hazam et al., 2017; Hazam et al., 2018). They reported the design of AMPs with distinct cationic and hydrophobic zones for interaction with

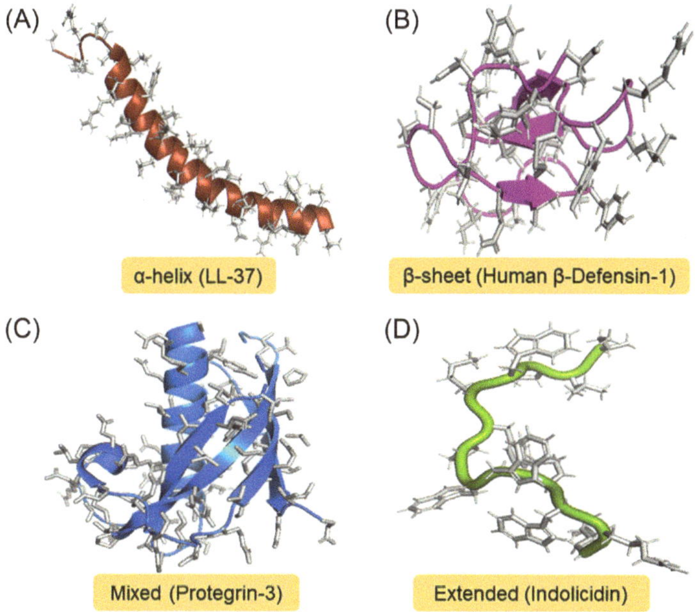

FIGURE 4.2 Secondary structures of AMPs. (A) LL-37 peptide has α-helical structure (Pdb Id: 2K6O). (B) Human β-defensin-1 peptide has β-sheet structure (Pdb Id: 1KJ5). (C) Protegrin-3 peptide having a combination of α-helix and β-sheet structures (Pdb Id: 1KWI). (D) Indolicidin peptide with no extended conformation (Pdb Id: 1G89). *AMPs*, Antimicrobial peptides.

membranes of Gram-positive, Gram-negative, and acid-fast bacteria (Hazam et al., 2021).

4.3.2 Cyclization

Peptides whose end terminals are linked by disulfide bonds, side chains, or head-to-tail bonds creating a ring-like structure are termed cyclic peptides. In nature, few AMPs like AS-48 and theta-defensins are circular in nature. Orlov et al. (2019) have reported that the main chain cyclization of arenicin-1 enhanced its activity against drug-resistant clinical isolates and possessed no cytotoxicity. Cyclic analogs of tachyplesins I, II, and III are known to have similar structure and activity as their parent peptide (Vernen et al., 2019), to form nanoassemblies by stacking up in a way to create a hollow tube-like nanochannel. These peptides are resistant to proteolytic cleavage and have broad-spectrum antimicrobial activity (Dartois et al., 2005; Fernandez-Lopez et al., 2001).

4.3.3 Chemical modifications

4.3.3.1 Lipidation

Lipopeptides like daptomycin, polymyxin B, and colistin are naturally lipidated antimicrobial agents. Lipidation refers to the addition of hydrophobic lipid tails by amidation, esterification, and disulfide bridging. It modulates peptide pharmacokinetic and pharmacodynamics properties by influencing peptide secondary structure, hydrophobicity, and self-assembly (Zhang & Bulaj, 2012). In addition, lipidation also confers enzymatic stability, bioavailability, membrane permeability, and bioactivity (Li et al., 2013; Zhang & Bulaj, 2012).

4.3.3.2 Acylation

Acylation refers to the addition of lipophilic acids to the N-terminus or amino acid side chain amine (N-acylation) or cysteine (S-acylation) residue of the peptide. Naturally occurring AMPs having fatty acid linked to the N-terminus of peptides are known as lipopeptaibols. Synthetically, many groups have generated acylated variants of lactoferricin B, magainin, LL-37, mastoparan-X, etc., and showed an enhancement in their antimicrobial properties upon acylation. Mak et al. have shown that the covalent conjugation of saturated fatty acids (C8 − C12 hydrocarbon chains) has improved the antibacterial potency (Scheme 4.1) of CG117−136 fragment of human cathepsin G (Mak et al., 2003).

4.3.3.3 Glycosylation

The covalent attachment of glycans to specific amino acid residues in peptides or proteins is glycosylation. It is an important posttranslational modification that led to the change in the physicochemical properties and consequently, biological functions like protein folding, target specificity, host − pathogen interactions, immunomodulation, etc. (Moradi, Hussein, Varamini, Simerska, & Toth, 2016; Opdenakker, Rudd, Wormald, Dwek, & Van Damme, 1995). Glycosylation is categorized into the following four types depending on the residue site: N-glycosylation (Asn), S-glycosylation (Cys), O-glycosylation (Ser, Thr, or Tyr), and C-glycosylation (Trp).

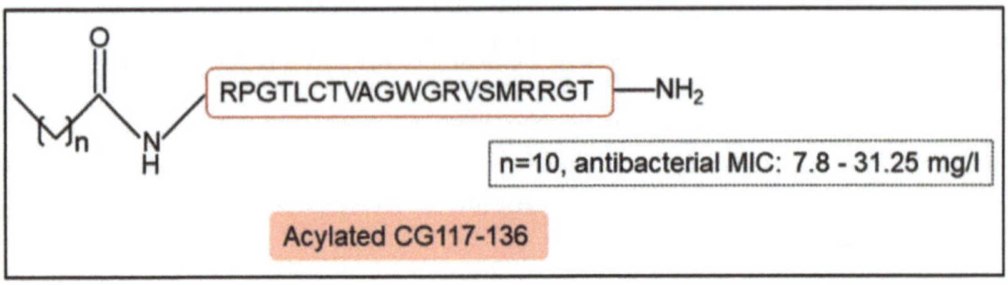

SCHEME 4.1 N-terminal acylation of CG 117−136 domain of human cathepsin G.

Proline-rich AMPs like drosocin, formaecin 1, and formaecin 2 have O-linked glycosylation (Bulet et al., 1993; Mackintosh, Veal, Beattie, & Gooley, 1998). Bulet and coworkers have shown that the activity of drosocin gets reduced 5–10 times against Escherichia *coli* without O-glycosylation (Bulet et al., 1993; Bulet, Urge, Ohresser, Hetru, & Otvos, 1996). Kaur et al. investigated the effect of sugar chain length on the activity of drosocin (Lele, Kaur, Thiruvikraman, & Kaur, 2017). They observed a twofold decrease in antibacterial activity upon increasing the sugar chain length. O-glycosylation also helps in receptor recognition without membrane disruption and enhances the antimicrobial activity of AMPs (Bulet et al., 1993; Mackintosh et al., 1998).

Guo et al. have synthesized glycol-analogs of tryrocidine A by adding monosaccharide, disaccharide, or lactose to the asparagine side chain, which affects its conformation and activity (Hu et al., 2009). In another study, cross-linking of nisin with chitosan in different ratios via genipin cross-linker has influenced its antibacterial properties (Butler, Ng, & Pudney, 2003). Nisin to chitosan ratio of 200:1 has led to 80% bacterial growth inhibition. It may be due to the disturbance of phospholipid assembly of the bacterial membrane by high local concentration of nisin, which led to its penetration in cells and hence, high bioactivity.

S-glycosylated peptides are rare in nature and found in human urine (Lote & Weiss, 1971). The prominent examples of antibacterial glycopeptides having sugar groups linked to the sulfur of cysteine residues are sublancin 168, ASM1, thurandacins A/B, and glycocin F. Donk et al. showed that S-glycosylated galactosyl, mannosyl, and GlcNAc sublancin analogs have similar activity to S-glucosylated sublancin but displayed much better bioactivity than nonglycosylated forms (Oman, Boettcher, Wang, Okalibe, & van der Donk, 2011). Even the cationic peptidopolysaccharides have high selectivity toward microorganisms. Li et al. have reported an ε-poly-L-lysine chitosan conjugate with antibacterial and antifungal properties (Su et al., 2017). S-glycosylated peptides have improved enzymatic stability than O/N-linked congeners against transglycosylases and glycosidases (Ludolph, Eisele, & Waldmann, 2002; Piazza et al., 1999).

C-glycosylated peptides have carbohydrate moieties attached to the peptide sequence. Copper(I)-catalyzed azide-alkyne cycloaddition (CuAAc) reaction via triazole ring formation has been employed to generate glycotriazole peptides. Verly et al. have developed more potent antifungal peptides with a monosaccharide coupled to the hylaseptin-P1 peptide (Junior et al., 2017). Similarly, the addition of carbohydrates to tyrocidine A by CuAAc reaction has improved its therapeutic index (Lin & Walsh, 2004). C-glycosylation assists in interaction with membranes and reduces toxicity against mammalian cells like other glycosylation strategies.

4.3.4 Conjugation to conventional antibiotics

This strategy involves the fusion of two distinct compounds to yield a potent antimicrobial agent. The combination of AMP with conventional antibiotics has proven to generate hybrid molecules with enhanced antimicrobial activity. For instance, Bishop et al. have conjugated levofloxacin (LVFX) to Pep-4 AMP, which showed $10-80$-fold more antibacterial activity under intermediate and high salt conditions compared to Pep-4 (Rodriguez, Papanastasiou, Juba, & Bishop, 2014). Similarly, Devocelle et al. have demonstrated the coupling of cephalosporin to D-Bac8c(Leu^{2,5}) AMP by CuAAc chemistry (Desgranges et al., 2012). A series of HLopt2 and LFcinB (truncated) AMP conjugates with combinations of ciprofloxacin, LVFX, and fluconazole have yielded hybrid molecules with enhanced antibacterial activities (Ptaszyńska, Gucwa et al., 2019; Ptaszyńska, Olkiewicz et al., 2019). In another study, LVFX conjugated to indolicidin AMP and transactivator of transcription (TAT) CPP improved its antibacterial activity against Gram-positive pathogens (Ghaffar et al., 2015). Kasko et al. have reported the conjugation of tobramycin to the penetratin-derived segment membrane-active antibiotic-peptide conjugates (MAAPC) for increasing the intracellular accumulation of tobramycin (Deshayes et al., 2017). These findings showed the potential of peptide-antibiotic hybrids for the future design of novel antimicrobials.

4.3.5 Multivalent approach

Peptides with repeat units linked by disulfide bridges, amide linkages, or other bond types led to highly branched structures like dendrimers, which have shown better biocompatibility, stability, selectivity, and protease resistance (Bracci et al., 2003). Dimeric lentivirus lytic peptides and magainin 2 were generated using disulfide bridging and showed $4-8$ times more Gram-positive bactericidal activity than the parent sequences (Mukai, Matsushita, Niidome, Hatekeyama, & Aoyagi, 2002; Tencza et al., 1999). Tetrameric structures have reported a higher cationic charge for better interaction with anionic bacterial membranes. The dimeric oxidation of bovine lactoferrin-derived peptide, LfcinB, by interdisulfide bridge has led to tetrameric LfcinB, which showed a 5–10-fold enhanced bioactivity (Vargas Casanova et al., 2017; Vega, Martínez, Chalá, Vargas, & Rosas, 2018).

Further, the development of multiple antigen peptide (MAP) by tethering branching peptides to divalent Lys core has been established by an amide linkage. Tam and coworkers have pioneered in generating MAP-like dendrimers of AMPs like protegrin, tachypleisin, and RTD-1 using Lys as the multivalent core unit (Tam, Lu, & Yang, 2002). Pini et al. have demonstrated that tetrameric M6 peptide displayed more effective bacterial inhibition and bactericidal activity against many Gram-negative bacteria than their monomeric counterparts (Pini et al., 2005). Similarly, vancomycin dimer formation by diamine linkers showed higher antibacterial activity even in vancomycin-resistant strains (Rao et al., 1999).

The polyvalency of peptides has also been reported using the carboxyl group of amino acids or other small molecules. The tetramers of cationic Lys-Leu-Ala-Arg (KLAR) or Arg-Leu-Ala-Arg (RLAR) peptides showed significant antibacterial activity against Gram-negative bacteria (Khrushchev, Kashparov, Klimenko, & Mitin, 2007). Cilli et al. synthesized two aurein 1.2 dimers [$(AU)_2K$ and $E(AU)_2$] linked with lysine or glutamic acid residues. These dimers did not only increase bacterial growth inhibition, but enhanced the pore formation ability of peptides and increased membrane permeability (Lorenzón, Sanches, Nogueira, Bauab, & Cilli, 2013). These reports showed that multimerization of peptides constitutes a promising strategy for improving the activity of AMPs.

4.3.6 Antimicrobial peptide mimics

Synthetic mimics of AMPs are AMP-like molecules created using the knowledge of peptidomimetics. Their backbones are composed of β-amino acid oligomers, phenylene ethynylenes, acrylamide oligomers, etc. (Michael Henderson & Lee, 2013). Similarly, peptoids are peptide-like molecules with poly-N-substituted glycines, where the side chain is linked to amide nitrogen instead of α-carbon of the backbone (Andreev et al., 2018; Simon et al., 1992). Peptoid's synthetic structure ensures sustained bioactivity by providing resistance to proteases. Sharma et al. have reported an SPO peptoid, a homolog of AMP SA4, with an inhibitory effect against planktonic and biofilm formation of *Acinetobacter baumannii* strains (Sharma, Choudhary, Vashistt, Shrivastava, & Bisht, 2019). Peptoid nanoassemblies showed tremendous potential in bioscience applications (Mannige et al., 2015; Wang et al., 2020; Zhu et al., 2017). The lipidation and end-chain modification of antimicrobial peptoids form nanostructure by self-assembly (Hasan et al., 2020).

4.4 Antimicrobial peptides: Strategy to combat antimicrobial resistance

AMPs are favored as new alternatives over antibiotics due to the emergence of antimicrobial resistance (AMR). The excessive use of antibiotics in the recent past has introduced a global public health challenge. According to a Centers for Disease Control and Prevention (CDC) report, around 2.8 million people in the United States develop an antibiotic-resistant infection, and more than 35,000 people die (Antibiotic resistance threats in the United States, 2019). O'Neill has produced a similar report on behalf of the UK government, which predicted that AMR would claim around 10 million lives

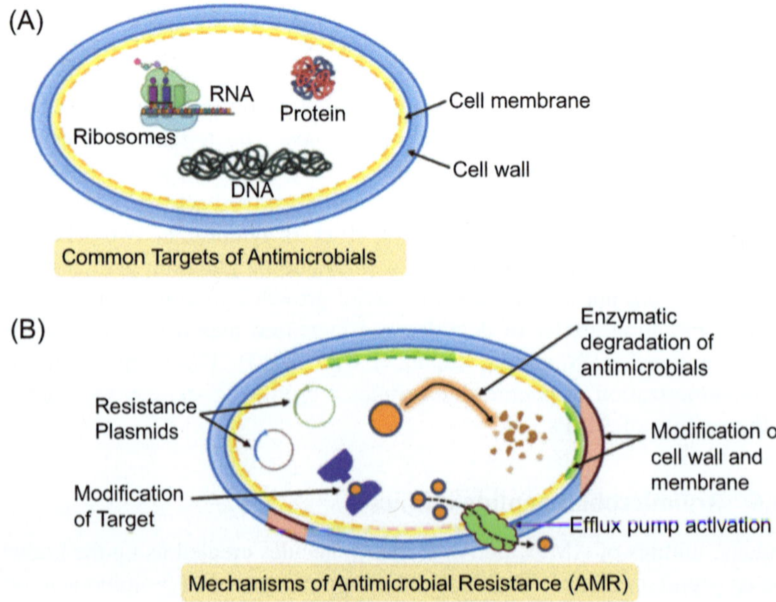

FIGURE 4.3 Antimicrobial targets and AMR mechanisms. (A) Common targets of antimicrobial agents. (B) Different mechanisms by which bacteria gain resistance to antimicrobials.

per year by 2050, overshooting the number of people dying from cancer (8.2 million each year) (http://www.amr-review.org). The present statistics on AMR and currently available treatment options demand more effective solutions.

The common bacterial targets of antimicrobials are cell walls, cell membranes, ribosomes, and biomolecules like DNA, RNA, and proteins (Fig. 4.3A). Bacterial species show AMR by surpassing the targeted pathways or by modifying the cellular components. The common ways of adapting AMR include modification of cell wall, cell membrane, target proteins, activation of efflux pumps, and enzymatic degradation of antimicrobials [Fig. 4.3B (Abdi, Mirkalantari, & Amirmozafari, 2019)].

AMPs gained more attraction to combat with AMR due to simple composition, shorter length, ease of synthesis, programmability, and relatively safe therapeutic usage being natural components of an innate defense system. The first few antimicrobial agents include lysozyme (Fleming & Wright, 1922) and gramicidin (Dubos, 1939a, 1939b). There is a constant rise in the number of AMPs followed, including defensins (Ganz, 2003), cathelicidins (Zanetti, Gennaro, & Romeo, 1995), cecropins (Steiner, 1982), sarcotoxins (Okada & Natori, 1985), and dermaseptins (Amiche, Seon, Pierre, & Nicolas, 1999). Many AMPs are in the different phases of research and clinical trials (Table 4.2), and some are already in the market. FDA-approved

TABLE 4.2 Antimicrobial peptides in Phase II or III clinical trials.

Peptide	Parent compound	Medical use	Phase
Omiganan	Indolicidin	Genital warts, seborrheic dermatitis	II
PAC 113	Histatin	Oral candidiasis	
1–11	Lactoferricin	Bacterial infections and mycoses	
Delmitide (RDP58)	HLA	Inflammatory bowel disease	
Lytixar (LTX-109)	Synthetic tripeptide	Atopic dermatitis, skin infections	
Brilacidin	Arylamide oligomer	Oral mucositis	
Novexatin (NP213)	Cyclic peptide	Fungal nail infection	
OP-145	LL-37	Chronic middle ear infection	
D2A21	Synthetic peptide	Burns and wound infection	III
Pexiganan	Magainin	Diabetic foot infection	
Iseganan (IB-367)	Protegrin I	Oral mucositis	
XMP 629	BPI protein	Impetigo	

AMPs include nisin (Hansen & Sandine, 1994), gramicidin (Kelkar & Chattopadhyay, 2007), vancomycin (Andersson, Hughes, & Kubicek-Sutherland, 2016), etc. (Table 4.3). In addition to this, the nanocarriers of AMPs are effective against highly virulent ESKAPE pathogens (like *Staphylococcus aureus*, *Pseudomonas aeruginosa*, and *Enterobacter spp.*) relevant to nosocomial infections (Mukhopadhyay, Bharath Prasad, Mehta, & Nayak, 2020).

4.5 Mechanism of action and selectivity of antimicrobial peptides

The antimicrobial activity and selectivity of AMPs for bacterial membranes are the consequence of electrostatic and hydrophobic interactions. Electrostatics plays a significant role in the selective interactions of cationic AMPs and anionic bacterial membranes. In general, bacterial membranes are rich in anionic lipids like cardiolipin and phosphatidylserine. They are

TABLE 4.3 FDA-approved antimicrobial peptides.

Peptide	Indications	Activity spectra	References
Nisin	Food preservative	Gram-positive and some Gram-negative bacteria	Hsu et al. (2004)
Dalbavancin	Acute bacterial skin infections	Gram-positive bacteria	Scott (2015)
Enfuvirtide	HIV-1 infection	Virus	Kitchen, Nuño, Kitchen, and Krogstad (2008)
Gramicidin D	Skin, eye, and wound infections	Gram-positive bacteria	Kessler, Schuhmann, Morneweg, Linne, and Marahiel (2004)
Gramicidin S	Wound infections and genital ulcers	Gram-positive, some Gram-negative bacteria, and fungi	Prenner, Lewis, and McElhaney (1999), Swierstra, Kapoerchan, Knijnenburg, van Belkum, and Overhand (2016)
Daptomycin	Skin infections, endocarditis, and bacteraemia	Gram-positive bacteria	Randall, Mariner, Chopra, and O'Neill (2013), Straus and Hancock (2006)
Telaprevir	Hepatitis C	Virus	Matthews and Lancaster (2012)
Polymyxin B	Pneumonia, meningitis, sepsis, UTI	Gram-negative bacteria	Tsubery, Ofek, Cohen, and Fridkin (2000), Velkov, Thompson, Nation, and Li (2010), Yu, Qin, Lin, Fang, and Qiu (2015)
Bacitracin	Skin, eye, and wound infections	Gram-positive bacteria	Ming and Epperson (2002), Stone and Strominger (1971)
Colistin	Intestinal infections and cystic fibrosis	Gram-negative bacteria	Poirel, Jayol, and Nordmann (2017)
Vancomycin	Skin infections, meningitis, bloodstream infections, endocarditis, bone, and joint infections.	Gram-positive bacteria	Alosaimy et al. (2020), Hawley and Gump (1973)

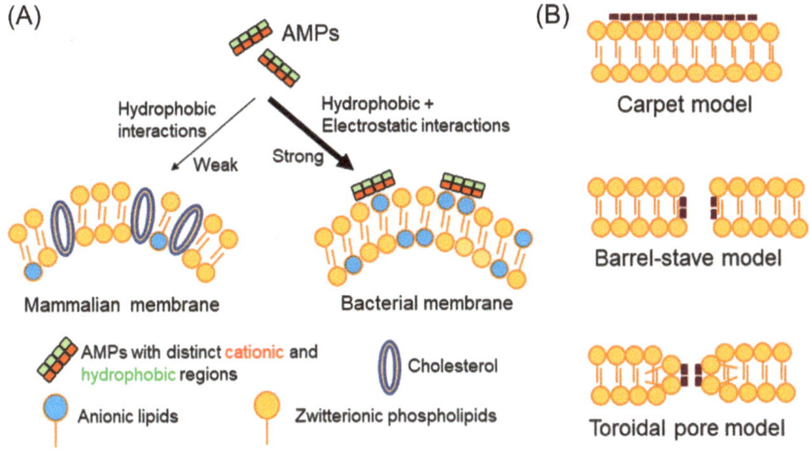

FIGURE 4.4 Selectivity and mechanism of AMPs. (A) Selectivity of AMPs for bacterial membranes over mammalian membranes. (B) Representative models of the known mechanism of action of AMPs. *AMPs*, Antimicrobial peptides.

exposed to the outer leaflet of the bacterial membrane, unlike mammalian cells (Fig. 4.4A).

Also, teichoic acid in Gram-positive bacterial membrane and lipopolysaccharide in Gram-negative bacterial membrane provides net negative charge in the outer membranes, respectively (Silhavy, Kahne, & Walker, 2010). This assists AMPs in selective targeting of bacteria. Also, lipid packing in bacterial membranes affects the interaction of AMPs and hence, modulates their bactericidal potency. Few proposed models like a barrel-stave model, carpet model, toroidal-pore model, aggregated channel model etc., are known to mediate the action of AMPs through membrane interactions [Fig. 4.4B; (Yeaman & Yount, 2003)]. In addition to this, few AMPs penetrate into bacterial cells and affect biochemical processes like DNA replication, RNA transcription, protein, and cell wall synthesis (Graf & Wilson, 2019; Moravej et al., 2018). For instance, β-defensins inhibit bacterial growth by inhibiting protein synthesis and blocking cell wall formation (Aisenbrey, Marquette, & Bechinger, 2019; Moravej et al., 2018).

4.5.1 Barrel-stave model

According to this model, the barrel-stave pore is formed due to the exact alignment of facial amphipathicity and hydrophobic distribution of AMPs with the bilayer(Hazam, Goyal, & Ramakrishnan, 2019; Sharma, Choudhary, Vashistt, Shrivastava, & Bisht, 2019) In other words, toroidal pores are formed when peptides attach to the membranes coaxially, leading to the merging of two hydrophobic regions of AMPs and membrane, respectively.

This structural arrangement appears like a hollow cylinder running parallel to the membrane. During the interaction process, the hydrophilic part of the peptide forms the interior part of the transmembrane pore.

4.5.2 Carpet model

This model proposed that AMPs scatter parallelly on the membranes upon interaction, disorienting the stacked lipid molecules of membranes. The structural arrangement of AMPs looks like a carpet over the membrane. During the interaction, the disoriented lipid molecules degenerate into small aggregates disintegrating the membrane, resulting in the membranolytic activity of AMPs with no pore formation.

4.5.3 Toroidal model

As the name suggests, the formation of a pore-like structure is the key feature of this model. In this model, each AMP molecule interacts independently with the membrane altering the local membrane curvature cooperatively (Hazam et al., 2019). They initiate inward bending of lipid molecules leading to the toroid formation, whose inner core is partially aligned by lipid heads and peptides (Yang, Harroun, Weiss, Ding, & Huang, 2001).

4.5.4 Disordered toroidal-pore model

The difference between the toroid pore and disordered toroidal-pore model is that the latter one has a hydrophilic inner core due to the polar head groups of phospholipids. According to this model, phospholipids' head groups line the inner lumen of the pore. The peptides with less rigid conformations follow this mechanism.

4.6 Strengths and weakness of antimicrobial peptides

4.6.1 Merits of antimicrobial peptides

AMPs are favored over conventional antibiotics due to their low toxicity, biocompatibility, and broad-spectrum activity against multidrug resistant (MDR) species and bacterial biofilms. AMPs like SMAP-29 were effective against MDR, vancomycin-resistant enterococcos faecium (VREF), and mutated *P. aeruginosa* (Skerlavaj, Benincasa, Risso, Zanetti, & Gennaro, 1999). AMPs show synergistic effects when used in combinations of known antibiotics like penams, nalidixic acid derivatives, and β-lactams (Giacometti et al., 2005). For instance, B2088 peptide, combined with antibiotics like chloramphenicol, gentamicin, tobramycin, etc., shows selective bactericidal activity to *P. aeruginosa* (Lakshminarayanan et al., 2016). The selectivity of

AMPs for anionic microbial membranes over neutral mammalian membranes encourages their design paradigm (Teixeira, Feio, & Bastos, 2012). AMPs like LL-37, lactoferrin, melamine, citropin, etc., showed broad-spectrum biofilm inhibiting activity when used synergistically with antibiotics like rifampicin and minocycline (Chung & Khanum, 2017).

4.6.2 Limitations of antimicrobial peptides

High production cost, reduced bioavailability, AMR, and proteolytic susceptibility are major limitations for AMP development as therapeutic options. However, these limitations are addressed by increasing the bioavailability of AMPs through modifications like terminal capping, N-methylation, cyclization, stapling, conjugation with large polymers, and fusion with proteins like albumin. (Di, 2015; Renukuntla, Vadlapudi, Patel, Boddu, & Mitra, 2013; Zorzi, Middendorp, Wilbs, Deyle, & Heinis, 2017). In addition, reduced sequestration of AMPs can be achieved by altering the membrane anionic charge using alanine and lysine residues in AMPs (Epand, Walker, Epand, & Magarvey, 2016).

4.7 Conclusion

Antimicrobial peptides are promising alternatives as novel antibacterial agents compared to conventional antibiotics. This is due to their high programmability, long stability, reduced side effects, and fair biocompatibility. This chapter discussed the recent advancement in their design using rational approaches, chemical modifications, and nanotization strategies, which showed their potential to combat AMR. FDA-approved AMPs and accelerated clinical trials suggest their translational benefits. Together with the latest investigations and development of AMPs using interdisciplinary research, AMP seems like a potent peptide-based antibacterial agent in the coming decades.

References

Abdi, M., Mirkalantari, S., & Amirmozafari, N. (2019). Bacterial resistance to antimicrobial peptides. *Journal of Peptide Science: An Official Publication of the European Peptide Society*, *25*(11), e3210. Available from https://doi.org/10.1002/psc.3210.

Aisenbrey, C., Marquette, A., & Bechinger, B. (2019). The mechanisms of action of cationic antimicrobial peptides refined by novel concepts from biophysical investigations. *Advances in Experimental Medicine and Biology*, *1117*, 33–64. Available from https://doi.org/10.1007/978-981-13-3588-4_4.

Alosaimy, S., Sabagha, N. L., Lagnf, A. M., Zasowski, E. J., Morrisette, T., Jorgensen, S. C. J., ... Rybak, M. J. (2020). Monotherapy with Vancomycin or Daptomycin vs combination therapy with β-lactams in the treatment of methicillin-resistant *Staphylococcus aureus* bloodstream infections: A retrospective cohort analysis. *Infectious Diseases and Therapy*, *9*(2), 325–339. Available from https://doi.org/10.1007/s40121-020-00292-8.

Amiche, M., Seon, A. A., Pierre, T. N., & Nicolas, P. (1999). The dermaseptin precursors: A protein family with a common preproregion and a variable C-terminal antimicrobial domain. *FEBS Letters*, *456*(3), 352–356. Available from https://doi.org/10.1016/S0014-5793(99)00964-3.

Andersson, D. I., Hughes, D., & Kubicek-Sutherland, J. Z. (2016). Mechanisms and consequences of bacterial resistance to antimicrobial peptides. *Drug Resistance Updates*, *26*, 43–57. Available from https://doi.org/10.1016/j.drup.2016.04.002.

Andreev, K., Martynowycz, M. W., Huang, M. L., Kuzmenko, I., Bu, W., Kirshenbaum, K., & Gidalevitz, D. (2018). Hydrophobic interactions modulate antimicrobial peptoid selectivity towards anionic lipid membranes. *Biochimica et Biophysica Acta (BBA)—Biomembranes*, *1860*(6), 1414–1423. Available from https://doi.org/10.1016/j.bbamem.2018.03.021.

Antibiotic resistance threats in the United States. (2019). Retrieved from https://stacks.cdc.gov/view/cdc/82532, Available from http://dx.doi.org/10.15620/cdc:82532.

Bracci, L., Falciani, C., Lelli, B., Lozzi, L., Runci, Y., Pini, A., … Neri, P. (2003). Synthetic peptides in the form of dendrimers become resistant to protease activity. *The Journal of Biological Chemistry*, *278*(47), 46590–46595. Available from https://doi.org/10.1074/jbc.M308615200.

Bulet, P., Dimarcq, J. L., Hetru, C., Lagueux, M., Charlet, M., Hegy, G., … Hoffmann, J. A. (1993). A novel inducible antibacterial peptide of Drosophila carries an O-glycosylated substitution. *The Journal of Biological Chemistry*, *268*(20), 14893–14897.

Bulet, P., Urge, L., Ohresser, S., Hetru, C., & Otvos, L., Jr. (1996). Enlarged scale chemical synthesis and range of activity of drosocin, an O-glycosylated antibacterial peptide of Drosophila. *European Journal of Biochemistry / FEBS*, *238*(1), 64–69. Available from https://doi.org/10.1111/j.1432-1033.1996.0064q.x.

Butler, M. F., Ng, Y.-F., & Pudney, P. D. A. (2003). Mechanism and kinetics of the crosslinking reaction between biopolymers containing primary amine groups and genipin. *Journal of Polymer Science Part A: Polymer Chemistry*, *41*(24), 3941–3953. Available from https://doi.org/10.1002/pola.10960.

Chung, P. Y., & Khanum, R. (2017). Antimicrobial peptides as potential anti-biofilm agents against multidrug-resistant bacteria. *Journal of Microbiology, Immunology and Infection*, *50*(4), 405–410. Available from https://doi.org/10.1016/j.jmii.2016.12.005.

Copolovici, D. M., Langel, K., Eriste, E., & Langel, U. (2014). Cell-penetrating peptides: Design, synthesis, and applications. *ACS nano*, *8*(3), 1972–1994.

Dartois, V., Sanchez-Quesada, J., Cabezas, E., Chi, E., Dubbelde, C., Dunn, C., … Parr, T. R., Jr. (2005). Systemic antibacterial activity of novel synthetic cyclic peptides. *Antimicrobial Agents and Chemotherapy*, *49*(8), 3302–3310. Available from https://doi.org/10.1128/aac.49.8.3302-3310.2005.

Desgranges, S., Ruddle, C. C., Burke, L. P., McFadden, T. M., O'Brien, J. E., Fitzgerald-Hughes, D., … Devocelle, M. (2012). β-lactam-host defence peptide conjugates as antibiotic prodrug candidates targeting resistant bacteria. *RSC Advances*, *2*(6), 2480–2492. Available from https://doi.org/10.1039/C2RA01351G.

Deshayes, S., Xian, W., Schmidt, N. W., Kordbacheh, S., Lieng, J., Wang, J., … Kasko, A. M. (2017). Designing hybrid antibiotic peptide conjugates to cross bacterial membranes. *Bioconjugate Chemistry*, *28*(3), 793–804. Available from https://doi.org/10.1021/acs.bioconjchem.6b00725.

Di, L. (2015). Strategic approaches to optimizing peptide ADME properties. *The AAPS Journal, 17*(1), 134–143. Available from https://doi.org/10.1208/s12248-014-9687-3.

Di Luca, M., Maccari, G., Maisetta, G., & Batoni, G. (2015). BaAMPs: The database of biofilm-active antimicrobial peptides. *Biofouling, 31*(2), 193–199. Available from https://doi.org/10.1080/08927014.2015.1021340.

Dubos, R. J. (1939a). Studies on a bactericidal agent extracted from a soil bacillus: I. Preparation of the agent. Its activity in vitro. *Journal of Experimental Medicine, 70*(1), 1–10. Available from https://doi.org/10.1084/jem.70.1.1.

Dubos, R. J. (1939b). Studies on a bactericidal agent extracted from a soil bacillus: II. Protective effect of the bactericidal agent against experimental pneumococcus infections in mice. *Journal of Experimental Medicine, 70*(1), 11–17. Available from https://doi.org/10.1084/jem.70.1.11.

Epand, R. M., Walker, C., Epand, R. F., & Magarvey, N. A. (2016). Molecular mechanisms of membrane targeting antibiotics. *Biochimica et Biophysica Acta (BBA)—Biomembranes, 1858*(5), 980–987. Available from https://doi.org/10.1016/j.bbamem.2015.10.018.

Fan, L., Sun, J., Zhou, M., Zhou, J., Lao, X., Zheng, H., & Xu, H. (2016). DRAMP: A comprehensive data repository of antimicrobial peptides. *Scientific Reports, 6*, 24482. Available from https://doi.org/10.1038/srep24482.

Fernandez-Lopez, S., Kim, H. S., Choi, E. C., Delgado, M., Granja, J. R., Khasanov, A., … Ghadiri, M. R. (2001). Antibacterial agents based on the cyclic D,L-alpha-peptide architecture. *Nature, 412*(6845), 452–455, 10.1038/35086601.

Fleming, A., & Wright, A. E. (1922). On a remarkable bacteriolytic element found in tissues and secretions. *Proceedings of the Royal Society of London. Series B, Containing Papers of a Biological Character, 93*(653), 306–317. Available from https://doi.org/10.1098/rspb.1922.0023.

Ganz, T. (2003). Defensins: Antimicrobial peptides of innate immunity. *Nature Reviews. Immunology, 3*(9), 710–720. Available from https://doi.org/10.1038/nri1180.

Ghaffar, K. A., Hussein, W. M., Khalil, Z. G., Capon, R. J., Skwarczynski, M., & Toth, I. (2015). Levofloxacin and indolicidin for combination antimicrobial therapy. *Current Drug Delivery, 12*(1), 108–114. Available from https://doi.org/10.2174/1567201811666140910094050.

Giacometti, A., Cirioni, O., Kamysz, W., Silvestri, C., Licci, A., Riva, A., … Scalise, G. (2005). In vitro activity of amphibian peptides alone and in combination with antimicrobial agents against multidrug-resistant pathogens isolated from surgical wound infection. *Peptides, 26*(11), 2111–2116. Available from https://doi.org/10.1016/j.peptides.2005.03.009.

Gogoladze, G., Grigolava, M., Vishnepolsky, B., Chubinidze, M., Duroux, P., Lefranc, M.-P., & Pirtskhalava, M. (2014). dbaasp: Database of antimicrobial activity and structure of peptides. *FEMS Microbiology Letters, 357*(1), 63–68. Available from https://doi.org/10.1111/1574-6968.12489.

Gómez, E. A., Giraldo, P., & Orduz, S. (2017). InverPep: A database of invertebrate antimicrobial peptides. *Journal of Global Antimicrobial Resistance, 8*, 13–17. Available from https://doi.org/10.1016/j.jgar.2016.10.003.

Goyal, R., Jerath, G., Akhil, R., Chandrasekharan, A., Puppala, E. R., Ponneganti, S., … Ramakrishnan, V. (2021). Geometry encoded functional programming of tumor homing peptides for targeted drug delivery. *Journal of Controlled Release, 333*, 16–27. Available from https://doi.org/10.1016/j.jconrel.2021.03.010.

Goyal, R., Jerath, G., Chandrasekharan, A., Christian, Y., Kumar, T. R. S., & Ramakrishnan, V. (2021). Correction to: Molecular hybridization combining tumor homing and penetrating peptide domains for cellular targeting. *Drug Delivery and Translational Research*. Available from https://doi.org/10.1007/s13346-021-01051-z.

Goyal, R., Jerath, G., Chandrasekharan, A., Kumar, T. R. S., & Ramakrishnan, V. (2020). Peptide-based delivery vectors with pre-defined geometrical locks. *RSC Medicinal Chemistry*. Available from https://doi.org/10.1039/D0MD00229A.

Goyal, R., & Ramakrishnan, V. (2019). Chapter 2—Peptide-based drug delivery systems. In S. S. Mohapatra, S. Ranjan, N. Dasgupta, R. K. Mishra, & S. Thomas (Eds.), *Characterization and Biology of Nanomaterials for Drug Delivery* (pp. 25−45). Elsevier.

Graf, M., & Wilson, D. N. (2019). Intracellular antimicrobial peptides targeting the protein synthesis machinery. *Advances in Experimental Medicine and Biology*, *1117*, 73−89. Available from https://doi.org/10.1007/978-981-13-3588-4_6.

Hammami, R., Ben Hamida, J., Vergoten, G., & Fliss, I. (2008). PhytAMP: A database dedicated to antimicrobial plant peptides. *Nucleic Acids Research*, *37*(suppl_1), D963−D968. Available from https://doi.org/10.1093/nar/gkn655.

Hammami, R., Zouhir, A., Ben Hamida, J., & Fliss, I. (2007). BACTIBASE: A new web-accessible database for bacteriocin characterization. *BMC Microbiology*, *7*(1), 89. Available from https://doi.org/10.1186/1471-2180-7-89.

Hammami, R., Zouhir, A., Le Lay, C., Ben Hamida, J., & Fliss, I. (2010). BACTIBASE second release: A database and tool platform for bacteriocin characterization. *BMC Microbiology*, *10*(1), 22. Available from https://doi.org/10.1186/1471-2180-10-22.

Hancock, R. E., & Diamond, G. (2000). The role of cationic antimicrobial peptides in innate host defences. *Trends in Microbiology*, *8*(9), 402−410.

Hancock, R. E., & Scott, M. G. (2000). The role of antimicrobial peptides in animal defenses. *Proceedings of the National Academy of Sciences*, *97*(16), 8856−8861.

Hansen, J. N., & Sandine, W. E. (1994). Nisin as a model food preservative. *Critical Reviews in Food Science and Nutrition*, *34*(1), 69−93. Available from https://doi.org/10.1080/10408399409527650.

Hasan, A., Saxena, V., Castelletto, V., Zimbitas, G., Seitsonen, J., Ruokolainen, J., ... Lau, K. H. A. (2020). Chain-end modifications and sequence arrangements of antimicrobial peptoids for mediating activity and nano-assembly. *Frontiers in Chemistry*, *8*(416). Available from https://doi.org/10.3389/fchem.2020.00416.

Hawley, H. B., & Gump, D. W. (1973). Vancomycin therapy of bacterial meningitis. *American Journal of Diseases of Children*, *126*(2), 261−264. Available from https://doi.org/10.1001/archpedi.1973.02110190231025.

Hazam, P. K., Akhil, R., Jerath, G., Saikia, J., & Ramakrishnan, V. (2019). Topological effects on the designability and bactericidal potency of antimicrobial peptides. *Biophysical Chemistry*, *248*, 1−8. Available from https://doi.org/10.1016/j.bpc.2019.02.005.

Hazam, P. K., Goyal, R., & Ramakrishnan, V. (2019). Peptide based antimicrobials: Design strategies and therapeutic potential. *Progress in Biophysics and Molecular Biology*, *142*, 10−22. Available from https://doi.org/10.1016/j.pbiomolbio.2018.08.006.

Hazam, P. K., Jerath, G., Chaudhary, N., & Ramakrishnan, V. (2018). Peptido-mimetic approach in the design of syndiotactic antimicrobial peptides. *International Journal of Peptide Research and Therapeutics*, *24*(2), 299−307.

Hazam, P. K., Jerath, G., Kumar, A., Chaudhary, N., & Ramakrishnan, V. (2017). Effect of tacticity-derived topological constraints in bactericidal peptides. *Biochimica et biophysica acta. Biomembranes*, *1859*(8), 1388−1395. Available from https://doi.org/10.1016/j.bbamem.2017.05.002.

Hazam, P. K., Phukan, C., Akhil, R., Singh, A., & Ramakrishnan, V. (2021). Antimicrobial effects of syndiotactic polypeptides. *Scientific Reports*, *11*(1), 1823. Available from https://doi.org/10.1038/s41598-021-81394-2.

Hazam, P. K., Singh, A., Chaudhary, N., & Ramakrishnan, V. (2019). Bactericidal potency and extended serum life of stereo-chemically engineered peptides against *Mycobacterium*. *International Journal of Peptide Research and Therapeutics*, *25*(2), 465–472. Available from https://doi.org/10.1007/s10989-018-9690-0.

Hsu, S. T., Breukink, E., Tischenko, E., Lutters, M. A., de Kruijff, B., Kaptein, R., ... van Nuland, N. A. (2004). The nisin-lipid II complex reveals a pyrophosphate cage that provides a blueprint for novel antibiotics. *Nature Structural & Molecular Biology*, *11*(10), 963–967. Available from https://doi.org/10.1038/nsmb830.

Hu, H., Xue, J., Swarts, B. M., Wang, Q., Wu, Q., & Guo, Z. (2009). Synthesis and antibacterial activities of N-glycosylated derivatives of Tyrocidine A, a macrocyclic peptide antibiotic. *Journal of Medicinal Chemistry*, *52*(7), 2052–2059. Available from https://doi.org/10.1021/jm801577r.

Huan, Y., Kong, Q., Mou, H., & Yi, H. (2020). Antimicrobial peptides: Classification, design, application and research progress in multiple fields. *Frontiers in Microbiology*, *11*(2559). Available from https://doi.org/10.3389/fmicb.2020.582779.

Jerath, G., Goyal, R., Trivedi, V., Santhoshkumar, T. R., & Ramakrishnan, V. (2019). Syndiotactic peptides for targeted delivery. *Acta Biomaterialia*, *87*, 130–139. Available from https://doi.org/10.1016/j.actbio.2019.01.036.

Jerath, G., Goyal, R., Trivedi, V., Santhoshkumar, T. R., & Ramakrishnan, V. (2020). Conformationally constrained peptides for drug delivery. *Journal of Peptide Science*, *26*(4-5), e3244. Available from https://doi.org/10.1002/psc.3244.

Junior, E. F. C., Guimarães, C., Franco, L. L., Alves, R. J., Kato, K. C., Martins, H. R., ... Verly, R. M. (2017). Glycotriazole-peptides derived from the peptide HSP1: Synergistic effect of triazole and saccharide rings on the antifungal activity. *Amino Acids*, *49*(8), 1389–1400. Available from https://doi.org/10.1007/s00726-017-2441-2.

Kelkar, D. A., & Chattopadhyay, A. (2007). The gramicidin ion channel: A model membrane protein. *Biochimica et Biophysica Acta (BBA)—Biomembranes*, *1768*(9), 2011–2025. Available from https://doi.org/10.1016/j.bbamem.2007.05.011.

Kessler, N., Schuhmann, H., Morneweg, S., Linne, U., & Marahiel, M. A. (2004). The linear pentadecapeptide gramicidin is assembled by four multimodular nonribosomal peptide synthetases that comprise 16 modules with 56 catalytic domains. *The Journal of Biological Chemistry*, *279*(9), 7413–7419. Available from https://doi.org/10.1074/jbc.M309658200.

Khrushchev, A. Y., Kashparov, I. A., Klimenko, L. V., & Mitin, Y. V. (2007). Branched antimicrobial peptides. *Russian Journal of Bioorganic Chemistry*, *33*(6), 544–548. Available from https://doi.org/10.1134/S1068162007060027.

Kitchen, C. M., Nuño, M., Kitchen, S. G., & Krogstad, P. (2008). Enfuvirtide antiretroviral therapy in HIV-1 infection. *Therapeutics and Clinical Risk Management*, *4*(2), 433–439. Available from https://doi.org/10.2147/tcrm.s1962.

Kumar, A., Ramakrishnan, V., Ranbhor, R., Patel, K., & Durani, S. (2009). Homochiral stereochemistry: The missing link of structure to energetics in protein folding. *The Journal of Physical Chemistry. B*, *113*(51), 16435–16442. Available from https://doi.org/10.1021/jp906811k.

Kumar, A., Ranbhor, R., Patel, K., Ramakrishnan, V., & Durani, S. (2017). Automated protein design: Landmarks and operational principles. *Progress in Biophysics and Molecular Biology*, *125*, 24–35. Available from https://doi.org/10.1016/j.pbiomolbio.2016.12.002.

Lakshminarayanan, R., Tan, W. X., Aung, T. T., Goh, E. T. L., Muruganantham, N., Li, J., ... Beuerman, R. W. (2016). Branched peptide, B2088, disrupts the supramolecular organization of lipopolysaccharides and sensitizes the Gram-negative bacteria. *Scientific Reports*, *6* (1), 25905. Available from https://doi.org/10.1038/srep25905.

Lele, D. S., Kaur, G., Thiruvikraman, M., & Kaur, K. J. (2017). Comparing naturally occurring glycosylated forms of proline rich antibacterial peptide, Drosocin. *Glycoconjugate Journal*, *34*(5), 613–624. Available from https://doi.org/10.1007/s10719-017-9781-8.

Li, J., Qu, X., He, X., Duan, L., Wu, G., Bi, D., ... Ou, H.-Y. (2012). ThioFinder: A web-based tool for the identification of thiopeptide gene clusters in DNA sequences. *PLoS One*, *7*(9), e45878. Available from https://doi.org/10.1371/journal.pone.0045878.

Li, Z., Yuan, P., Xing, M., He, Z., Dong, C., Cao, Y., & Liu, Q. (2013). Fatty acid conjugation enhances the activities of antimicrobial peptides. *Recent Patents on Food, Nutrition & Agriculture*, *5*(1), 52–56. Available from https://doi.org/10.2174/2212798411305010008.

Lin, H., & Walsh, C. T. (2004). A chemoenzymatic approach to glycopeptide antibiotics. *Journal of the American Chemical Society*, *126*(43), 13998–14003. Available from https://doi.org/10.1021/ja045147v.

Lorenzón, E. N., Sanches, P. R., Nogueira, L. G., Bauab, T. M., & Cilli, E. M. (2013). Dimerization of aurein 1.2: Effects in structure, antimicrobial activity and aggregation of *Cândida albicans* cells. *Amino Acids*, *44*(6), 1521–1528. Available from https://doi.org/10.1007/s00726-013-1475-3.

Lote, C. J., & Weiss, J. B. (1971). Identification of digalactosylcysteine in a glycopeptide isolated from urine by a new preparative technique. *FEBS Letters*, *16*(2), 81–85. Available from https://doi.org/10.1016/0014-5793(71)80337-X.

Ludolph, B., Eisele, F., & Waldmann, H. (2002). Solid-phase synthesis of lipidated peptides. *Journal of the American Chemical Society*, *124*(21), 5954–5955. Available from https://doi.org/10.1021/ja025768t.

Mackintosh, J. A., Veal, D. A., Beattie, A. J., & Gooley, A. A. (1998). Isolation from an ant *Myrmecia gulosa* of two inducible O-glycosylated proline-rich antibacterial peptides. *The Journal of Biological Chemistry*, *273*(11), 6139–6143. Available from https://doi.org/10.1074/jbc.273.11.6139.

Mak, P., Pohl, J., Dubin, A., Reed, M. S., Bowers, S. E., Fallon, M. T., & Shafer, W. M. (2003). The increased bactericidal activity of a fatty acid-modified synthetic antimicrobial peptide of human cathepsin G correlates with its enhanced capacity to interact with model membranes. *International Journal of Antimicrobial Agents*, *21*(1), 13–19. Available from https://doi.org/10.1016/s0924-8579(02)00245-5.

Mannige, R. V., Haxton, T. K., Proulx, C., Robertson, E. J., Battigelli, A., Butterfoss, G. L., ... Whitelam, S. (2015). Peptoid nanosheets exhibit a new secondary-structure motif. *Nature*, *526*(7573), 415–420. Available from https://doi.org/10.1038/nature15363.

Matthews, S. J., & Lancaster, J. W. (2012). Telaprevir: A hepatitis C NS3/4A protease inhibitor. *Clinical Therapeutics*, *34*(9), 1857–1882. Available from https://doi.org/10.1016/j.clinthera.2012.07.011.

Michael Henderson, J., & Lee, K. Y. C. (2013). Promising antimicrobial agents designed from natural peptide templates. *Current Opinion in Solid State and Materials Science*, *17*(4), 175–192. Available from https://doi.org/10.1016/j.cossms.2013.08.003.

Ming, L. J., & Epperson, J. D. (2002). Metal binding and structure-activity relationship of the metalloantibiotic peptide bacitracin. *Journal of Inorganic Biochemistry*, *91*(1), 46–58. Available from https://doi.org/10.1016/s0162-0134(02)00464-6.

Mishra, B., & Wang, G. (2012). Ab initio design of potent anti-MRSA peptides based on database filtering technology. *Journal of the American Chemical Society*, *134*(30), 12426–12429. Available from https://doi.org/10.1021/ja305644e.

Moradi, S. V., Hussein, W. M., Varamini, P., Simerska, P., & Toth, I. (2016). Glycosylation, an effective synthetic strategy to improve the bioavailability of therapeutic peptides. *Chemical Science*, *7*(4), 2492–2500. Available from https://doi.org/10.1039/C5SC04392A.

Moravej, H., Moravej, Z., Yazdanparast, M., Heiat, M., Mirhosseini, A., Moosazadeh Moghaddam, M., & Mirnejad, R. (2018). Antimicrobial peptides: Features, action, and their resistance mechanisms in bacteria. *Microbial Drug Resistance (Larchmont, N.Y.)*, *24*(6), 747–767. Available from https://doi.org/10.1089/mdr.2017.0392.

Mukai, Y., Matsushita, Y., Niidome, T., Hatekeyama, T., & Aoyagi, H. (2002). Parallel and antiparallel dimers of magainin 2: Their interaction with phospholipid membrane and antibacterial activity. *Journal of Peptide Science*, *8*(10), 570–577. Available from https://doi.org/10.1002/psc.416.

Mukhopadhyay, S., Bharath Prasad, A. S., Mehta, C. H., & Nayak, U. Y. (2020). Antimicrobial peptide polymers: No escape to ESKAPE pathogens—A review. *World Journal of Microbiology and Biotechnology*, *36*(9), 131. Available from https://doi.org/10.1007/s11274-020-02907-1.

Muttenthaler, M., King, G. F., Adams, D. J., & Alewood, P. F. (2021). Trends in peptide drug discovery. *Nature Reviews. Drug Discovery*, *20*(4), 309–325. Available from https://doi.org/10.1038/s41573-020-00135-8.

Novković, M., Simunić, J., Bojović, V., Tossi, A., & Juretić, D. (2012). DADP: The database of anuran defense peptides. *Bioinformatics (Oxford, England)*, *28*(10), 1406–1407. Available from https://doi.org/10.1093/bioinformatics/bts141.

Okada, M., & Natori, S. (1985). Primary structure of sarcotoxin I, an antibacterial protein induced in the hemolymph of *Sarcophaga peregrina* (flesh fly) larvae. *Journal of Biological Chemistry*, *260*(12), 7174–7177. Available from https://doi.org/10.1016/S0021-9258(17)39590-X.

Oman, T. J., Boettcher, J. M., Wang, H., Okalibe, X. N., & van der Donk, W. A. (2011). Sublancin is not a lantibiotic but an S-linked glycopeptide. *Nature Chemical Biology*, *7*(2), 78–80. Available from https://doi.org/10.1038/nchembio.509.

Opdenakker, G., Rudd, P. M., Wormald, M., Dwek, R. A., & Van Damme, J. (1995). Cells regulate the activities of cytokines by glycosylation. *The FASEB Journal*, *9*(5), 453–457. Available from https://doi.org/10.1096/fasebj.9.5.7896019.

Orlov, D. S., Shamova, O. V., Eliseev, I. E., Zharkova, M. S., Chakchir, O. B., Antcheva, N., ... Tossi, A. (2019). Redesigning Arenicin-1, an antimicrobial peptide from the marine polychaeta *Arenicola marina*, by strand rearrangement or branching, substitution of specific residues, and backbone linearization or cyclization. *Marine Drugs*, *17*(6), 376. Available from https://doi.org/10.3390/md17060376.

Piazza, C., Formaggio, F., Crisma, M., Toniolo, C., Kamphuis, J., Kaptein, B., & Broxterman, Q. B. (1999). Total synthesis and membrane modifying properties of the lipopeptaibol trikoningin KB II and its analogues with acyl chains of different length at the N- and C-termini. *Journal of Peptide Science: An Official Publication of the European Peptide Society*, *5*(2), 96–102, 10.1002/(sici)1099-1387(199902)5:2 < 96::Aid-psc185 > 3.0.Co;2-g.

Pini, A., Giuliani, A., Falciani, C., Runci, Y., Ricci, C., Lelli, B., ... Bracci, L. (2005). Antimicrobial activity of novel dendrimeric peptides obtained by phage display selection and rational modification. *Antimicrobial Agents and Chemotherapy*, *49*(7), 2665–2672. Available from https://doi.org/10.1128/aac.49.7.2665-2672.2005.

Piotto, S. P., Sessa, L., Concilio, S., & Iannelli, P. (2012). YADAMP: Yet another database of antimicrobial peptides. *International Journal of Antimicrobial Agents*, *39*(4), 346–351. Available from https://doi.org/10.1016/j.ijantimicag.2011.12.003.

Poirel, L., Jayol, A., & Nordmann, P. (2017). Polymyxins: Antibacterial activity, susceptibility testing, and resistance mechanisms encoded by plasmids or chromosomes. *Clinical Microbiology Reviews*, *30*(2), 557–596. Available from https://doi.org/10.1128/cmr.00064-16.

Prakash, V., Ranbhor, R., & Ramakrishnan, V. (2020). De novo designed heterochiral blue fluorescent protein. *ACS Omega*, *5*(41), 26382–26388. Available from https://doi.org/10.1021/acsomega.0c02574.

Prenner, E. J., Lewis, R. N., & McElhaney, R. N. (1999). The interaction of the antimicrobial peptide gramicidin S with lipid bilayer model and biological membranes. *Biochimica et Biophysica Acta*, *1462*(1-2), 201–221. Available from https://doi.org/10.1016/s0005-2736(99)00207-2.

Ptaszyńska, N., Gucwa, K., Olkiewicz, K., Łęgowska, A., Okońska, J., Ruczyński, J., … Rolka, K. (2019). Antibiotic-based conjugates containing antimicrobial HLopt2 peptide: Design, synthesis, antimicrobial and cytotoxic activities. *ACS Chemical Biology*, *14*(10), 2233–2242. Available from https://doi.org/10.1021/acschembio.9b00538.

Ptaszyńska, N., Olkiewicz, K., Okońska, J., Gucwa, K., Łęgowska, A., Gitlin-Domagalska, A., … Rolka, K. (2019). Peptide conjugates of lactoferricin analogues and antimicrobials—Design, chemical synthesis, and evaluation of antimicrobial activity and mammalian cytotoxicity. *Peptides*, *117*, 170079. Available from https://doi.org/10.1016/j.peptides.2019.04.006.

Ramakrishnan, V., Ranbhor, R., & Durani, S. (2004). Existence of specific "folds" in polyproline II ensembles of an "unfolded" alanine peptide detected by molecular dynamics. *Journal of the American Chemical Society*, *126*(50), 16332–16333.

Ramakrishnan, V., Ranbhor, R., & Durani, S. (2005). Simulated folding in polypeptides of diversified molecular tacticity: Implications for protein folding and de novo design. *Biopolymers*, *78*(2), 96–105.

Ramakrishnan, V., Ranbhor, R., Kumar, A., & Durani, S. (2006). The link between sequence and conformation in protein structures appears to be stereochemically established. *The Journal of Physical Chemistry. B*, *110*(18), 9314–9323. Available from https://doi.org/10.1021/jp056417e.

Ranbhor, R., Kumar, A., Patel, K., Ramakrishnan, V., & Durani, S. (2018). Automated design evolution of stereochemically randomized protein foldamers. *Physical Biology*, *15*(3), 036001. Available from https://doi.org/10.1088/1478-3975/aaac9a. (Accession No. 29393061).

Ranbhor, R., Kumar, A., Tendulkar, A., Patel, K., Ramakrishnan, V., & Durani, S. (2018). IDeAS: Automated design tool for hetero-chiral protein folds. *Physical Biology*, *15*(6), 066005. Available from https://doi.org/10.1088/1478-3975/aacdc3.

Ranbhor, R., Ramakrishnan, V., Kumar, A., & Durani, S. (2006). The interplay of sequence and stereochemistry in defining conformation in proteins and polypeptides. *Biopolymers*, *83*(5), 537–545. Available from https://doi.org/10.1002/bip.20584.

Randall, C. P., Mariner, K. R., Chopra, I., & O'Neill, A. J. (2013). The target of daptomycin is absent from *Escherichia coli* and other gram-negative pathogens. *Antimicrobial Agents and Chemotherapy*, *57*(1), 637–639. Available from https://doi.org/10.1128/aac.02005-12.

Rao, J., Yan, L., Lahiri, J., Whitesides, G. M., Weis, R. M., & Warren, H. S. (1999). Binding of a dimeric derivative of vancomycin to L-Lys-D-Ala-D-lactate in solution and at a surface. *Chemistry & Biology*, *6*(6), 353–359. Available from https://doi.org/10.1016/s1074-5521(99)80047-7.

Reddy, K., Yedery, R., & Aranha, C. (2004). Antimicrobial peptides: Premises and promises. *International Journal of Antimicrobial Agents*, *24*(6), 536–547.

Renukuntla, J., Vadlapudi, A. D., Patel, A., Boddu, S. H. S., & Mitra, A. K. (2013). Approaches for enhancing oral bioavailability of peptides and proteins. *International Journal of Pharmaceutics*, *447*(1), 75–93. Available from https://doi.org/10.1016/j.ijpharm.2013.02.030.

Rodriguez, C. A., Papanastasiou, E. A., Juba, M., & Bishop, B. (2014). Covalent modification of a ten-residue cationic antimicrobial peptide with levofloxacin. *Frontiers in Chemistry*, *2*(71). Available from https://doi.org/10.3389/fchem.2014.00071.

Saikia, K., Sravani, Y. D., Ramakrishnan, V., & Chaudhary, N. (2017). Highly potent antimicrobial peptides from N-terminal membrane-binding region of *E. coli* MreB. *Scientific Reports*, *7*, 42994. Available from https://doi.org/10.1038/srep42994.

Scott, L. J. (2015). Dalbavancin: A review in acute bacterial skin and skin structure infections. *Drugs*, *75*(11), 1281–1291. Available from https://doi.org/10.1007/s40265-015-0430-x.

Seebah, S., Suresh, A., Zhuo, S., Choong, Y. H., Chua, H., Chuon, D., ... Verma, C. (2006). Defensins knowledgebase: A manually curated database and information source focused on the defensins family of antimicrobial peptides. *Nucleic Acids Research*, *35*(1), D265–D268. Available from https://doi.org/10.1093/nar/gkl866.

Seshadri Sundararajan, V., Gabere, M. N., Pretorius, A., Adam, S., Christoffels, A., Lehväslaiho, M., ... Bajic, V. B. (2011). DAMPD: A manually curated antimicrobial peptide database. *Nucleic Acids Research*, *40*(D1), D1108–D1112. Available from https://doi.org/10.1093/nar/gkr1063.

Sharma, D., Choudhary, M., Vashistt, J., Shrivastava, R., & Bisht, G. S. (2019). Cationic antimicrobial peptide and its poly-N-substituted glycine congener: Antibacterial and antibiofilm potential against *A. baumannii*. *Biochemical and Biophysical Research Communications*, *518*(3), 472–478. Available from https://doi.org/10.1016/j.bbrc.2019.08.062.

Silhavy, T. J., Kahne, D., & Walker, S. (2010). The bacterial cell envelope. *Cold Spring Harbor Perspectives in Biology*, *2*(5), a000414.

Simon, R. J., Kania, R. S., Zuckermann, R. N., Huebner, V. D., Jewell, D. A., Banville, S., et al. (1992). Peptoids: A modular approach to drug discovery. *Proceedings of the National Academy of Sciences of the United States of America*, *89*(20), 9367–9371. Available from https://doi.org/10.1073/pnas.89.20.9367.

Skerlavaj, B., Benincasa, M., Risso, A., Zanetti, M., & Gennaro, R. (1999). SMAP-29: A potent antibacterial and antifungal peptide from sheep leukocytes. *FEBS Letters*, *463*(1), 58–62. Available from https://doi.org/10.1016/S0014-5793(99)01600-2.

Steiner, H. (1982). Secondary structure of the cecropins: Antibacterial peptides from the moth *Hyalophora cecropia*. *FEBS Letters*, *137*(2), 283–287. Available from https://doi.org/10.1016/0014-5793(82)80368-2.

Stone, K. J., & Strominger, J. L. (1971). Mechanism of action of bacitracin: complexation with metal ion and C 55-isoprenyl pyrophosphate. *Proceedings of the National Academy of Sciences of the United States of America*, *68*(12), 3223–3227. Available from https://doi.org/10.1073/pnas.68.12.3223.

Straus, S. K., & Hancock, R. E. (2006). Mode of action of the new antibiotic for Gram-positive pathogens daptomycin: Comparison with cationic antimicrobial peptides and lipopeptides. *Biochimica et Biophysica Acta*, *1758*(9), 1215–1223. Available from https://doi.org/10.1016/j.bbamem.2006.02.009.

Su, Y., Tian, L., Yu, M., Gao, Q., Wang, D., Xi, Y., ... Li, P. (2017). Cationic peptidopolysaccharides synthesized by 'click' chemistry with enhanced broad-spectrum antimicrobial activities. *Polymer Chemistry*, *8*(25), 3788–3800. Available from https://doi.org/10.1039/C7PY00528H.

Swierstra, J., Kapoerchan, V., Knijnenburg, A., van Belkum, A., & Overhand, M. (2016). Structure, toxicity and antibiotic activity of gramicidin S and derivatives. *European Journal of Clinical Microbiology & Infectious Diseases*, *35*(5), 763–769. Available from https://doi.org/10.1007/s10096-016-2595-y.

Tam, J. P., Lu, Y. A., & Yang, J. L. (2002). Antimicrobial dendrimeric peptides. *European Journal of Biochemistry / FEBS*, *269*(3), 923–932. Available from https://doi.org/10.1046/j.0014-2956.2001.02728.x.

Teixeira, V., Feio, M. J., & Bastos, M. (2012). Role of lipids in the interaction of antimicrobial peptides with membranes. *Progress in Lipid Research*, *51*(2), 149–177. Available from https://doi.org/10.1016/j.plipres.2011.12.005.

Tencza, S. B., Creighton, D. J., Yuan, T., Vogel, H. J., Montelaro, R. C., & Mietzner, T. A. (1999). Lentivirus-derived antimicrobial peptides: Increased potency by sequence engineering and dimerization. *The Journal of Antimicrobial Chemotherapy*, *44*(1), 33–41. Available from https://doi.org/10.1093/jac/44.1.33.

Théolier, J., Fliss, I., Jean, J., & Hammami, R. (2014). MilkAMP: A comprehensive database of antimicrobial peptides of dairy origin. *Dairy Science & Technology*, *94*(2), 181–193. Available from https://doi.org/10.1007/s13594-013-0153-2.

Thomas, S., Karnik, S., Barai, R. S., Jayaraman, V. K., & Idicula-Thomas, S. (2009). CAMP: A useful resource for research on antimicrobial peptides. *Nucleic Acids Research*, *38*(1), D774–D780. Available from https://doi.org/10.1093/nar/gkp1021.

Tsubery, H., Ofek, I., Cohen, S., & Fridkin, M. (2000). The functional association of polymyxin B with bacterial lipopolysaccharide is stereospecific: Studies on polymyxin B nonapeptide. *Biochemistry*, *39*(39), 11837–11844. Available from https://doi.org/10.1021/bi000386q.

Vargas Casanova, Y., Rodríguez Guerra, J. A., Umaña Pérez, Y. A., Leal Castro, A. L., Almanzar Reina, G., García Castañeda, J. E., & Rivera Monroy, Z. J. (2017). Antibacterial synthetic peptides derived from bovine lactoferricin exhibit cytotoxic effect against MDA-MB-468 and MDA-MB-231 breast cancer cell lines. *Molecules (Basel, Switzerland)*, *22*(10), 1641, Retrieved from. Available from https://www.mdpi.com/1420-3049/22/10/1641.

Vega, S. C., Martínez, D. A., Chalá, M. D. S., Vargas, H. A., & Rosas, J. E. (2018). Design, synthesis and evaluation of branched RRWQWR-based peptides as antibacterial agents against clinically relevant Gram-positive and Gram-negative pathogens. *Frontiers in Microbiology*, *9*, 329. Available from https://doi.org/10.3389/fmicb.2018.00329.

Velkov, T., Thompson, P. E., Nation, R. L., & Li, J. (2010). Structure–activity relationships of polymyxin antibiotics. *Journal of Medicinal Chemistry*, *53*(5), 1898–1916. Available from https://doi.org/10.1021/jm900999h.

Vernen, F., Harvey, P. J., Dias, S. A., Veiga, A. S., Huang, Y. H., Craik, D. J., … Troeira Henriques, S. (2019). Characterization of Tachyplesin peptides and their cyclized analogues to improve antimicrobial and anticancer properties. *International Journal of Molecular Sciences*, *20*(17). Available from https://doi.org/10.3390/ijms20174184.

Wade, D., Boman, A., Wåhlin, B., Drain, C. M., Andreu, D., Boman, H. G., & Merrifield, R. B. (1990). All-D amino acid-containing channel-forming antibiotic peptides. *Proceedings of the National Academy of Sciences of the United States of America*, *87*(12), 4761–4765. Available from https://doi.org/10.1073/pnas.87.12.4761.

Wang, C. K., Kaas, Q., Chiche, L., & Craik, D. J. (2008). CyBase: A database of cyclic protein sequences and structures, with applications in protein discovery and engineering. *Nucleic*

Acids Research, *36*, D206–D210. Available from https://doi.org/10.1093/nar/gkm953. (Database issue).

Wang, G., Li, X., & Wang, Z. (2009). APD2: The updated antimicrobial peptide database and its application in peptide design. *Nucleic Acids Research*, *37*, D933–D937. Available from https://doi.org/10.1093/nar/gkn823. (Database issue).

Wang, G., Li, X., & Wang, Z. (2015). APD3: The antimicrobial peptide database as a tool for research and education. *Nucleic Acids Research*, *44*(D1), D1087–D1093. Available from https://doi.org/10.1093/nar/gkv1278.

Wang, S.-T., Gray, M. A., Xuan, S., Lin, Y., Byrnes, J., Nguyen, A. I., ... Gang, O. (2020). DNA origami protection and molecular interfacing through engineered sequence-defined peptoids. *Proceedings of the National Academy of Sciences*, *117*(12), 6339–6348. Available from https://doi.org/10.1073/pnas.1919749117.

Wang, Z., & Wang, G. (2004). APD: The antimicrobial peptide database. *Nucleic Acids Research*, *32*, D590–D592. Available from https://doi.org/10.1093/nar/gkh025. (Database issue).

Whitmore, L., & Wallace, B. A. (2004). The Peptaibol database: A database for sequences and structures of naturally occurring peptaibols. *Nucleic Acids Research*, *32*, D593–D594. Available from https://doi.org/10.1093/nar/gkh077. (Database issue).

Wimley, W. C. (2010). Describing the mechanism of antimicrobial peptide action with the interfacial activity model. *ACS Chemical Biology*, *5*(10), 905–917. Available from https://doi.org/10.1021/cb1001558.

Wu, H., Lu, H., Huang, J., Li, G., & Huang, Q. (2012). EnzyBase: A novel database for enzybiotic studies. *BMC Microbiology*, *12*(1), 54. Available from https://doi.org/10.1186/1471-2180-12-54.

Yang, L., Harroun, T. A., Weiss, T. M., Ding, L., & Huang, H. W. (2001). Barrel-Stave model or toroidal model? A case study on melittin pores. *Biophysical Journal*, *81*(3), 1475–1485. Available from https://doi.org/10.1016/S0006-3495(01)75802-X.

Yeaman, M. R., & Yount, N. Y. (2003). Mechanisms of antimicrobial peptide action and resistance. *Pharmacological Reviews*, *55*(1), 27. Available from https://doi.org/10.1124/pr.55.1.2.

Yu, Z., Qin, W., Lin, J., Fang, S., & Qiu, J. (2015). Antibacterial mechanisms of polymyxin and bacterial resistance. *BioMed Research International*, *2015*, 679109. Available from https://doi.org/10.1155/2015/679109.

Zanetti, M., Gennaro, R., & Romeo, D. (1995). Cathelicidins: A novel protein family with a common proregion and a variable C-terminal antimicrobial domain. *FEBS Letters*, *374*(1), 1–5. Available from https://doi.org/10.1016/0014-5793(95)01050-O.

Zhang, L., & Bulaj, G. (2012). Converting peptides into drug leads by lipidation. *Current Medicinal Chemistry*, *19*(11), 1602–1618. Available from https://doi.org/10.2174/092986712799945003.

Zhao, X., Wu, H., Lu, H., Li, G., & Huang, Q. (2013). LAMP: A database linking antimicrobial peptides. *PLoS One*, *8*(6), e66557. Available from https://doi.org/10.1371/journal.pone.0066557.

Zhu, L., Zhao, Z., Cheng, P., He, Z., Cheng, Z., Peng, J., ... Hu, Z. (2017). Antibody-mimetic peptoid nanosheet for label-free serum-based diagnosis of Alzheimer's disease. *Advanced Materials*, *29*(30). Available from https://doi.org/10.1002/adma.201700057.

Zorzi, A., Middendorp, S. J., Wilbs, J., Deyle, K., & Heinis, C. (2017). Acylated heptapeptide binds albumin with high affinity and application as tag furnishes long-acting peptides. *Nature Communications*, *8*(1), 16092. Available from https://doi.org/10.1038/ncomms16092.

Chapter 5

Cell-penetrating peptides

Aparna Rai[1] and Gaurav Jerath[1,2,*]
[1]*Pepthera Laboratories Pvt. Ltd., Technology Incubation Centre, Indian Institute of Technology Guwahati, Guwahati, Assam, India,* [2]*Department of Medical Biotechnology, Gujarat Biotechnology University, Gandhinagar, Gujarat, India*

Peptides are oligomers of amino acids with associated chemical and biological functions. These functions are the results of the peptide components such as the amino acid sequence, peptide conformation, and spatial physicochemical properties. This chapter will focus on the peptides that can traverse the cell membrane of mammalian cells. Such peptides are often referred to as cell-penetrating peptides (CPPs), Trojan peptides, Protein Transduction Domains (PTDs), etc. The discovery of this class of peptides is attributed to the identification of an amino acid sequence in the trans-activator protein of the Human Immunodeficiency Virus (HIV), which is responsible for the penetration of the virus inside host cells.

CPPs are short oligomeric peptides with a chain length of 5–30 amino acids, which traverse through cell membranes and can deliver a variety of molecules including proteins, nucleic acids, small molecules, etc. in the cell (Heitz, Morris, & Divita, 2009). Naturally, given their ability to crossover biological membranes, CPPs have been extensively studied for their drug delivery applications. CPPSite 2.0 (a database) has records of over 1800 peptides with cell-penetrative activity (Agrawal et al., 2016). In the last 30 years or so, CPPs have been used from the basic research to preclinical studies for treatment of multiple diseases. Most often, the drug delivery applications of CPPs have been studied in the field of oncology (Feni & Neundorf, 2017).

Following the general conception of understanding the product prior to "designing" the product, this chapter will focus on the various what's and how's of CPPs for a better understanding of the de novo design of CPPs. The design of a peptide for the intended purpose of cell penetration requires the

* Equal Contribution.

De Novo Peptide Design. DOI: https://doi.org/10.1016/B978-0-323-99917-5.00005-6

understanding of what are CPPs; what are their physicochemical properties; how CPPs work (mechanism); what are the strengths, limitations, and opportunities available for CPPs; what are their applications and how can we design them.

5.1 Cell-penetrating peptides: a brief history

Membranes are integral components of biological systems, as they help defining a system's boundary at the physical, chemical, and biological levels. An important function of membranes other than defining a system is to regulate the transfer of molecules to and from the system and its surroundings. Cell membranes, therefore, regulate the flow of molecules to and from the intracellular locale. Cell membranes restrict the transport of molecules to and from the intracellular locale. The membranes are selectively permeable to molecules recognized by various cell surface receptors. This barrier acts as an obstacle for intracellular delivery of various therapeutic molecules. Most therapeutic molecules (drugs) are chemical moieties with various levels of hydrophobicity and charge. Majority of these drugs are mimics of a natural ligand for cell surface receptors, for example, methotrexate is the chemical mimic of folic acid and recognized by the folate receptors on the cell surface (Antony, 1992; Jansen, 1999). The uptake of such chemi-mimics is mostly through receptor-mediated endocytosis (Jansen, 1999). In other cases, they may require a carrier molecule to cross the cell membranes. The use of carriers for drug encapsulation provides a variety of advantages like better drug solubility, enhanced drug stability, decreased side effects, targeted delivery, etc. in comparison to the native drug (Gelperina, Kisich, Iseman, & Heifets, 2005; Sastry, Nyshadham, & Fix, 2000; Singh & Lillard, 2009; Tiwari et al., 2012).

The discovery of protein transduction properties of HIV TAT protein in late-eighties contradicted the established impression of the scientific community that the cell membrane was impermeable to hydrophilic molecules (Frankel & Pabo, 1988). Further, a transcription factor of *Drosophila melanogaster*, Antennapedia homeodomain, was also reported to penetrate live nerve cells in 1991 (Joliot, Pernelle, Deagostini-Bazin, & Prochiantz, 1991). Together, the transduction properties of the two proteins presented an exciting opportunity to identify the least amino acid sequence required for cellular uptake, leading to the discovery of Tat and Penetratin, the first CPPs (Bechara & Sagan, 2013). In 1997 MPG peptides were designed to transport nucleic acids, followed by Pep-1 for delivery of peptides and proteins (Kang et al., 2011; Morris, Depollier, Mery, Heitz, & Divita, 2001; Morris, Vidal, Chaloin, Heitz, & Divita, 1997). A chimeric peptide derived from the neuropeptide galanin (Transportan) was used to transport small peptides and large proteins as a proof of concept for the in vivo application of CPPs (Pooga, Hällbrink, Zorko, Uuml, & Langel, 1998). A list of commonly known CPPs is given in Table 5.1.

TABLE 5.1 Common cell-penetrating peptides and their sequence.

Name	Sequence	Source	References
TAT (48–60)	GRKKRRQRRRPPQ	Human Immunodeficiency Virus	Frankel and Pabo (1988)
Penetratin	RQIKIWFQNRRMKWKK	*Drosophila melanogaster*	Joliot et al. (1991)
Transportan	GWTLNSLKALAALAKKIL	Chimeric (galanin and mastoporan)	Pooga et al. (1998)
VP22	NAKTRRHERRRKLAIER	Herpes Simplex Virus Type 1	Elliott and O'Hare (1997)
MPG	GALFLGFLGAAGSTMGA	Simian Virus 40	Morris et al. (1997)
Pep-1	KETWWETWWTEWSQPKKKRKV	Simian Virus 40	Morris et al. (2001)
pVEC	LLIILRRRIRKQAHAHSK	Murine Vascular Endothelial-Cadherin	Elmquist, Lindgren, Bartfai, and Langel (2001)
YTA2	YTAIAWVKAFIRKLRK	Phage Library	Lindgren et al. (2006)
M918	MVTVLFRRLRIRRACGPPRVRV	Human	El-Andaloussi, Johansson, Holm, and Langel (2007)
CADY	GLWRALWRLLRSLWRLLWRA	Synthetic	Morris et al. (2001)

5.2 Classification of cell-penetrating peptides

There is no unified classification for CPP and they are classified on different basis. The first classification of CPPs was on the basis of their origin (protein-derived, chimeric, or synthetic). Like Tat and Penetratin, most CPPs have been identified from different protein sequences. Such CPPs have also been termed as PTDs and Trojan horses due their functional characterization. Nona- and octa-arginines were among the first synthetic CPPs reported (Futaki et al., 2001; Wender et al., 2000).

CPPs, since their discovery have been classified on different basis as shown in Fig. 5.1.

For the scope of this chapter, we will discuss the classification of CPPs based on their physical and chemical properties.

5.2.1 Cationic cell-penetrating peptides

The first CPPs reported (TAT and Penetratin) were cationic peptides (Frankel & Pabo, 1988; Joliot et al., 1991). Based on this criterion synthetic poly-arginine constructs were tested for their cellular uptake and it was discovered that an octa-arginine is the minimum sequence required for cellular uptake (Tunnemann et al., 2008; Wender et al., 2000). Lower uptake of poly-lysine in comparison to poly-arginine suggests that charge alone is not responsible for cellular uptake (Tunnemann et al., 2008). This could be explained due to the guanidium group in arginine, which facilitates the formation of hydrogen bonds (Herce, Garcia, & Cardoso, 2014). The uptake of Penetratin, on the other hand, is diminished by a single W14F mutation, suggesting that hydrophobic interactions also play an important role for cellular uptake of cationic peptides (Prochiantz, 1996).

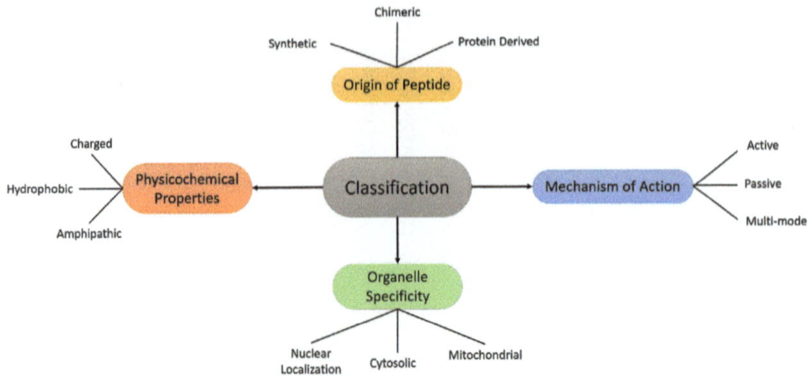

FIGURE 5.1 CPP classification. CPPs can be classified on different basis of their origins, mechanisms of action, physicochemical properties, and cell organelle specificity. *CPPs*, Cell-penetrating peptides.

5.2.2 Amphipathic cell-penetrating peptides

Peptides consisting of specific distribution of polar and nonpolar (hydrophobic) residues are termed amphipathic peptides. Amphipathicity can exist in the amino acid sequence (primary) or the secondary structure (secondary) for a given peptide (Fig. 5.2).

5.2.2.1 Primary amphipathic

Primary amphipathic peptides, for example, MPG (GLAFLGFLGAAGSTMGAWSQPKKKRKV) and Pep-1 (KETWWETWWTEWSQPKKRKV) are chimeras of a hydrophobic and hydrophilic (KKRKV) sequences (Morris et al., 1997; Morris et al., 2001). Here, the KKRKV sequence is a nuclear localizing sequence, which otherwise has poor cell-penetration capability. WSQP is a linker for the two CPP chimeras. The hydrophobic part of the sequences improves the cellular uptake of the nuclear localizing sequence (Oglecka et al., 2008). Other primary amphipathic of note is pVEC, which was derived from VE-cadherin (Elmquist et al., 2001).

5.2.2.2 Secondary amphipathic α-helical cell-penetrating peptides

α-Helical conformation is the most common structural motif reportedly adopted by amphipathic CPPs (Agrawal et al., 2016). Secondary amphipathic peptides have distinct regions of hydrophobic and hydrophilic amino acid clusters along their surface. Helical wheels are an important tool for visualizing the structural amphipathicity in helical peptides. However, given the

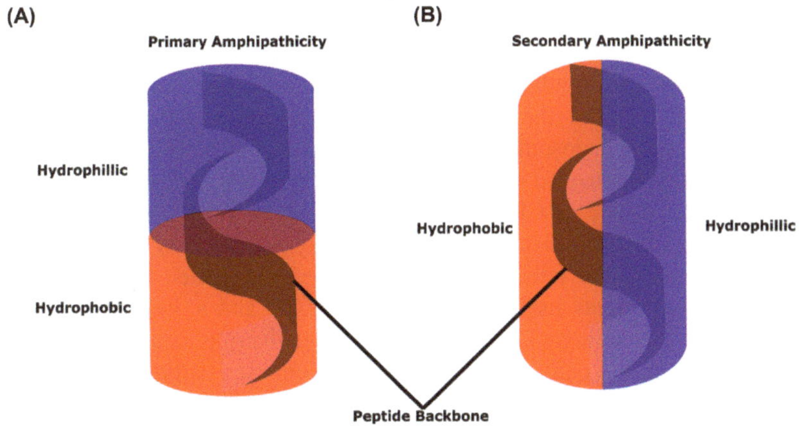

FIGURE 5.2 **Amphipathicity in peptides.** Amphipathicity in peptides can either exist due to the distinct regions of hydrophilic and hydrophobic amino acids in the sequence (A) or arrangement of such residues resulting distinct hydrophilic and hydrophobic planes in the secondary structure of the peptide (B).

probabilistic nature of an amino acid sequence adopting a helical conformation, a helical wheel may not always hold true in depicting the structural amphipathicity of peptides. The importance of amphipathicity for membrane transduction was established through studies on the model amphipathic peptide, MAP (KLALKLALKALKAALKLA) (Oehlke et al., 1998). The substitution of lysine residues with glutamine brings the overall charge of the peptide to zero but conserves its amphipathicity. The mutant MAP peptide was also cell penetrating, thereby affirming the significance of amphipathicity on cell penetration (Oehlke et al., 1999; Wolf et al., 2006). Even though many CPPs reported to date are amphipathic, it is not necessary for all amphipathic peptides to be CPPs. This has been illustrated through single point mutations of transportan and MAP peptides, which severely affected the uptake of these peptides (Soomets et al., 2000; Wolf et al., 2006).

5.3 Mechanism of uptake

The exact mechanism(s) of uptake for CPPs vary with peptide sequence and composition (Guidotti, Brambilla, & Rossi, 2017; Guo, Peng, Kang, & Sun, 2016; Kauffman, Fuselier, He, & Wimley, 2015; Richard et al., 2003). The choice of mechanism is further dependent on the physicochemical properties of the cargo to be delivered like size, hydrophobicity, charge, etc. However, multiple mechanisms of uptake for CPP are suggested in existing literature (Guidotti et al., 2017; Guo et al., 2016; Kauffman et al., 2015; Richard et al., 2003). The understanding of mechanism of penetration is paramount when considering the intracellular interactions and localization of CPPs for cargo delivery applications. The fate of a CPP is dependent on the uptake mechanism it follows; for example, if a CPP enters the cell through direct penetration, it is highly probable that it may be able to deliver the cargo directly to the cytosol and then finally to other intracellular compartments. On the other hand, a CPP entering the cell through endocytosis is more likely to be degraded in a late endosome or lysosome and may not be able to deliver the cargo to another intended target, unless it escapes the endosome/lysosome. Therefore the choice of CPP for delivering cargoes is highly dependent on the proposed mechanism of delivery to be followed. Majority of CPPs adopt multiple pathways of internalization simultaneously (Patel, Zaro, & Shen, 2007), which presents itself as both an advantage and a disadvantage. The advantage is that a single CPP can be used for delivering cargoes to multiple locations within the cells. The disadvantage is that the cargo is also distributed to other nonintended organelles, thereby decreasing the overall delivery efficiency.

The cellular entry of a CPP can be categorized into the following steps: (1) Interaction with cell surface, (2) crossing the cell membrane, (3) intracellular interactions and localizations, and (4) intracellular degradation.

5.3.1 Interaction with cell surface

The first contact between a CPP and the cell surface is the initiation step for cell penetration. The CPP can interact with either the proteoglycans at the cell surface, extracellular receptors, or the membrane phospholipids. The driving force for such biological contacts is generally attributed to electrostatic interactions.

5.3.1.1 Role of proteoglycans

Glycosaminoglycans (GAGs) like heparin sulfate, chondroitin sulfate, sialic acid, etc. are negatively charged and omnipresent on cell membranes. Cationic CPPs bind to these GAGs, which leads to activation of small GTPase RhoA and/or GTPase Rac1 and actin network remodeling [105.88]. This leads to an increase in membrane fluidity (Conner & Schmid, 2003; Eitzen, 2003) and finally the engulfment of the peptide through endocytosis. The process is similar to the uptake of other cationic molecules by the cell like poly-lysines (Frankel & Pabo, 1988; Green & Loewenstein, 1988; Ryser & Shen, 1978; Shen & Ryser, 1978). The proteoglycan − CPP interaction is the primary step of cellular entry for majority of peptides with L-isoforms and poly-D-isoforms of amino acids. The D-isoform amino acid–containing sequences are suggested to employ other modes of first contact (Verdurmen et al., 2011).

5.3.1.2 Role of cellular receptors

Many cellular receptors are involved in the binding of natural ligands, inducing cascades of associated pathways and ligand ingestion. Similarly, peptides bind to the extracellular domain of the membrane-bound receptors, leading to their ingestion. The most commonly used motif for tumor homing and penetration, RGD, binds to the interleukin receptor, which activates the receptor-mediated uptake, leading to peptide intake (Aoki, Hosaka, Kawa, & Kiyosawa, 2001; Arap, Pasqualini, & Ruoslahti, 1998; Pasqualini, Koivunen, & Ruoslahti, 1997; Ruoslahti, 2000).

5.3.1.3 Interaction with membrane phospholipids

Plasma membranes of cells are composed of lipid bilayers, consisting of zwitterionic and negatively charged phospholipids. The composition of negatively charged lipids, Palmitoyloleoylphosphatidylglycerol and POPS (Palmitoyloleoylphosphatidylserine) is higher in bacterial cells (50%) than mammalian cells (10%) (Copolovici, Langel, Eriste, & Langel, 2014; Fleming, Maharaj, Chen, Nelson, & Elmore, 2008; Klahn & Zacharias, 2013). The distribution of negatively charged lipids is asymmetrical between the two layers of the plasma membrane in mammalian cells, with a higher concentration in the cytosolic part of the bilayer (Klahn & Zacharias, 2013). This asymmetric distribution of anionic lipids in mammalian cells is highly regulated and known to fail during disease conditions like cancer. This loss of regulation in cancer cells brings out a higher concentration of POPS in the outer layer, thereby making the cell surface more negatively charged and reduces the membrane potential (Klahn & Zacharias, 2013). The reduction in membrane potential acts as the driving force for higher accumulation of cationic CPPs on the cancer cell surface than in normal cells. As seen in the case of Antimicrobial Peptides (AMPs), cationic CPPs interact with these negatively charged phospholipids (Hazam, Jerath, Kumar, Chaudhary, & Ramakrishnan, 2017). Therefore the relative cationic charge and hydrophobicity of CPPs, just like AMPs, play an important role in their membrane activity and hints for their cancer-targeting ability.

The cationic amino acid residues in CPPs attract the negatively charged headgroups of anionic lipids, establishing strong electrostatic contacts. This further leads to the formation of a dent on the membrane surface, leading to the internalization of CPPs. The detailed models of membrane-based penetration are discussed in the following sections.

5.3.2 Cellular internalization of cell-penetrating peptides

Most of the reported CPPs enter the cells through either endocytosis or direct penetration or a mix of both (Guidotti et al., 2017; Guo et al., 2016; Kauffman et al., 2015; Richard et al., 2003). The choice of pathway for internalization of CPPs is dependent on the type of their cell surface interactions. The peptides interacting with proteoglycans or receptors follow the endocytosis route of entry. On the other hand, peptides interacting with the membrane lipids tend to penetrate through either micropinocytosis (a form of endocytosis) or various models of direct penetration. CPPs can adopt multiple pathways of cellular entry simultaneously, which are dependent on multiple factors like, temperature, pH, cargo size (Guidotti et al., 2017; Guo et al., 2016; Kauffman et al., 2015; Richard et al., 2003), CPP structure, and others. Almost all CPPs partially penetrate cells through membrane-based mechanisms with higher dependency on endocytic routes of entry. As per status quo, the most prevalent form of cellular uptake for CPPs is endocytosis. The studies relating to the characterization of the cellular entry pathway are based mostly on the inhibition of certain mechanisms pertaining to endocytosis (Copolovici et al., 2014). Though these studies provide immense insights into the possible pathway(s) of entry, they discount the possible up-regulation of another pathway of entry due to the down-regulation of another. Various modes of endocytosis and direct penetration reported for CPPs (Fig. 5.3) to date have been explained in the following subsections.

5.3.2.1 Endocytosis

Endocytosis is the collective term for all energy-dependent cellular uptake processes, characterized by vesicle formation (Copolovici et al., 2014;

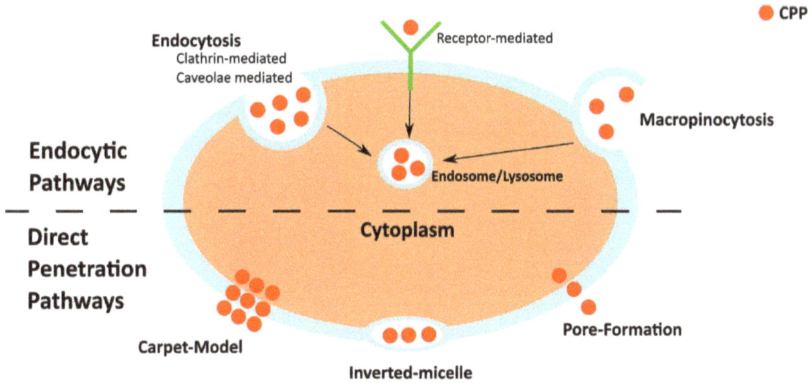

FIGURE 5.3 Mechanism of cellular internalization. CPPs enter cells through either of the two paths, direct penetration through the cell membrane or through endocytosis. The fate of the peptide is highly dependent on the path of uptake. *CPPs*, Cell-penetrating peptides.

Kauffman et al., 2015). It can be broadly classified as phagocytosis and pinocytosis. Phagocytosis is majorly limited to specialized cells (macrophages and leukocytes). Pinocytosis occurs in all cell types and can further be subclassified into receptor-mediated and nonreceptor-mediated endocytosis. Receptor-mediated endocytosis is either clathrin- or caveolin-mediated, while nonreceptor-mediated are clathrin and caveolin-independent endocytosis and macropinocytosis. The choice of clathrin/caveolin-mediated and independent mechanisms is dependent on the size of a molecule (Fittipaldi et al., 2003). Vesicles formed from clathrin-mediated mechanisms are roughly 120 nm in diameter, from caveolin mediation are 70–80 nm, clathrin/caveolin-independent mechanism vesicles are ~90 nm, and macropinocytosis formed vesicles are 1–5 μM. Therefore the choice of cargo is an important attribute for the choice of endocytic entry.

CPPs binding to proteoglycans can enter the cells during the recycling of proteoglycans, which is a continuous process or via clustering of proteoglycans, which initiates endocytosis through activation of intracellular pathways and actin remodeling. The clathrin- and caveolin-mediated pathways are also involved in the cellular uptake of peptides bound to other cellular receptors as well.

Another major form of endocytosis prevalent for cellular uptake of CPPs is macropinocytosis (Komin, Russell, Hristova, & Searson, 2017). Macropinocytosis differs from other forms of endocytosis in terms of the size of vesicles formed (macropinosomes) and the ruffling of membrane. The plasma membrane of cells is not a smooth surface; rather it is ubiquitous with dents (pits) and outward protrusions. Macropinosomes are formed due to the collapse of such membrane ruffles, which are actin-rich protrusions of the cell surface. The size of macropinosomes is three order of magnitude more than other endocytic vesicles and thus the choice of uptake process for relatively large protein structures and other cargoes. Liposomes coated with high concentrations of oligoarginines have been demonstrated to penetrate cells through macropinocytosis (Kaplan, Wadia, & Dowdy, 2005; Wadia, Stan, & Dowdy, 2004). In addition to the big molecules like proteins and liposomes, macropinocytosis is also a mode of cell penetration for the Tat peptide (without cargo).

5.3.2.2 Direct penetration

Peptides interacting with membrane phospholipids follow the direct penetration mode of entry. Direct penetration through plasma membrane is an energy and temperature-independent mechanism of uptake (Reissmann, 2014; Ye et al., 2016). Interestingly, it was the first reported mechanism of entry for CPPs but was disregarded as an artifact arising from cell fixation (Richard et al., 2003). Thereafter, all studies on the uptake mechanisms of CPPs are carried out on live cells. The membrane-based uptake mechanism of CPPs is quite similar to that of AMPs, owing to similar characteristics of the two classes of membrane-active peptides. The membrane interaction of these peptides is foremost dependent on the spatial electrostatic potential distribution on their surface, which is complemented by the anionic lipids on membrane surface. Multiple models of direct penetration have been theorized for both CPPs and AMPs (Jobin & Alves, 2014).

Carpet-like model, as the name suggests, involves the lateral positioning of the cationic residues of peptides over the negatively charged phospholipids on the outer membrane. The accumulation of negative charge at surface and at threshold concentration of peptides, the peptides rotate about their principal axis, leading to phospholipid redirection. This leads to the interaction of the hydrophobic residues in peptide sequences with hydrophobic tails of the phospholipids and finally cell penetration in a manner similar to a detergent-like effect (Raucher & Ryu, 2015; Shin et al., 2014; Ye et al., 2016).

The inverted micelle model has been proposed following the findings of Alain Prochiantz's group, using Nuclear Magnetic Resonance imaging for the penetration of phospholipid membranes by Penetratin (Derossi et al., 1996). This model is useful in explaining the uptake of amphipathic CPPs but not others like oligoarginines, which do not contain any hydrophobic residues (Alves et al., 2008). The primary step of this model differs from that of the carpet-like model in the nonparallel positioning of the peptides on the cell surface. The interaction of cationic peptide residues and anionic lipid headgroups leads to the formation of a micelle through the invagination of the membrane. The micelle then opens toward the cytosolic end of the membrane, thereby causing membrane permeation (Joanne et al., 2009; Kawamoto et al., 2011).

Barrel stave model is one of the two pore-formation models proposed for CPP entry (Guidotti et al., 2017; Jarver, Mager, & Langel, 2010; Lopez-Meza & Ochoa-Zarzosa, 2015). The peptides align perpendicular to bilayer plane, insert, associate and, as a result of peptide amphipathicity, form a pore. It can be best understood as a drilling process arising from a mix of electrostatic interactions at the membrane surface and the hydrophobic interactions with the bilayer core. This model is similar to the mechanism of action proposed for the antibacterial peptide, Gramicidin (Pomès & Roux, 2002). However, Gramicidin requires two end-to-end stacked monomers, for forming a water channel and relates to the toxicity of the peptide. In case of CPPs, the pore results from a cumulative barrel-like positioning of individual peptide units in parallel.

The Toroidal pore-formation model deviates from the barrel stave model in the relative positioning of peptides in the pore formed (Bechara & Sagan, 2013). Instead of parallel positioning of peptide molecules, the peptides induce a membrane curvature local to the site of pore formed. This leads to a continuity of the outer and inner layers, such that, the two layers are indistinguishable at the site of the pore. However, the pore–formation models discussed are highly dependent on the helical structure of the peptides and their amphipathicity, in summary, their spatial electrostatic distributions.

Adaptive penetration model is among the most recently proposed models for cell penetration (Rothbard, Jessop, Lewis, Murray, & Wender, 2004; Wender, Galliher, Goun, Jones, & Pillow, 2008). Identical to all the models discussed earlier, it relies on the electrostatic interactions of positively charged amino acids with anionic lipid head groups. However, it is only applicable to guanidium group containing CPPs, that is, arginine-containing CPPs. The guanidium group of arginine forms bidentate hydrogen bonds with lipid head groups, leading to peptide charge neutralization or inversion, which is dependent upon the counterion present. This results in the phase transition of CPP from hydrophilic to more lipophilic solvability and its further insertion into the bilayer. The presence of different counterions on the cytosolic end of the bilayer reinstates the original physical properties of the peptide, leading to its release into the cytosol. The diffusion of the peptides across the bilayer is a function of membrane potential and requires a net positive charge for a CPP. The partition constant for peptides is defined as the ratio of concentrations of peptide in lipid and aqueous phases. The partition constant is indicative of the membrane disruption efficiency of AMPs but no studies have been reported on correlating it with membrane permeation efficiency of CPPs yet.

5.3.3 Cellular localization

The intracellular localization of CPPs is highly dependent on the pathway(s) of cell penetration, which are further dependent on the type of peptide – cell surface binding interactions. As illustrated in Fig. 5.3, peptides entering the cell through either of the endocytic pathways are primarily confined to vesicles, thereby primarily localizing in the endosomes/macropinosomes. The next barrier for such peptide molecules is to escape the vesicles for intracellular delivery of cargo to other cell organelles. Therefore endosomal escaping ability is a vital attribute for any CPP to be considered as a drug delivery vector, unless it is required to deliver the cargo to endosomes or lysosomes. After endosomal escape, the peptide localizes to the cytosol and/or further to other cellular organelles.

On the other hand, CPPs entering cells through direct translocation localize to the cytosol first and thereafter can localize to other organelles, including endosomes. Moreover, a CPP can also localize to multiple cell compartments. Though favorable in many situations, the peptide concentration distribution decreases its bioavailability to a specific organelle for effecting a cargo function, leading to higher dose requirements. However, a few CPPs in literature are reported to target mitochondria for site-directed delivery (Horton, Stewart, Fonseca, Guo, & Kelley, 2008).

5.3.4 Transcellular transport and degradation

Premature degradation of CPPs hinder their applicability as drug delivery vectors; however, it is also required that the vector can be metabolized and excreted from the system. There are major barriers where a CPP may be degraded within an in vivo system. However, in the confines of mechanism of penetration, only intracellular degradation is discussed in this section. Major degradation of peptides inside cells is either in lysosomes or in cytosol. In the event that a CPP is unable to escape endosomes after penetration, it will encounter the lysosomal proteolytic degradation pathways. Lysosomal proteases comprise of more than 40 lytic enzymes.

The cytosol also comprises of numerous proteases and acts as a natural proteolytic site for cytosol localizing peptides. The ubiquitin–proteasome degradation pathway, 26S proteasome-mediated degradation, and cytosolic tripeptidases are among the most well-documented degradation pathways in cytosol. The presence of D-amino acids has been reported to protect the CPP from premature proteolytic cleavage; however, a protein cargo would effectively be prone to such degradation too. Therefore the degradation or stability of CPP in cytosol is of lesser relevance when transporting a protein-based cargo (Komin et al., 2017).

Though CPPs can be degraded in both lysosomes and cytosol, the ability of escaping endosomes is more vital than stability in cytosol, for effective transport, unless the site intended is lysosomes, as discussed earlier.

The penetration of cell by CPPs also opens the debate as to whether the CPPs can permeate to outside the cell too. The Tat protein is known to be released from HIV-infected cells to other noninfected cells. Therefore it is understandable that other CPPs may as well have such a property of transcellular transport. The applications of such a property possessing CPPs can be multifold ranging from blood–brain barrier, epithelial barrier (skin, intestine, pulmonary) to 3D tumor penetration.

5.4 Strengths, limitations, and opportunities

The use of peptides as drug delivery vectors offers its own sets of advantages and disadvantages as advocated through various reports (Fig. 5.4). Given their biological nature, peptides are comparatively safer and more tolerable for biological systems than chemi-mimics. Peptides can be designed for higher selectivity, efficacy, and potency. The metabolism of peptides is more predictable than other chemical moieties. Moreover, peptides can be synthesized through standard synthetic protocols and can be conjugated with any molecule of interest through simple chemical reactions. However, peptides may sometimes require modifications due to poor chemical and physical stability in biological fluids, which may otherwise hinder their use for clinical applications. Most peptides have short plasma life leading to premature degradation in blood and, thus, fast elimination. Further, the mode of administration for peptides is presently limited to intravenous routes. Many reports have described alternate modes for administration but the consensus is the same (Fosgerau & Hoffmann, 2015).

The weaknesses of peptide-based drug delivery vectors provide new avenues and opportunities for the development of novel peptide sequences to address these issues. Using the 20 naturally occurring peptides as a set of variables in a peptide sequence of sequence length 10–15, which is the length of most CPPs, a total of 20^{10}–20^{15} combinations (sequences) are possible. This signifies the vast sequence universe available for designing peptides with tailored features. The use of unnatural amino acids as described by Kumar et al. can increase this available sequence universe by twofolds (Kumar & Ramakrishnan, 2010). Further, the conjugation of peptides with different therapeutic molecules increases the number of sequential combinations available for peptide design.

Strengths: • High potency and efficacy • Cell-type selectivity • Predictable metabolism • Shorter time for lab to clinic • Better safety and tolerability	Opportunities: • De novo design • Alternate delivery routes • Multifunctional peptides • Focussed libraries • Formulation development
Weaknesses: • Instability (physical and chemical) • Pre-mature degradation • Prone to aggregation • Non-oral modes of delivery • Hydrolysis and oxidation	Threats: • Immunogenicity • Cost of production • Stricter safety and efficacy norms • Advancements in personalized and Network medicine • Patent expiries

FIGURE 5.4 SWOT analysis of peptide therapeutics. The Strengths, Weaknesses, Opportunities, and Threats for the use of peptides in development of novel therapies are shown.

Moreover, peptides with multiple features can also be designed, for example, many AMPs share similar properties with CPPs. Overall, in the given purview, peptides do offer a great potential for development of future therapies and therapeutics (Fosgerau & Hoffmann, 2015).

5.5 Cell-type specificity: to be or not to be

CPPs can deliver a variety of cargoes including small molecules, proteins, nucleic acids, etc. inside any cell type (Gupta, Levchenko, & Torchilin, 2005; Milletti, 2012; Temsamani & Vidal, 2004). This property makes them highly efficient in an in vitro system. However, the same property of CPP to deliver cargoes to other normal tissues in an in vivo system hinders their clinical applications (Huang et al., 2013; Koren & Torchilin, 2012; Shi, Qi, Xiang, & Zhang, 2014). In case of an antineoplasic cargo, the peptides can cause systemic toxicity, which in lower levels may well be acceptable but beyond a threshold may cause severe side effects (Huang et al., 2013; Koren & Torchilin, 2012; Shi et al., 2014). The positive charge, a common feature of most CPPs, though advantageous for cell penetration may lead to drug delivery inefficacy due to interactions with negatively charged serum proteins, resulting in their premature degradation in blood (Buyens et al., 2012).

Some research groups have demonstrated that some CPPs can selectively permeate cancer cells instead of noncancerous or normal cells (Bechara et al., 2013; Farkhani et al., 2014; Nakase, Takeuchi, Tanaka, & Futaki, 2008; Papo et al., 2006; Schröder-Borm, Bakalova, & Andrä, 2005). This could be partly due to the differences in membrane composition of normal and cancer cells (Jobin & Alves, 2014). Cancer cells have more numbers of GAGs and anionic phosphatidylserine on their surface than normal cells, which attract more cationic CPPs than normal cells (Åmand et al., 2012; Jobin & Alves, 2014; Ziegler & Seelig, 2011). Jobin et al. have previously demonstrated that the RW16 peptide derived from penetratin interacts more with anionic liposomes than zwitterionic ones (Jobin et al., 2013). Further, tumor cells express higher levels of different receptors, which can be utilized for "tumor homing" applications (Jobin & Alves, 2014). Apart from the membrane-based differences between cancer and healthy cells, the tumor macroenvironment factors like angiogenesis, hypoxia, and inflammation can also be used for differentiating between cancer and healthy cells (Danhier, Feron, & Préat, 2010; Quail & Joyce, 2013). Therefore in the purview of anticancer therapies and targeted delivery, a CPP should have some form of cell type specificity.

5.6 Cell-penetrating peptides for anticancer drug delivery

Cancer is an epigenetic disease characterized by unregulated cell proliferation and a major cause of death worldwide (Farkhani et al., 2014; Hanahan, & Weinberg, 2000, 2011; Pavlova & Thompson, 2016; Pietras & Östman,

2010). Present chemotherapeutic strategies are synonymous with the term "a Pyrrhic victory" due to a variety of side effects. Further, the drugs sometimes are unable to penetrate the vasculature of cancer cells leading to inaction. The delivery of anticancer drugs through CPPs offers avenues for targeted delivery of drug molecules to intended site, increasing drug efficacy and reducing side effects (Ruoslahti, 2017). Multiple reports in literature describe the use of peptides for cancer therapy. Some of these findings are discussed in the following sections.

5.6.1 Cell-penetrating peptides for targeted delivery

The heterogeneity of cancer and healthy cells discussed in the previous section provides avenues for targeted delivery exploiting these differences in cell types. Peptides reported for cancer-specific targeting can further be classified as (1) Homing Peptides (HP), (2) CPPs with homing domains, and (3) Cell-penetrating and Tumor Homing peptides (CPTHP) (Svensen, Walton, & Bradley, 2012). HP do not penetrate cells and deliver the cargo to the surface. HP conjugated with a CPP can penetrate cell membranes for intracellular drug delivery. CPTHP peptides are peptides that selectively penetrate tumor cells without an externally conjugated homing domain.

5.6.2 Delivery of anticancer drugs using cell-penetrating peptides

Conjugation of taxol (Dubikovskaya, Thorne, Pillow, Contag, & Wender, 2008), methotrexate (MTX) (Lindgren et al., 2006), doxorubicin (Dox) (Aroui, Brahim, De Waard, Breard, & Kenani, 2009; Dawidczyk, Russell, & Searson, 2015) to CPPs reportedly increase their membrane permeability. The linking of taxol to octa-arginine through disulfide linkers increases its solubility in aqueous systems and pharmacokinetics. The delivery of MTX and Dox has shown to increase the cytotoxicity of the drugs and overcome multidrug resistance (Jerath, Goyal, Trivedi, Santhoshkumar, & Ramakrishnan, 2019; Jerath, Goyal, Trivedi, Santhoshkumar, & Ramakrishnan, 2020). In addition, CPPs have also been used to deliver nucleotides and therapeutic proteins to cancer cells. A detailed account of the same is beyond the scope of this study and can be found in the following reference (Borrelli, Tornesello, Tornesello, & Buonaguro, 2018).

5.7 Cell-penetrating peptide prediction

Predicting the cell-penetrative capability of a given amino acid sequence is a tedious task, which is further complicated by the fact that the exact mechanism of CPP uptake is still debatable (Copolovici et al., 2014). As discussed in the previous section, endocytosis is the major contributor for cellular uptake of CPPs conjugated with a cargo; membrane transduction has also been indicated to be the choice of pathway adopted by specific peptides (Ezzat et al., 2012; Guterstam et al., 2009;

Rangel et al., 2012; Rydström, Deshayes, Konate, Crombez, & Padari, 2011). Further, more challenging is the question as to what is more critical for the penetrative action of an amino acid sequence. Is it the properties of the peptide or the membrane potential or the membrane composition or a combination of all or none? Given that CPPs penetrate plant, yeast, and bacterial cells in addition to mammalian cells, one may assume that the properties of the peptide sequence is more important (Good, Awasthi, Dryselius, Larsson, & Nielsen, 2001; Holm, Netzereab, Hansen, Langel, & Hällbrink, 2005; Mäe et al., 2005; Nekhotiaeva, Awasthi, Nielsen, & Good, 2004; Nekhotiaeva, Elmquist, et al., 2004).

The elucidation of properties of peptides responsible for cell penetration is no less than a Gordian knot. Majority of CPPs are cationic, which can be attributed to arginine, lysine, or histidine. As discussed in section 5.2, arginine residues contribute more toward the cellular uptake of CPPs than lysine residues. This may be due to the two hydrogen bond donor groups available in the guanidium group of arginine against a single hydrogen bond donor provided by the amino side-chain group in lysine. Further, the hydrophobicity and its distribution (sequential and structural) can affect the uptake of CPPs. The maintenance of structural amphipathicity in various solvents and buffers is rarely replicated under physiological conditions. The poly-L-amino acid sequences exist as an ensemble of disordered structures in solutions and may or may not reconform to helical or sheeted structures as intended in the design of their sequence (Medina et al., 2016). The length of the peptide is also an important factor. CPPs reported in literature vary in sequence length of 5–30 amino acid residues and in the absence of a smaller range of peptide length critical for cell penetration, it is almost impossible to identify a cell-penetrating sequence from a peptide sequence library (Agrawal et al., 2016). However, despite the roadblocks mentioned earlier, it may still be possible to predict the cell-penetrative capability of an amino acid sequence. One mode of prediction is heuristic approach also regarded as "an educated guess" and the other is a completely predictive approach, which relies on a variety of physicochemical properties of peptides (descriptors).

The heuristic approach is responsible for the majority of CPPs reported in literature. It is majorly based on a mechanistic sequence-activity relationship study, which includes the identification of a cationic fragment augmented by hydrophobic residues in a protein sequence. The method is laborious and time-intensive given that it requires the synthesis, characterization, and testing of multiple overlapping sequences. However, a number of guidelines are available for identifying a cell-penetrating sequence. These include:

1. Cationic residues are preferred and anionic residues are suggestively avoided, even when the sequence possesses more positive residues.
2. Sequences with more arginine are preferred over lysine-rich sequences.
3. Helical wheel plotting is required to check for possible structural amphipathicity in the given sequence.

4. Shorter peptides are preferred over longer sequences (> 20) due to ease of synthesis and purification.
5. Highly hydrophobic sequences should be avoided due to poor water solubility, which may complicate the use of the peptide.

The predictive approach relies on a set of peptidic properties (descriptors). The use of Sandberg's z-descriptors has been demonstrated to predict CPPs from protein sequences; however, it does not take into account the amino acid sequence but only the composition (Hansen, Kilk, & Langel, 2008; Lindgren & Langel, 2011; Sandberg, Eriksson, Jonsson, Sjöström, & Wold, 1998). Another approach is the use of support vector machines. Different groups have reported the use of support vector machines (A Dobchev et al., 2010; Sanders, Johnston, Bridges, Burgess, & Willeford, 2011; Suhorutsenko, Eriste, Copolovici, & Langel, 2012; Yamada et al., 2013). Among these, a noteworthy effort is the availability of a support vector machine based web-server, CELLPDD (Gautam, Chaudhary, Kumar, & Raghava, 2015).

5.8 Designing cell-penetrating peptides

In terms of protein folding, CPPs with a poly-L backbone can adopt α-helical, β-sheet, or disordered conformations in solution. The disordered CPPs can reconform to either of the two secondary structures. In this section, we will discuss the different avenues available for the design of novel peptides with cell-penetrative activity based on amino acid sequence and structural parameters.

5.8.1 Amino acid sequence

Proteins and peptides are polymers of different amino acid monomers. Each amino acid monomer consists of a main chain composed of backbone atoms (N, C-α, C, and O) and a variable side-chain. The spatial orientation of the different side-chain array combinations provides uniqueness to each peptide molecule, which is responsible for its function (Jerath, Hazam, & Ramakrishnan, 2014; Jerath, Hazam, Shekhar, & Ramakrishnan, 2016). In the following subsections, we will discuss the effects of various amino acid properties on cell penetration.

5.8.1.1 Chirality

Most CPPs derived from natural proteins are composed of L-amino acids (Agrawal et al., 2016). This makes them more susceptible to proteolytic degradation, which may lead to their inaction in an in vivo system. The use of D-stereoisomers increases the stability of peptides against proteolytic cleavage (Pujals, Sabido, Tarrago, & Giralt, 2007). Poly-D peptides have more stability but often have lower cellular uptake than the poly-L sequences

(Pujals et al., 2007; Verdurmen et al., 2011; Yamada, Signorelli, Cannistraro, Beattie, & Bizzarri, 2015). The internalization efficiency of peptides is also reported to be affected by the number of D-amino acids in a sequence (Ma et al., 2012; Yamada et al., 2015). Multiple substitutions of L-amino acids to D-amino acids can alter the α-helix to β-sheet ratio of CPPs, thereby affecting their cellular uptake (Yamada et al., 2015). Interestingly, the use of D-Pro to conform the disordered structure of a CPP provided proteolytic stability and a nonendosomal pathway of uptake (Medina et al., 2016). The contribution of chirality toward cell penetration remains unclear in the absence of a detailed study.

5.8.1.2 Cationic charge

Cationic charged residues are known to promote cell penetration of CPPs (Futaki et al., 2001; Liu et al., 2016; Takayama et al., 2009). The first CPP, TAT, contains 11 cationic residues (Frankel & Pabo, 1988; Joliot et al., 1991). Poly-arginines were among the first synthetic CPPs due to their cationicity and the ability of guanidium groups to form bidentate hydrogen bonds, critical for membrane permeation (Futaki et al., 2001). However, the length of sequence, position, and number of arginines in sequence affect their cellular uptake (Kawaguchi et al., 2016; Mishra et al., 2011). Kelley and coworkers demonstrated an interesting example of designing cationic CPPs (Horton et al., 2008). The designed peptides were imparted with positive charge through arginine and lysine along with a lipophilic character, which is due to the presence of phenylalanine and cyclohexylalanine. The combination of these two characters gave the peptides the ability to selectively penetrate the mitochondria (Horton et al., 2008).

5.8.1.3 Hydrophobicity and aromaticity

Hydrophobicity in peptide sequences can be accomplished by the use of aliphatic and aromatic amino acids. Aromatic amino acids are more hydrophobic and possess favorable free energy of interaction with the lipid bilayer. Aromatic rings can form $\Pi - \Pi$ stack with membrane proteins and may contribute to stabilization/promotion of CPPs with membranes (deRonde, Birke, & Tew, 2015). The uses of polymeric aromatic groups as peptide modifications have been reported to increase the uptake of poly-arginines (Perret et al., 2005).

5.8.2 Secondary structure folding and its effects on cellular uptake

Two interconnected parameters should be considered to study the effect of peptide folding on cellular uptake: first, the affinity of the CPP for membranes and second, the folding capacity of proteins in presence of membranes. Electrostatic forces, hydrophobic interactions, and hydrogen bonding characterize the membrane − CPP interaction. Most CPPs adopt α-helical conformations upon membrane interaction, which provides a template for the interplay of different interaction forces (Kalafatovic & Giralt, 2017). Therefore constricting the geometry of a known CPP to a helical conformation may provide the ideal structural template for redesign of novel CPPs with tailored structural amphipathicity, which in turn may help increase their cellular uptake.

Ramakrishnan and colleagues have previously described the role of syndiotactic backbone on the conformational locking of peptidic structures (Goyal et al., 2021; Goyal, Jerath, Chandrasekharan, Kumar, & Ramakrishnan, 2020; Hazam, Akhil, Jerath, Saikia, & Ramakrishnan, 2019; Hazam, Jerath, Chaudhary, & Ramakrishnan, 2018; Hazam, Jerath, Kumar, Chaudhary, & Ramakrishnan, 2017; Jerath, Goyal, Trivedi, Santhoshkumar, & Ramakrishnan, 2019, 2020; Ramakrishnan, Ranbhor, Kumar, & Durani, 2006). Syndiotactic peptides are capable of forming a Π(L,D) helix, which is thermodynamically more stable than the α-helix. Further, the use of unnatural amino acids can help in conforming the peptide structure in a stable configuration (Ramakrishnan et al., 2006; Urry, 1971). Together, these design inputs available in literature along with other specific methods in literature can be exploited for the design of novel peptides with cell-penetrative ability.

5.9 Conclusion

In the present chapter, we discussed the use of peptide-based drug delivery vectors for transporting different cargoes inside the cell. The use of peptides as drug delivery agents presents its own set of pros and cons. The peptides termed as PTDs, Trojan horse, and CPPs have the ability to traverse live cell membranes. The classification of CPPs on the basis of their physicochemical properties provides important insights into their mechanism of action (cellular uptake). The fate of peptides inside the cell is dependent on the path of cellular uptake. Moreover, we learnt from different reports in literature that cell-type specificity is required for the targeted delivery of anticancer agents inside the cell for future in vivo applications. This may be achieved by the inclusion of a homing domain with the cell-penetrating domain of the designed peptides. Alternately, some peptides possess the dualistic nature of CPP and HP. The design of CPPs based on various sequence and structural parameters offers new directions for the de novo design of novel CPPs with tailored features and functions.

References

Dobchev, A. D., Mager, I., Tulp, I., Karelson, G., Tamm, T., Tamm, K., et al. (2010). Prediction of cell-penetrating peptides using artificial neural networks. *Current Computer-aided Drug Design*, *6*(2), 79–89.

Agrawal, P., Bhalla, S., Usmani, S. S., Singh, S., Chaudhary, K., Raghava., Gajendra, P. S., et al. (2016). CPPsite 2.0: A repository of experimentally validated cell-penetrating peptides. *Nucleic Acids Research*, *44*(D1), D1098–D1103.

Alves, I. D., Goasdoué, N., Correia, I., Aubry, S., Galanth, C., Sagan, S., et al. (2008). Membrane interaction and perturbation mechanisms induced by two cationic cell penetrating peptides with distinct charge distribution. *Biochimica et Biophysica Acta (BBA)—General Subjects*, *1780*(7), 948–959.

Åmand, H. L., Rydberg, H. A., Fornander, L. H., Lincoln, P., Nordén, B., & Esbjörner, E. K. (2012). Cell surface binding and uptake of arginine-and lysine-rich penetratin peptides in absence and presence of proteoglycans. *Biochimica et Biophysica Acta (BBA)—Biomembranes*, *1818*(11), 2669–2678.

Antony, A. C. (1992). The biological chemistry of folate receptors. *Blood*, *79*(11), 2807–2820.

Aoki, Y., Hosaka, S., Kawa, S., & Kiyosawa, K. (2001). Potential tumor-targeting peptide vector of histidylated oligolysine conjugated to a tumor-homing RGD motif. *Cancer Gene Therapy*, *8*(10), 783.

Arap, W., Pasqualini, R., & Ruoslahti, E. (1998). Cancer treatment by targeted drug delivery to tumor vasculature in a mouse model. *Science (New York, N.Y.)*, *279*(5349), 377–380.

Aroui, S., Brahim, S., De Waard, M., Breard, J., & Kenani, A. (2009). Efficient induction of apoptosis by doxorubicin coupled to cell-penetrating peptides compared to unconjugated doxorubicin in the human breast cancer cell line MDA-MB 231. *Cancer Letters*, *285*(1), 28–38.

Bechara, C., Pallerla, M., Zaltsman, Y., Burlina, F., Alves, I. D., Lequin, O., et al. (2013). Tryptophan within basic peptide sequences triggers glycosaminoglycan-dependent endocytosis. *The FASEB Journal*, *27*(2), 738–749.

Bechara, C., & Sagan, S. (2013). Cell-penetrating peptides: 20 years later, where do we stand? *FEBS Letters*, *587*(12), 1693–1702.

Borrelli, A., Tornesello, A. L., Tornesello, M. L., & Buonaguro, F. M. (2018). Cell penetrating peptides as molecular carriers for anti-cancer agents. *Molecules (Basel, Switzerland)*, *23*(2), 295.

Buyens, K., De Smedt, S. C., Braeckmans, K., Demeester, J., Peeters, L., van Grunsven, L. A., et al. (2012). Liposome based systems for systemic siRNA delivery: Stability in blood sets the requirements for optimal carrier design. *Journal of Controlled Release*, *158*(3), 362–370.

Conner, S. D., & Schmid, S. L. (2003). Regulated portals of entry into the cell. *Nature*, *422* (6927), 37–44.

Copolovici, D. M., Langel, K., Eriste, E., & Langel, Ü. (2014). Cell-penetrating peptides: Design, synthesis, and applications. *ACS Nano*, *8*(3), 1972–1994.

Danhier, F., Feron, O., & Préat, V. (2010). To exploit the tumor microenvironment: Passive and active tumor targeting of nanocarriers for anti-cancer drug delivery. *Journal of Controlled Release*, *148*(2), 135–146.

Dawidczyk, C. M., Russell, L. M., & Searson, P. C. (2015). Recommendations for benchmarking preclinical studies of nanomedicines. *Cancer Research*, *75*(19), 4016–4020.

deRonde, B. M., Birke, A., & Tew, G. N. (2015). Design of aromatic-containing cell-penetrating peptide mimics with structurally modified pi electronics. *Chemistry (Weinheim an der Bergstrasse, Germany)*, *21*(7), 3013–3019.

Derossi, D., Calvet, S., Trembleau, A., Brunissen, A., Chassaing, G., & Prochiantz, A. (1996). Cell internalization of the third helix of the antennapedia homeodomain is receptor-independent. *Journal of Biological Chemistry*, *271*(30), 18188–18193.

Dubikovskaya, E. A., Thorne, S. H., Pillow, T. H., Contag, C. H., & Wender, P. A. (2008). Overcoming multidrug resistance of small-molecule therapeutics through conjugation with releasable octaarginine transporters. *Proceedings of the National Academy of Sciences of the United States of America*, *105*(34), 12128–12133.

Eitzen, G. (2003). Actin remodeling to facilitate membrane fusion. *Biochimica et Biophysica Acta*, *1641*(2–3), 175–181.

El-Andaloussi, S., Johansson, H. J., Holm, T., & Langel, Ü. (2007). A novel cell-penetrating peptide, M918, for efficient delivery of proteins and peptide nucleic acids. *Molecular Therapy*, *15*(10), 1820–1826.

Elliott, G., & O'Hare, P. (1997). Intercellular trafficking and protein delivery by a herpes virus structural protein. *Cell*, *88*(2), 223–233.

Elmquist, A., Lindgren, M., Bartfai, T., & Langel, Ü. (2001). VE-cadherin-derived cell-penetrating peptide, pVEC, with carrier functions. *Experimental Cell Research*, *269*(2), 237–244.

Ezzat, K., Helmfors, H., Tudoran, O., Juks, C., Lindberg, S., Padari, K., et al. (2012). Scavenger receptor-mediated uptake of cell-penetrating peptide nanocomplexes with oligonucleotides. *The FASEB Journal*, *26*(3), 1172–1180.

Farkhani, S. M., Valizadeh, A., Karami, H., Mohammadi, S., Sohrabi, N., & Badrzadeh, F. (2014). Cell penetrating peptides: Efficient vectors for delivery of nanoparticles, nanocarriers, therapeutic and diagnostic molecules. *Peptides*, *57*, 78–94.

Feni, L., & Neundorf, I. (2017). The current role of cell-penetrating peptides in cancer therapy. *Advances in Experimental Medicine and Biology*, *1030*, 279–295.

Fittipaldi, A., Ferrari, A., Zoppe, M., Arcangeli, C., Pellegrini, V., Beltram, F., et al. (2003). Cell membrane lipid rafts mediate caveolar endocytosis of HIV-1 Tat fusion proteins. *The Journal of Biological Chemistry*, *278*(36), 34141–34149.

Fleming, E., Maharaj, N. P., Chen, J. L., Nelson, R. B., & Elmore, D. E. (2008). Effect of lipid composition on buforin II structure and membrane entry. *Proteins: Structure, Function, and Bioinformatics*, *73*(2), 480–491.

Fosgerau, K., & Hoffmann, T. (2015). Peptide therapeutics: current status and future directions. *Drug Discovery Today*, *20*(1), 122–128.

Frankel, A. D., & Pabo, C. O. (1988). Cellular uptake of the tat protein from human immunodeficiency virus. *Cell*, *55*(6), 1189–1193.

Futaki, S., Suzuki, T., Ohashi, W., Yagami, T., Tanaka, S., Ueda, K., et al. (2001). Arginine-rich peptides. An abundant source of membrane-permeable peptides having potential as carriers for intracellular protein delivery. *The Journal of Biological Chemistry*, *276*(8), 5836–5840.

Gautam, A., Chaudhary, K., Kumar, R., & Raghava, G. P. S. (2015). Computer-aided virtual screening and designing of cell-penetrating peptides. *Methods in Molecular Biology*, *1324*, 59–69.

Gelperina, S., Kisich, K., Iseman, M. D., & Heifets, L. (2005). The potential advantages of nanoparticle drug delivery systems in chemotherapy of tuberculosis. *American Journal of Respiratory and Critical Care Medicine*, *172*(12), 1487–1490.

Good, L., Awasthi, S. K., Dryselius, R., Larsson, O., & Nielsen, P. E. (2001). Bactericidal antisense effects of peptide–PNA conjugates. *Nature Biotechnology*, *19*(4), 360.

Goyal, R., Jerath, G., Akhil, R., Chandrasekharan, A., Puppala, E. R., Ponneganti, S., et al. (2021). Geometry encoded functional programming of tumor homing peptides for targeted drug delivery. *Journal of Controlled Release*, *333*, 16–27.

Goyal, R., Jerath, G., Chandrasekharan, A., Kumar, T. S., & Ramakrishnan, V. (2020). Peptide-based delivery vectors with pre-defined geometrical locks. *RSC Medicinal Chemistry*, *11*(11), 1303–1313.

Green, M., & Loewenstein, P. M. (1988). Autonomous functional domains of chemically synthesized human immunodeficiency virus TAT trans-activator protein. *Cell*, *55*(6), 1179–1188.

Guidotti, G., Brambilla, L., & Rossi, D. (2017). Cell-penetrating peptides: from basic research to clinics. *Trends in Pharmacological Sciences*, *38*(4), 406–424.

Guo, Z., Peng, H., Kang, J., & Sun, D. (2016). Cell-penetrating peptides: Possible transduction mechanisms and therapeutic applications. *Biomedical Reports*, *4*(5), 528–534.

Gupta, B., Levchenko, T. S., & Torchilin, V. P. (2005). Intracellular delivery of large molecules and small particles by cell-penetrating proteins and peptides. *Advanced Drug Delivery Reviews*, *57*(4), 637–651.

Guterstam, P., Madani, F., Hirose, H., Takeuchi, T., Futaki, S., Andaloussi, S. E., et al. (2009). Elucidating cell-penetrating peptide mechanisms of action for membrane interaction, cellular uptake, and translocation utilizing the hydrophobic counter-anion pyrenebutyrate. *Biochimica et Biophysica Acta (BBA)—Biomembranes*, *1788*(12), 2509–2517.

Hanahan, D., & Weinberg, R. A. (2000). The hallmarks of cancer. *Cell*, *100*(1), 57–70.

Hanahan, D., & Weinberg, R. A. (2011). Hallmarks of cancer: The next generation. *Cell*, *144*(5), 646–674.

Hansen, M., Kilk, K., & Langel, Ü. (2008). Predicting cell-penetrating peptides. *Advanced Drug Delivery Reviews*, *60*(4–5), 572–579.

Hazam, P. K., Akhil, R., Jerath, G., Saikia, J., & Ramakrishnan, V. (2019). Topological effects on the designability and bactericidal potency of antimicrobial peptides. *Biophysical Chemistry*, *248*, 1–8.

Hazam, P. K., Jerath, G., Chaudhary, N., & Ramakrishnan, V. (2018). Peptido-mimetic approach in the design of syndiotactic antimicrobial peptides. *International Journal of Peptide Research and Therapeutics*, *24*(2), 299–307.

Hazam, P. K., Jerath, G., Kumar, A., Chaudhary, N., & Ramakrishnan, V. (2017). Effect of tacticity-derived topological constraints in bactericidal peptides. *Biochimica et Biophysica Acta (BBA)—Biomembranes*, *1859*(8), 1388–1395.

Heitz, F., Morris, M. C., & Divita, G. (2009). Twenty years of cell-penetrating peptides: From molecular mechanisms to therapeutics. *British Journal of Pharmacology*, *157*(2), 195–206.

Herce, H. D., Garcia, A. E., & Cardoso, M. C. (2014). Fundamental molecular mechanism for the cellular uptake of guanidinium-rich molecules. *Journal of the American Chemical Society*, *136*(50), 17459–17467.

Holm, T., Netzereab, S., Hansen, M., Langel, Ü., & Hällbrink, M. (2005). Uptake of cell-penetrating peptides in yeasts. *FEBS Letters*, *579*(23), 5217–5222.

Horton, K. L., Stewart, K. M., Fonseca, S. B., Guo, Q., & Kelley, S. O. (2008). Mitochondria-penetrating peptides. *Chemistry & Biology*, *15*(4), 375–382.

Huang, Y., Jiang, Y., Wang, H., Wang, J., Shin, M. C., Byun, Y., et al. (2013). Curb challenges of the "Trojan Horse" approach: Smart strategies in achieving effective yet safe cell-penetrating peptide-based drug delivery. *Advanced Drug Delivery Reviews*, *65*(10), 1299–1315.

Jansen, G. (1999). *Receptor-and carrier-mediated transport systems for folates and antifolates. Antifolate Drugs in Cancer Therapy* (pp. 293–321). Springer.

Jarver, P., Mager, I., & Langel, U. (2010). In vivo biodistribution and efficacy of peptide mediated delivery. *Trends in Pharmacological Sciences*, *31*(11), 528–535.

Jerath, G., Goyal, R., Trivedi, V., Santhoshkumar, T. R., & Ramakrishnan, V. (2019). Syndiotactic peptides for targeted delivery. *Acta Biomaterialia*, *87*, 130–139.

Jerath, G., Goyal, R., Trivedi, V., Santhoshkumar, T. R., & Ramakrishnan, V. (2020). Conformationally constrained peptides for drug delivery. *Journal of Peptide Science*, *26*(4–5), e3244.

Jerath, G., Hazam, P. K., & Ramakrishnan, V. (2014). bPE toolkit: Toolkit for computational protein engineering. *Systems and Synthetic Biology*, *8*(4), 337–341.

Jerath, G., Hazam, P. K., Shekhar, S., & Ramakrishnan, V. (2016). Mapping the geometric evolution of protein folding motor. *PLoS One*, *11*(10), e0163993.

Joanne, P., Galanth, C., Goasdoué, N., Nicolas, P., Sagan, S., Lavielle, S., et al. (2009). Lipid reorganization induced by membrane-active peptides probed using differential scanning calorimetry. *Biochimica et Biophysica Acta (BBA)—Biomembranes*, *1788*(9), 1772–1781.

Jobin, M.-L., & Alves, I. D. (2014). On the importance of electrostatic interactions between cell penetrating peptides and membranes: A pathway toward tumor cell selectivity? *Biochimie*, *107*, 154–159.

Jobin, M.-L., Bonnafous, P., Temsamani, H., Dole, F., Grélard, A., Dufourc, E. J., et al. (2013). The enhanced membrane interaction and perturbation of a cell penetrating peptide in the presence of anionic lipids: Toward an understanding of its selectivity for cancer cells. *Biochimica et Biophysica Acta (BBA)—Biomembranes*, *1828*(6), 1457–1470.

Joliot, A., Pernelle, C., Deagostini-Bazin, H., & Prochiantz, A. (1991). Antennapedia homeobox peptide regulates neural morphogenesis. *Proceedings of the National Academy of Sciences of the United States of America*, *88*(5), 1864–1868.

Kalafatovic, D., & Giralt, E. (2017). Cell-penetrating peptides: Design strategies beyond primary structure and amphipathicity. *Molecules (Basel, Switzerland)*, *22*, 11.

Kang, M. J., Kim, B. G., Eum, J. Y., Park, S. H., Choi, S. E., An, J. J., et al. (2011). Design of a Pep-1 peptide-modified liposomal nanocarrier system for intracellular drug delivery: Conformational characterization and cellular uptake evaluation. *Journal of Drug Targeting*, *19*(7), 497–505.

Kaplan, I. M., Wadia, J. S., & Dowdy, S. F. (2005). Cationic TAT peptide transduction domain enters cells by macropinocytosis. *Journal of Controlled Release: Official Journal of the Controlled Release Society*, *102*(1), 247–253.

Kauffman, W. B., Fuselier, T., He, J., & Wimley, W. C. (2015). Mechanism matters: A taxonomy of cell penetrating peptides. *Trends in Biochemical Sciences*, *40*(12), 749–764.

Kawaguchi, Y., Takeuchi, T., Kuwata, K., Chiba, J., Hatanaka, Y., Nakase, I., et al. (2016). Syndecan-4 is a receptor for clathrin-mediated endocytosis of arginine-rich cell-penetrating peptides. *Bioconjugate Chemistry*, *27*(4), 1119–1130.

Kawamoto, S., Takasu, M., Miyakawa, T., Morikawa, R., Oda, T., Futaki, S., et al. (2011). Inverted micelle formation of cell-penetrating peptide studied by coarse-grained simulation: Importance of attractive force between cell-penetrating peptides and lipid head group. *The Journal of Chemical Physics*, *134*(9), 095103.

Klahn, M., & Zacharias, M. (2013). Transformations in plasma membranes of cancerous cells and resulting consequences for cation insertion studied with molecular dynamics, [10.1039/C3CP52085D] *Physical Chemistry Chemical Physics*, *15*(34), 14427–14441.

Komin, A., Russell, L. M., Hristova, K. A., & Searson, P. C. (2017). Peptide-based strategies for enhanced cell uptake, transcellular transport, and circulation: Mechanisms and challenges. *Advanced Drug Delivery Reviews*, *110–111*, 52–64.

Koren, E., & Torchilin, V. P. (2012). Cell-penetrating peptides: Breaking through to the other side. *Trends in Molecular Medicine*, *18*(7), 385–393.

Kumar, A., & Ramakrishnan, V. (2010). Creating novel protein scripts beyond natural alphabets. *Systems and Synthetic Biology*, *4*(4), 247–256.

Lindgren, M., & Langel, Ü. (2011). *Classes and prediction of cell-penetrating peptides. Cell-penetrating Peptides* (pp. 3–19). Springer.

Lindgren, M., Rosenthal-Aizman, K., Saar, K., Eiríksdóttir, E., Jiang, Y., Sassian, M., et al. (2006). Overcoming methotrexate resistance in breast cancer tumour cells by the use of a new cell-penetrating peptide. *Biochemical Pharmacology*, *71*(4), 416–425.

Liu, Y., Mei, L., Xu, C., Yu, Q., Shi, K., Zhang, L., et al. (2016). Dual receptor recognizing cell penetrating peptide for selective targeting, efficient intratumoral diffusion and synthesized anti-glioma therapy. *Theranostics*, *6*(2), 177–191.

Lopez-Meza, J. E., & Ochoa-Zarzosa, A. (2015). Antimicrobial peptides: Current and potential applications in biomedical therapies. *BioMed Research International*, *2015*, 367243.

Ma, Y., Gong, C., Ma, Y., Fan, F., Luo, M., Yang, F., et al. (2012). Direct cytosolic delivery of cargoes in vivo by a chimera consisting of D- and L-arginine residues. *Journal of Controlled Release: Official Journal of the Controlled Release Society*, *162*(2), 286–294.

Mäe, M., Myrberg, H., Jiang, Y., Paves, H., Valkna, A., & Langel, Ü. (2005). Internalisation of cell-penetrating peptides into tobacco protoplasts. *Biochimica et Biophysica Acta (BBA)—Biomembranes*, *1669*(2), 101–107.

Medina, S. H., Miller, S. E., Keim, A. I., Gorka, A. P., Schnermann, M. J., & Schneider, J. P. (2016). An intrinsically disordered peptide facilitates non-endosomal cell entry. *Angewandte Chemie International Edition*, *55*(10), 3369–3372.

Milletti, F. (2012). Cell-penetrating peptides: Classes, origin, and current landscape. *Drug Discovery Today*, *17*(15), 850–860.

Mishra, A., Lai, G. H., Schmidt, N. W., Sun, V. Z., Rodriguez, A. R., Tong, R., et al. (2011). Translocation of HIV TAT peptide and analogues induced by multiplexed membrane and cytoskeletal interactions. *Proceedings of the National Academy of Sciences of the United States of America*, *108*(41), 16883–16888.

Morris, M. C., Depollier, J., Mery, J., Heitz, F., & Divita, G. (2001). A peptide carrier for the delivery of biologically active proteins into mammalian cells. *Nature Biotechnology*, *19*, 1173.

Morris, M. C., Vidal, P., Chaloin, L., Heitz, F., & Divita, G. (1997). A new peptide vector for efficient delivery of oligonucleotides into mammalian cells. *Nucleic Acids Research*, *25*(14), 2730–2736.

Nakase, I., Takeuchi, T., Tanaka, G., & Futaki, S. (2008). Methodological and cellular aspects that govern the internalization mechanisms of arginine-rich cell-penetrating peptides. *Advanced Drug Delivery Reviews*, *60*(4–5), 598–607.

Nekhotiaeva, N., Awasthi, S. K., Nielsen, P. E., & Good, L. (2004). Inhibition of *Staphylococcus aureus* gene expression and growth using antisense peptide nucleic acids. *Molecular Therapy*, *10*(4), 652–659.

Nekhotiaeva, N., Elmquist, A., Rajarao, G. K., Hallbrink, M., Langel, U., & Good, L. (2004). Cell entry and antimicrobial properties of eukaryotic cell-penetrating peptides. *The FASEB Journal*, *18*(2), 394–396.

Oehlke, J., Scheller, A., Janek, K., Wiesner, B., Krause, E., Beyermann, M., et al. (1999). *Rapid translocation of amphipathic β-helical and β-sheet-forming peptides through plasma membranes of endothelial cells. Peptide Science—Present and Future* (pp. 782–783). Springer.

Oehlke, J., Scheller, A., Wiesner, B., Krause, E., Beyermann, M., Klauschenz, E., et al. (1998). Cellular uptake of an alpha-helical amphipathic model peptide with the potential to deliver polar compounds into the cell interior non-endocytically. *Biochimica et Biophysica Acta*, *1414*(1–2), 127–139.

Oglecka, K., Lundberg, P., Magzoub, M., Goran Eriksson, L. E., Langel, U., & Graslund, A. (2008). Relevance of the N-terminal NLS-like sequence of the prion protein for membrane perturbation effects. *Biochimica et Biophysica Acta*, *1778*(1), 206–213.

Papo, N., Seger, D., Makovitzki, A., Kalchenko, V., Eshhar, Z., Degani, H., et al. (2006). Inhibition of tumor growth and elimination of multiple metastases in human prostate and breast xenografts by systemic inoculation of a host defense–like lytic peptide. *Cancer Research*, *66*(10), 5371–5378.

Pasqualini, R., Koivunen, E., & Ruoslahti, E. (1997). αv integrins as receptors for tumor targeting by circulating ligands. *Nature Biotechnology*, *15*(6), 542.

Patel, L. N., Zaro, J. L., & Shen, W.-C. (2007). Cell penetrating peptides: intracellular pathways and pharmaceutical perspectives. *Pharmaceutical Research, 24*(11), 1977–1992.

Pavlova, N. N., & Thompson, C. B. (2016). The emerging hallmarks of cancer metabolism. *Cell Metabolism, 23*(1), 27–47.

Perret, F., Nishihara, M., Takeuchi, T., Futaki, S., Lazar, A. N., Coleman, A. W., et al. (2005). Anionic fullerenes, calixarenes, coronenes, and pyrenes as activators of oligo/polyarginines in model membranes and live cells. *Journal of the American Chemical Society, 127*(4), 1114–1115.

Pietras, K., & Östman, A. (2010). Hallmarks of cancer: Interactions with the tumor stroma. *Experimental Cell Research, 316*(8), 1324–1331.

Pomès, R., & Roux, B. (2002). Molecular mechanism of H + conduction in the single-file water chain of the gramicidin channel. *Biophysical Journal, 82*(5), 2304–2316.

Pooga, M., Hällbrink, M., Zorko, M., Uuml., & Langel, L. (1998). Cell penetration by transportan. *The FASEB Journal, 12*(1), 67–77.

Prochiantz, A. (1996). Getting hydrophilic compounds into cells: Lessons from homeopeptides. *Current Opinion in Neurobiology, 6*(5), 629–634.

Pujals, S., Sabido, E., Tarrago, T., & Giralt, E. (2007). all-D proline-rich cell-penetrating peptides: A preliminary in vivo internalization study. *Biochemical Society Transactions, 35*(Pt 4), 794–796.

Quail, D. F., & Joyce, J. A. (2013). Microenvironmental regulation of tumor progression and metastasis. *Nature Medicine, 19*(11), 1423.

Ramakrishnan, V., Ranbhor, R., Kumar, A., & Durani, S. (2006). The link between sequence and conformation in protein structures appears to be stereochemically established. *The Journal of Physical Chemistry. B, 110*(18), 9314–9323.

Rangel, R., Guzman-Rojas, L., Le Roux, L. G., Staquicini, F. I., Hosoya, H., Barbu, E. M., et al. (2012). Combinatorial targeting and discovery of ligand-receptors in organelles of mammalian cells. *Nature Communications, 3*, 788.

Raucher, D., & Ryu, J. S. (2015). Cell-penetrating peptides: Strategies for anticancer treatment. *Trends in Molecular Medicine, 21*(9), 560–570.

Reissmann, S. (2014). Cell penetration: Scope and limitations by the application of cell-penetrating peptides. *Journal of Peptide Science: An Official Publication of the European Peptide Society, 20*(10), 760–784.

Richard, J. P., Melikov, K., Vives, E., Ramos, C., Verbeure, B., Gait, M. J., et al. (2003). Cell-penetrating peptides. A reevaluation of the mechanism of cellular uptake. *Journal of Biological Chemistry, 278*(1), 585–590.

Rothbard, J. B., Jessop, T. C., Lewis, R. S., Murray, B. A., & Wender, P. A. (2004). Role of membrane potential and hydrogen bonding in the mechanism of translocation of guanidinium-rich peptides into cells. *Journal of the American Chemical Society, 126*(31), 9506–9507.

Ruoslahti, E. (2000). *Targeting tumor vasculature with homing peptides from phage display*. In: Paper presented at the Seminars in Cancer Biology.

Ruoslahti, E. (2017). Tumor penetrating peptides for improved drug delivery. *Advanced Drug Delivery Reviews, 110–111*, 3–12.

Rydström, A., Deshayes, S., Konate, K., Crombez, L., & Padari, K. (2011). Direct translocation as major cellular uptake for CADY self-assembling peptide. *PLoS One*.

Ryser, H. J., & Shen, W. C. (1978). Conjugation of methotrexate to poly(L-lysine) increases drug transport and overcomes drug resistance in cultured cells. *Proceedings of the National Academy of Sciences of the United States of America, 75*(8), 3867–3870.

Sandberg, M., Eriksson, L., Jonsson, J., Sjöström, M., & Wold, S. (1998). New chemical descriptors relevant for the design of biologically active peptides. A multivariate characterization of 87 amino acids. *Journal of Medicinal Chemistry*, *41*(14), 2481–2491.

Sanders, W. S., Johnston, C. I., Bridges, S. M., Burgess, S. C., & Willeford, K. O. (2011). Prediction of cell penetrating peptides by support vector machines. *PLoS Computational Biology*, *7*(7), e1002101.

Sastry, S. V., Nyshadham, J. R., & Fix, J. A. (2000). Recent technological advances in oral drug delivery—A review. *Pharmaceutical Science & Technology Today*, *3*(4), 138–145.

Schröder-Borm, H., Bakalova, R., & Andrä, J. (2005). The NK-lysin derived peptide NK-2 preferentially kills cancer cells with increased surface levels of negatively charged phosphatidylserine. *FEBS Letters*, *579*(27), 6128–6134.

Shen, W. C., & Ryser, H. J. (1978). Conjugation of poly-L-lysine to albumin and horseradish peroxidase: A novel method of enhancing the cellular uptake of proteins. *Proceedings of the National Academy of Sciences of the United States of America*, *75*(4), 1872–1876.

Shi, N.-Q., Qi, X.-R., Xiang, B., & Zhang, Y. (2014). A survey on "Trojan Horse" peptides: Opportunities, issues and controlled entry to "Troy.". *Journal of Controlled Release*, *194*, 53–70.

Shin, M. C., Zhang, J., Ah Min, K., Lee, K., Moon, C., Balthasar, J. P., et al. (2014). Combination of antibody targeting and PTD-mediated intracellular toxin delivery for colorectal cancer therapy. *Journal of Controlled Release: Official Journal of the Controlled Release Society*, *194*, 197–210.

Singh, R., & Lillard, J. W. (2009). Nanoparticle-based targeted drug delivery. *Experimental and Molecular Pathology*, *86*(3), 215–223.

Soomets, U., Lindgren, M., Gallet, X., Hallbrink, M., Elmquist, A., Balaspiri, L., et al. (2000). Deletion analogues of transportan. *Biochimica et Biophysica Acta*, *1467*(1), 165–176.

Suhorutsenko, J., Eriste, E., Copolovici, D.-M., & Langel, Ü. (2012). Human Protein 53-derived cell-penetrating peptides. *International Journal of Peptide Research and Therapeutics*, *18* (4), 291–297.

Svensen, N., Walton, J. G., & Bradley, M. (2012). Peptides for cell-selective drug delivery. *Trends in Pharmacological Sciences*, *33*(4), 186–192.

Takayama, K., Nakase, I., Michiue, H., Takeuchi, T., Tomizawa, K., Matsui, H., et al. (2009). Enhanced intracellular delivery using arginine-rich peptides by the addition of penetration accelerating sequences (Pas). *Journal of Controlled Release: Official Journal of the Controlled Release Society*, *138*(2), 128–133.

Temsamani, J., & Vidal, P. (2004). The use of cell-penetrating peptides for drug delivery. *Drug Discovery Today*, *9*(23), 1012–1019.

Tiwari, G., Tiwari, R., Sriwastawa, B., Bhati, L., Pandey, S., Pandey, P., et al. (2012). Drug delivery systems: An updated review. *International Journal of Pharmaceutical Investigation*, *2*(1), 2.

Tunnemann, G., Ter-Avetisyan, G., Martin, R. M., Stockl, M., Herrmann, A., & Cardoso, M. C. (2008). Live-cell analysis of cell penetration ability and toxicity of oligo-arginines. *Journal of Peptide Science: An Official Publication of the European Peptide Society*, *14*(4), 469–476.

Urry, D. W. (1971). The gramicidin A transmembrane channel: A proposed π(L,D) helix. *Proceedings of the National Academy of Sciences*, *68*(3), 672–676.

Verdurmen, W. P. R., Bovee-Geurts., Petra, H., Wadhwani, P., Ulrich., Anne, S., Hällbrink, M., van Kuppevelt., Toin, H., et al. (2011). Preferential uptake of L- vs D-amino acid cell-penetrating peptides in a cell type-dependent manner. *Chemistry & Biology*, *18*(8), 1000–1010.

Wadia, J. S., Stan, R. V., & Dowdy, S. F. (2004). Transducible TAT-HA fusogenic peptide enhances escape of TAT-fusion proteins after lipid raft macropinocytosis. *Nature Medicine*, *10*(3), 310–315.

Wender, P. A., Galliher, W. C., Goun, E. A., Jones, L. R., & Pillow, T. H. (2008). The design of guanidinium-rich transporters and their internalization mechanisms. *Advanced Drug Delivery Reviews*, *60*(4), 452–472.

Wender, P. A., Mitchell, D. J., Pattabiraman, K., Pelkey, E. T., Steinman, L., & Rothbard, J. B. (2000). The design, synthesis, and evaluation of molecules that enable or enhance cellular uptake: Peptoid molecular transporters. *Proceedings of the National Academy of Sciences of the United States of America*, *97*(24), 13003–13008.

Wolf, Y., Pritz, S., Abes, S., Bienert, M., Lebleu, B., & Oehlke, J. (2006). Structural requirements for cellular uptake and antisense activity of peptide nucleic acids conjugated with various peptides. *Biochemistry*, *45*(50), 14944–14954.

Yamada, T., Christov, K., Shilkaitis, A., Bratescu, L., Green, A., Santini, S., et al. (2013). p28, a first in class peptide inhibitor of cop1 binding to p53. *British Journal of Cancer*, *108*(12), 2495.

Yamada, T., Signorelli, S., Cannistraro, S., Beattie, C. W., & Bizzarri, A. R. (2015). Chirality switching within an anionic cell-penetrating peptide inhibits translocation without affecting preferential entry. *Molecular Pharmaceutics*, *12*(1), 140–149.

Ye, J., Liu, E., Yu, Z., Pei, X., Chen, S., Zhang, P., et al. (2016). CPP-assisted intracellular drug delivery, what is next? *International Journal of Molecular Sciences*, *17*(11).

Ziegler, A., & Seelig, J. (2011). Contributions of glycosaminoglycan binding and clustering to the biological uptake of the nonamphipathic cell-penetrating peptide WR9. *Biochemistry*, *50* (21), 4650–4664.

Chapter 6

Peptide-based nanomaterials: applications and challenges

Gaurav Pandey[1,2] and Debika Datta[3]

[1]*Department of Chemistry, University of California Riverside, Riverside, CA, United States,* [2]*Present address: Scripps Research, Department of Molecular Medicine, La Jolla, CA, United States,* [3]*Department of Nanoengineering, University of California San Diego, La Jolla, CA, United States*

6.1 Introduction to molecular self-assembly

Molecular self-assembly is central to life. In biological systems, simpler building blocks associate themselves in solution through molecular recognition and self-assembly to give rise to rather complex structures. For example, the self-assembly of lipid molecules in water to give rise to a lipid bilayer is perhaps one of the most well-known examples of self-assembly in biology (Fig. 6.1). The amphipathic nature of the phospholipids causes the hydrophobic tail to be shielded from the water inside the hydrophobic core, whereas

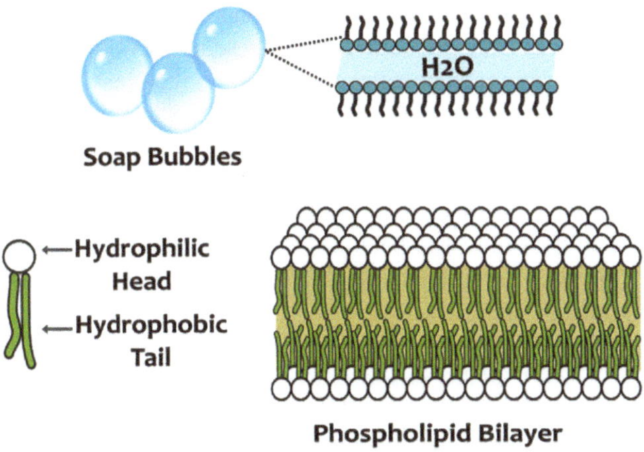

FIGURE 6.1 A diagrammatic representation depicting self-assembly of soap molecules forming the bubbles (top) and phospholipid molecules forming bilayer and micelles (bottom).

De Novo Peptide Design. DOI: https://doi.org/10.1016/B978-0-323-99917-5.00001-9

the hydrophilic polar headgroup is exposed to the aqueous environment (Schnur, 1993; Spector et al., 1996).

Other examples include assembly of nucleic acids through H-bonding and stacking interactions between the nitrogenous bases, the self-assembly of single protein/glycoprotein in the S-layers in bacterial cell surface (all archaea and many bacteria) (Schuster, Györvary, Pum, & Sleytr, 2005), spontaneous formation of template-assisted virus shell (like tobacco mosaic virus) (Fraenkel-Conrat, Singer, & Williams, 1957), the folding of the polypeptide chain to form the tertiary or quaternary protein structure (Sigler et al., 1998), silk fibroin protein fibrillar assembly (Feltwell, 1990; Winkler et al., 1999), and amyloid fibrils (Chapman et al., 2002; Elliot et al., 2003; Jarrett & Lansbury Jr, 1993; King et al., 1997).

Any physical system has the natural propensity to exchange energy with its surroundings to attain a low energy state. To bring the constituent particles to a stable configuration, they assume structures with a reduced free energy state. This spontaneous, self-assembly process happens without any external intervention. In the language of a chemist or biologist, the term self-assembly refers to the spontaneous organization of components of a system into an ordered functional unit. This phenomenon is reversible and can result from direct (specific interactions between the individual components) or indirect (environment mediated, for example, temperature, pH) interactions. It is driven by minimization of the free energy of the system, achieved by minimization of repulsive and maximization of attractive molecular interactions.

The intermolecular forces like van der Waals interactions, hydrogen bonding, hydrophobic interactions, weak polar interactions, $\pi - \pi$ interactions, as well as electrostatic interactions, hold the components into well-defined stable structures, such that the assembled structure is in equilibrium with the environment as well as the individual components (Pauling, 1960; Wang, Liu, Xing, & Yan, 2016). While all these noncovalent interactions are rather weak in isolation, when combined together, they play the decisive role in the conformation a molecule can take and the subsequent self-assembly (Ball, 1994; Lehn, 1993; Mandal, Shirazi, & Parang, 2014; Whitesides, Mathias, & Seto, 1991). Through these interactions, small building blocks associate to produce large superstructures with distinct physical and chemical properties. This assembly approach has been commonly called the "bottom-up" technique for designing and fabrication of nano and microstructures (Gazit, 2007a).

There are two main types of self-assembly: static and dynamic (Rughani & Schneider, 2008). Static self-assembly comprises of systems that are at global or local equilibrium, do not dissipate energy, and may require energy (for example, globular proteins are formed by static self-assembly) to form the assembly. In contrast, dynamic self-assembly involves energy dissipation to create ordered structures.

Molecular self-assembly has shown promise to be a powerful tool in the fabrication of sophisticated nanomaterials. Self-assembly of molecules in biological systems is highly regulated and carried out with very high fidelity. This is necessary to ensure proper function and avoid deleterious effects. Designing the molecules that can self-assemble in a controlled way and can respond to stimuli in a predicted manner is an emerging area in material science. Inspired by the tight control of assembly and precise emerging properties of the biological self-assembly, researchers have looked toward the molecules of biological origin to design such self-assembling molecules. Researchers have utilized molecules that include, but are not limited to, lipids, nucleic acid complexes, amino acids, peptides, proteins, and peptidomimetics (Aggeli et al., 2001; Alivisatos et al., 1996; Bong, Clark, Granja, & Ghadiri, 2001; Ghadiri, Granja, & Buehler, 1994; Mirkin, Letsinger, Mucic, & Storhoff, 1996; Schnur, 1993; Winfree, Liu, Wenzler, & Seeman, 1998; Zhang et al., 1995).

6.2 Protein and peptide self-assembly

Proteins are the workhorses of the cell, performing essentially every activity of life. Our body harbors over a million distinct types of proteins, and more than half the dry weight of a cell is constituted by proteins (Dobson, 2004). They serve a broad range of functions, central to the maintenance of life and have always attracted the attention of researchers.

Polymers of amino acids with roughly 2–50 amino acids are generally referred to as peptides. Proteins are heteropolymers of amino acids linked by peptide bonds (Brändén & Tooze, 1999). The linear chain of amino acids constituting a polypeptide chain is referred to as the primary structure of a protein. These polypeptide chains, driven by hydrogen bonding, organize themselves to have varying spatial arrangements like α-helices or β-sheets, forming the secondary structure of a protein. The secondary structures are driven by nonspecific hydrophobic interactions, further arranging themselves to create the functional three-dimensional domain, called the tertiary structure of a protein. The domains comprise the tertiary structure associated by ionic, salt bridges, disulfide and other noncovalent interactions like van der Waals forces between their amino acid residues, thus forming the quaternary structure of a protein (Fig. 6.2). The structural organization is directed by the amino acid chains constituting the primary structure.

The presence of the 20 natural amino acids along with the unnatural amino acids offers numerous possible combinations of sequences with intriguing physicochemical properties. The chemical space of a polymer made up of "X" different monomers is given by X^N, where N is the length of the polymer. Thus the larger the number of monomers available, the more diverse the polymers are. For example, for a tetrapeptide made up of naturally occurring 20 amino acids, there are 20^4, that is, 160,000 different

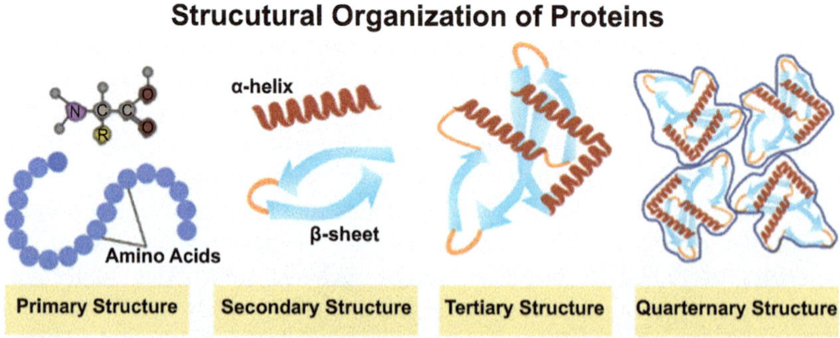

FIGURE 6.2 A schematic representation of four levels of protein structure organization.

sequences possible. An oligonucleotide of the same length, on the other hand, would have only 4^4, that is, 256 different sequences. Considering the possibility of including the D-enantiomers expands the diversity of a tetrapeptide to 40^4, that is, 1,560,000, a number four orders of magnitude larger than that for an oligonucleotide. Due to their inherent chemical versatility, ease of synthesis, physical and chemical stability, and well-established principles of folding, the peptides stand out as an excellent choice of starting material. Due to their properties such as biocompatibility and biodegradability, peptides, therefore, turn out to be ideal for making nanomaterials for various applications.

6.3 Peptide-based nanomaterials

Over the past three decades, the focus of nanotechnology has been toward the development of novel nanostructures and exploration of their potential applications such as materials for nanoelectronics, catalysts for bioprocessing, and sensors for biomedical technology (Chen & Rosi, 2010; Gazit, 2008).

The designed peptidic sequences can be either chemically synthesized or produced from microorganisms. The sequence-specific self-assembly makes the designed peptides take the desired conformations (α-helix, β-sheets), which could be fine-tuned to form three-dimensional building blocks for supramolecular structures (Fig. 6.3) (Mart, Osborne, Stevens, & Ulijn, 2006; Zhao & Zhang, 2006).

Peptides have different physicochemical properties depending on their conformation and stereochemical configuration of the amino acids. This makes them sensitive to environmental changes like pH, temperature, salt content, external stimulus, etc. Hence, they can be used as versatile biocompatible smart building blocks for the de novo design of functional materials (Briggs & Knecht, 2012; Ulijn & Woolfson, 2010). Some examples of peptide-based self-assembled structures include tubes, wires, tapes, fibrils,

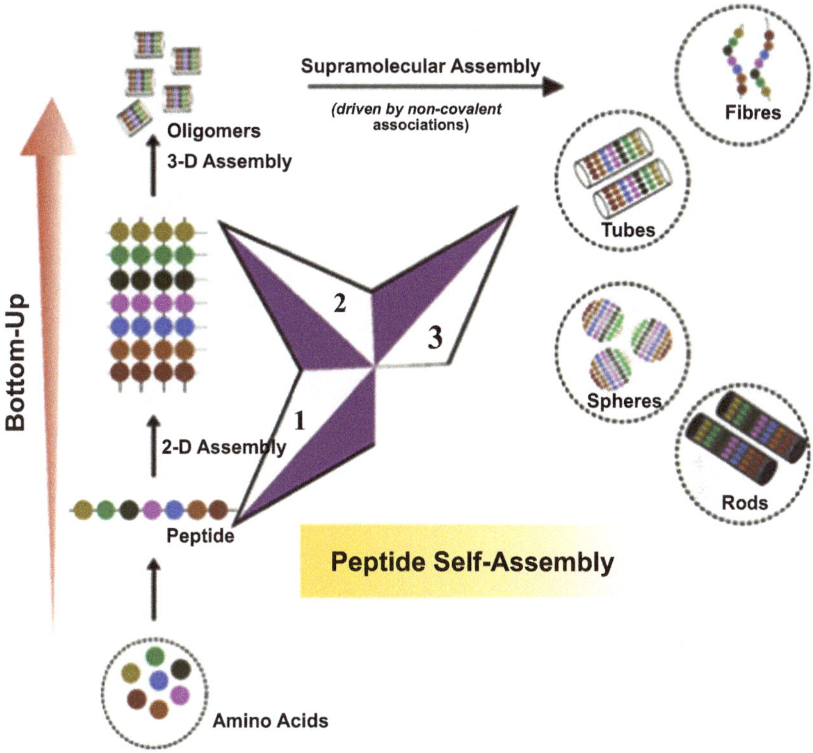

FIGURE 6.3 Diagrammatic illustration of the bottom-up approach of peptide self-assembly.

ribbons, hydrogels, vesicles, spheres (Aggeli et al., 1997; Cohen & Calkins, 1959; Del Mercato et al., 2007; Eanes & Glenner, 1968; Ghadiri, Granja, Milligan, McRee, & Khazanovich, 1993; Goux et al., 2004; Koley & Pramanik, 2011; Kretsinger, Haines, Ozbas, Pochan, & Schneider, 2005; Krysmann et al., 2008; Letchford & Burt, 2007; Lu, Jacob, Thiyagarajan, Conticello, & Lynn, 2003; Ozbas, Kretsinger, Rajagopal, Schneider, & Pochan, 2004; Ozbas, Rajagopal, Schneider, & Pochan, 2004; Reches & Gazit, 2003, 2004; Scheibel et al., 2003; Xie et al., 2015; Yokoi, Kinoshita, & Zhang, 2005).

Moreover, peptide-based supramolecular structures have been conjugated with various other molecules to produce hybrid compounds, peptidomimetics which find applications in material science (Sarikaya, Tamerler, Jen, Schulten, & Baneyx, 2003).

These materials have shown interesting electrical, optical, mechanical, and biological properties. Some of the exciting reports on self-assembled structures are listed and discussed in Table 6.1.

TABLE 6.1 Some of the important literature about self-assembly of the peptides, amino acids, and their analogs.

References	Work done
Amyloidogenic peptides	
Rudolf Virchow	Found that corpora amylacea in ependyme (the thin lining of the ventricular system of the brain) and choroid plexus (an area in the brain that produces cerebrospinal fluid) stained violet with iodine. He firmly believed that the stained material was starch and named it "amyloid."
Friedreich and Kekulé (1859)	Showed that protein was the primary substance present in the amyloid-rich segments isolated from the spleen.
Eanes and Glenner (1968)	The first report on the structural organization of proteins/peptides in amyloid fibrils (X-ray fiber diffraction). The first proof of the cross-β structure of amyloid fibrils.
Petkova et al. (2002)	Structural model of $A\beta_{1-40}$ from solid-state NMR.
Török et al. (2002)	Structural features of amyloid fibrils of Alzheimer's Aβ peptide using electron paramagnetic resonance spectroscopy.
Lu et al. (2003)	Formation of highly ordered nanotubes by the amyloidogenic fragment of β-amyloid, $A\beta_{16-22}$ at acidic pH.
Scheibel et al. (2003)	Conducting nanowires from amyloid fiber yeast Sup35p self-assembly.
Nelson et al. (2005)	The structure of the cross-β spine of a seven-residue amyloidogenic fragment of Sup35. The first atomic-resolution structure of an amyloid fibril.
Heise et al. (2005)	Structure and dynamics of full-length α-synuclein fibrils by solid-state NMR.
Smith, Knowles, Dobson, MacPhee, and Welland (2006)	Studied mechanical properties of insulin fibrils. The fibrils were found to have strength comparable to that of steel and mechanical stiffness similar to that of silk.
Del Mercato et al. (2007)	Intrinsically conducting and fluorescent amyloid-like fibrils of poly (VGGLG).
Krysmann et al. (2008)	Hydrogel from an amyloidogenic fragment of β-amyloid, KLVFF.

(Continued)

TABLE 6.1 (Continued)

References	Work done
Bhak, Lee, Park, Cho, and Paik (2010)	Accumulation of curly protein fibrils of α-synuclein produced amyloid hydrogel.
Guerrero-Ferreira et al. (2018)	Structure of α-synuclein fibrils using Cryo-EM.
Peptide nanotubes	
Ghadiri et al. (1993)	Formation of nanotubes from cyclic syndiotactic peptides. Extensive H-bonding network stabilizes the self-assembled structures.
Ghadiri et al. (1994)	Artificial ion channels using nanotubes made up of cyclic syndiotactic peptides having membrane-interacting properties.
Reches and Gazit (2003)	Nanotubes from diphenylalanine. Casting wires using nanotubes as the molds.
Reches and Gazit (2004)	Nanosphere formation from diphenylglycine derivatives.
Horne, Ashkenasy, and Ghadiri (2005)	Charge transfer through naphthalenetetracarboxylic acid diimide side chain in cyclic peptide self-assembly.
Carny, Shalev, and Gazit (2006)	Coaxial wires using diphenylalanine nanotubes.
Görbitz (2006)	Structure of the diphenylalanine nanotubes.
Liang et al. (2008)	Amyloid fibrils as the scaffold for light harvesting.
Kholkin, Amdursky, Bdikin, Gazit, and Rosenman (2010)	Reported strong shear piezoelectric activity in the diphenylalanine nanotubes.
Yan, Li, and Möhwald (2011)	Hexagonal crystalline microtubes through higher order organization of the diphenylalanine nanotubes. Show linear birefringence and optical waveguiding.
Tarabout et al. (2011)	Modulation of the peptide tube diameters by chemical modifications of the aromatic residues in close contacts.
Nonamyloid peptide–based fibers, ribbons, tapes, and hydrogels	
Aggeli et al. (1997)	Short oligopeptides form polymeric β-sheet nanotapes in water and methanol.

(Continued)

TABLE 6.1 (Continued)

References	Work done
Niece, Hartgerink, Donners, and Stupp (2003)	Nanofibers from peptide-amphiphile molecules. Assembly through electrostatic attraction and hydrophobic interaction.
Ozbas, Kretsinger, et al. (2004)	β-Hairpin peptide self-assembles to form hydrogel triggered by the presence of salt.
Yokoi et al. (2005)	Self-complementary peptide RADA16 forms nanofibers and hydrogel.
Schneider et al. (2002), Kretsinger et al. (2005)	MAX1 peptide folds into β-hairpin conformation and thereby forms the hydrogel.
Letchford and Burt (2007)	Formation of micelles, nanospheres, nanocapsules, and polymerosomes from amphiphilic copolymers of polyethylene glycol derivatives.
Mahler, Reches, Rechter, Cohen, and Gazit (2006)	Aromatic phenylalanine dipeptides forming hydrogels.
Zhang, Gu, Yang, and Xu (2003), Smith et al. (2008)	Hydrogel formation from Fmoc-diphenylalanine and other Fmoc-dipeptides.
Yang, Xu, Guo, Guo, and Xu (2007)	Intracellular formation of nanofibers (catalyzed by enzymes) that leads to hydrogelation and cell death.
Hirst, Escuder, Miravet, and Smith (2008)	Supramolecular gel-based nanostructures find application as biomaterials, regenerative medicine, hydrogels, metallogels, optical technologies, etc.
Lim et al. (2008)	A cyclic RGD-coated peptide self-assembles to form nanoribbons structure.
Rajagopal, Lamm, Haines-Butterick, Pochan, and Schneider (2009)	MAX1 peptide variant [MAX1(K15E)] undergoes thermally triggered hydrogelation at physiological buffer conditions.
Cui, Muraoka, Cheetham, and Stupp (2009)	A peptide amphiphile consisting of four amino acids and an alkyl tail self-assembles to form giant peptide nanobelts.
Ma et al. (2010)	Nanofibers and hydrogel formation from glycine-rich pentapeptide derivatives.
Huang et al. (2011)	Photoresponsive hydrogels from azobenzene substituted short peptides.
Fletcher, Lockett, and Dexter (2011)	A pH-responsive reversible hydrogel formed from a 21-residue α-helical peptide.

(*Continued*)

TABLE 6.1 (Continued)

References	Work done
Castelletto, Hamley, Adamcik, Mezzenga, and Gummel (2012)	A peptide amphiphile (C16-KTTKS) forms tapes, twisted ribbons, and fibrils on the addition of the oppositely charged anionic surfactant.
Ghosh, Adler-Abramovich, Gazit, and Verma (2013)	Derivatives of diphenylalanine conjugates form fibers.
Xie et al. (2015)	Helical nanoribbons from Fmoc-tripeptides.
Amino acid–based structures	
Gortner and Hoffman (1921)	One of the first examples of an amino acid–based hydrogel: the dibenzoyl cysteine formed a hydrogel.
Vegners, Shestakova, Kalvinsh, Ezzell, and Janmey (1995)	First report of a Fmoc dipeptide hydrogelator (Fmoc-Leu-Asp).
Yang, Gu, Zhang, Wang, and Xu (2004)	Hydrogel formation by N_α-Fmoc-Tyr and N_ε-Fmoc-Lys.
Toledano, Williams, Jayawarna, and Ulijn (2006)	Reversed hydrolysis by proteases formed potent peptide hydrogelators from Fmoc-tripeptides.
Jayawarna et al. (2006)	Fmoc dipeptides form hydrogels capable of supporting 3D cell culture.
Yang, Liang, Ma, Gao, and Xu (2007)	A β-amino acid derivative forms supramolecular hydrogel, in vivo and in vitro, triggered by phosphatase enzyme.
Gao et al. (2009)	Enzymatic dephosphorylation triggered the self-assembly and hydrogelation of a C-terminal methyl ester derivative of Fmoc-Tyr.
Liang et al. (2009)	Controlled drug release from D-amino acid dipeptide hydrogels.
Sutton et al. (2009)	Controlled dye-release from Fmoc-amino acid hydrogels.
Ryan, Anderson, Senguen, Youngman, and Nilsson (2010)	Compared to Fmoc-Tyr that forms hydrogels in 30 min with a rigidity, G′ of ~1000 Pa, Fmoc-Phe(F_5) forms optically transparent hydrogels within 5 min with a G′ of ~3000 Pa. Perfluorination of the phenyl side chain results in a significant electronic perturbation of the phenyl side chain (the side chain is much more electron-deficient than that of Tyr) and an increase in hydrophobicity.

(*Continued*)

TABLE 6.1 (Continued)

References	Work done
Koley and Pramanik (2011)	Formation of nanorods, nanofibrils, nanotubes, and nanovesicles from single hydrophobic amino acid- (Phe, Trp, Tyr, and Leu) based molecules.
Cao, Yuan, Zhu, Zhao, and Liu (2012)	Controlled self-assembly from achiral amino acid derivatives to form chiral twisted nanostructures.
Irwansyah et al. (2015)	Antimicrobial activity against Gram-positive bacteria, from coassembly of Fmoc-Phe and Fmoc-Leu hydrogels.

6.3.1 Amyloid peptides

Amyloid fibrils constitute a particularly important class of self-assembling proteins and peptides. Amyloid fibrils are long, fibrous aggregates that are composed of long β-sheets and are characterized by a meridional reflection at 4.7–4.8 Å and an equatorial reflection at ~10 Å. They are mostly known as the aggregates of misfolded proteins that lead to various neurodegenerative disease conditions like Alzheimer's, Parkinson's, type II diabetes, Huntington's disease, etc. (Glabe, 2006). However, studies in the past decade have identified various amyloidogenic proteins that are native functional forms in various living forms, including humans. The disease-unrelated amyloids have been found in aerial hyphae of filamentous bacteria, bacterial biofilm formation, memory consolidation in Drosophila, as well as in melanin synthesis in humans (Chapman et al., 2002; Fowler et al., 2005; Hervás et al., 2016; Maury, 2009). The growing evidence suggesting nontoxicity of amyloid fibrils eased their way for being used as functional materials in technological applications.

Amyloid fibrils have drawn the attention of the biological, chemistry, and material science community alike, owing to their impressive material properties. Amyloid fibers could have unusually high mechanical properties; insulin fibrils, for example, show mechanical strength comparable to that of steel, stiffness similar to that of silk (Smith et al., 2006). Owing to their stability and high aspect ratio (can go up to hundreds of micrometers), amyloid fibrils have been utilized to make nanowires by metal deposition (Reches & Gazit, 2003). Amyloid protein fiber from *Saccharomyces cerevisiae* Sup35 self-assembly has been used to form conducting nanowires (Scheibel et al., 2003). An investigation by Del Mercato et al. (2007) showed the electrical conductivity in intrinsically fluorescent amyloid-like fibril of polypeptide poly(ValGlyGlyLeuGly).

It is always advantageous to have short peptide sequences as monomers for self-assembly. Short peptides are not only economical and easy to synthesize but also self-assemble through a limited set of interactions, thereby providing better control over self-assembly. One of the most widely studied molecular systems is diphenylalanine (FF) self-assembly, which is a part of the core fragment of Alzheimer's amyloid-beta (Aβ) protein. Görbitz and Gazit showed the formation of nanotubes from the dipeptide diphenylalanine (Görbitz, 2006; Reches & Gazit, 2003).

FF self-assembles to form ordered nanotubes, microtubes, nanowires, nanorods, and nanocrystals under different conditions (Adler-Abramovich et al., 2009; Pandey et al., 2017; Reches & Gazit, 2003, 2006; Ryu & Park, 2008; Wang, Du, Wu, Xiong, & Chu, 2011; Zhu, Yan, Su, Yang, & Li, 2010). Reches and Gazit showed casting metal wires using diphenylalanine nanotubes as molds (Reches & Gazit, 2003). Diphenylalanine peptide nanotubes have also been fabricated to form trilayered coaxial nanocables using metal nanoparticles and can have unique electromagnetic properties (Carny et al., 2006).

Several short (4–10 residues) peptide sequences have been shown to form ordered amyloid-like fibrils with characteristic structural properties, which can serve as model systems for understanding the self-assembly of amyloid fibrils (Goux et al., 2004; Reches, Porat, & Gazit, 2002; Tenidis et al., 2000; Tjernberg, Hosia, Bark, Thyberg, & Johansson, 2002; Wagoner, Cheon, Chang, & Hall, 2014). The 16–22-residue stretch of β-amyloid ($A\beta_{16-22}$: Ac-KLVFFAE-am) has been known to assemble into fibers, nanotubes, and annular rings under different conditions (Balbach et al., 2000; Liu et al., 2012; Lu et al., 2003; Pachahara, Chaudhary, Subbalakshmi, & Nagaraj, 2012). Like Phe-Phe, highly ordered self-assembled nanotubes were obtained from the $A\beta_{16-22}$ fragment at acidic pH (Lu et al., 2003). Krysmann et al. showed the formation of hydrogels from an amyloidogenic fragment of β-amyloid, KLVFF (Krysmann et al., 2008). The aromatic cassette Phe-Phe is believed to be critical in modulating self-assembly (Chaudhary & Nagaraj, 2011; Inouye, Gleason, Zhang, Decatur, & Kirschner, 2010; Krone et al., 2008; Pachahara & Nagaraj, 2015). Nagaraj and coworkers investigated various aromatic analogs of $A\beta_{16-22}$, substituting Phe with Tyr and Trp residues and obtained highly amyloidogenic variants (Pachahara & Nagaraj, 2015). In a recent study, Chaudhary and coworkers investigated the hydrogelation propensity of end-capped $A\beta_{16-22}$ and its aromatic analog $A\beta_{16-22}$(F20Y)(Ac-KLVFYAE-am). The $A\beta_{16-22}$(F20Y) is reported to form hydrogels at concentrations ≥ 0.18 wt% (Datta, Kumar, Kumar, Nagaraj, & Chaudhary, 2019a).

In another remarkable study, Buchanan et al. investigated the mechanism of human islet amyloid polypeptide (IAPP) associated with type II diabetes

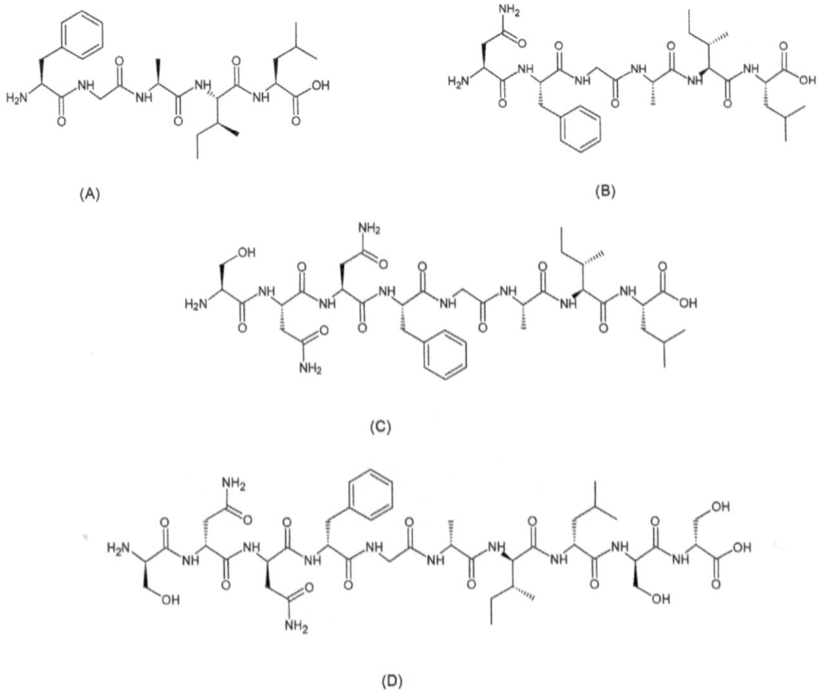

FIGURE 6.4 IAPP sequences forming amyloid-like fibrils identified by Kapurniotu and coworkers. (A) IAPP$_{23-37}$ (FGAIL), (B) IAPP$_{22-27}$, (C) (NFGAIL), and (D) IAPP$_{20-27}$ (SNNFGAIL), and IAPP$_{20-29}$ (SNNFGAILSS) (Tenidis et al., 2000).

using spectroscopic methods (Buchanan et al., 2013). Kapurniotu and coworkers identified short amyloidogenic stretches from human IAPP. The stretches IAPP$_{23-37}$ (FGAIL), IAPP$_{22-27}$ (NFGAIL), and IAPP$_{20-27}$ (SNNFGAIL), and IAPP$_{20-29}$ (SNNFGAILSS) are reported to form characteristic amyloid-like fibrils in vitro (Tenidis et al., 2000) (Fig. 6.4). Johansson and coworkers have reported typical amyloid-like fibrils from de novo designed tetrapeptides. The charge complementary in peptides like KFFE and KVVE is reported to play a critical role in the assembly (Tjernberg et al., 2002). Goux and coworkers identified short amyloidogenic stretches ($\leq$6 residues) from the tau protein (Goux et al., 2004).

Research in the past 20 years has shed light on the possible toxic protein species in amyloid diseases; it turns out that the toxicity is restricted chiefly to prefibrillar aggregates, and fibrils are the innocent bystanders (Hardy & Selkoe, 2002; Hartley et al., 1999; Lambert et al., 1998). This feature could make amyloid fibrils potential candidates as tissue engineering and drug delivery vehicles. The recognition principles involving the key use of aromatic interactions in driving the self-assembly of amyloid fibrils could thus be used in the design and fabrication of nanomaterials.

6.3.2 Cyclic peptides

A seminal work by Ghadiri and coworkers in 1993 gave the first peptide nanotubes from syndiotactic cyclic peptides (Ghadiri et al., 1993). The unique design principle involved the incorporation of amino acids with alternate Cα stereochemistry (alternate L and D-amino acids). In addition, the peptides were cyclized, thereby putting large-scale conformational constraints. The peptide molecules stack up, giving rise to hollow nanotubular structures. The stacked ring subunits are stabilized by the intermolecular H-bonding to form a β-sheet conformation (Bong et al., 2001). Artificial transmembrane ion channels were subsequently derived through the self-assembly of similar cyclic peptides (Ghadiri et al., 1994). Self-assembly of the syndiotactic cyclic peptides tagged with 1,4,5,8-naphthalenetetracarboxylic acid diimide (NDI) resulted in NDI excimer formation (Horne et al., 2005). This study shows that self-assembling cyclic syndiotactic peptides could be used to develop nanomaterials with interesting electronic and optical properties. A tryptophan-rich cationic peptide designed on the same principles has been shown to self-assemble into tubular structures that disrupt bacterial membranes, thereby acting as an antimicrobial agent (Fernandez-Lopez et al., 2001). The Lanreotide, a synthetic analog of somatostatin, is used as a growth hormone inhibitor. This cyclic octapeptide has been shown to form tubular structures in an aqueous solution (Valéry et al., 2003). The nanotubes have been reported to have a viral capsid-like dimension. An alternating pattern of aromatic and aliphatic amino acid residues resulted in antiparallel β-sheet conformation in the peptide building blocks. Furthermore, the self-assembly process for several cyclic peptides formed from 3–5 repeats of dipeptides having hydrophobic and charged amino acid residues have also been investigated (Mandal et al., 2013). The self-assembly of several cyclic syndiotactic peptides has been described in an excellent review by Perrier and coworkers (Chapman, Danial, Koh, Jolliffe, & Perrier, 2012).

6.3.3 Peptide amphiphiles

A peptide amphiphile (PA) is a peptide or peptide conjugate having distinct polar and nonpolar regions. An amphiphilic peptide can be made by attaching a stretch of nonpolar amino acid residues (forming the hydrophobic tail) to polar or charged amino acids (forming the hydrophilic head). Besides, the amphipathicity could be conferred through folding and self-assembly. The peptides may lack distinct hydrophilic and hydrophobic regions in their primary structure but can fold/assemble to attain an amphipathic character (Castelletto et al., 2012; Fletcher et al., 2011; Yokoi et al., 2005).

6.3.3.1 Lipopeptides

An important class of PAs is a peptide chain conjugated to a fatty acid/lipid/surfactant (Chu-Kung et al., 2004; Dasgupta, Mondal, & Das, 2013; Hamley, 2011; Hartgerink, Beniash, & Stupp, 2002; Zhao et al., 2010; Zhou, Yang, Zhao, Ning, & Xu, 2015). Surfactin, a cyclic lipopeptide of the *Bacillus subtilis* origin, is an example of naturally occurring lipopeptides (Heerklotz & Seelig, 2001). The generalized structure of lipopeptides includes an acyl carbon chain linked to one of the peptide termini. Stupp and coworkers made an extensive contribution to this class of self-assembling molecules by investigating several such PAs. They synthesized several amphiphilic molecules by attaching a long hydrophobic alkyl tail with a hydrophilic peptide chain head, mimicking a phospholipid molecule (Cui et al., 2009; Greenfield, Hoffman, Olvera de la Cruz, & Stupp, 2009; Guler et al., 2006; Hartgerink et al., 2002; Hartgerink, Beniash, & Stupp, 2001). One such example is that of the formation of giant nanobelts of length up to 0.1 mm formed from the self-assembly of $C_{16}H_{31}$OVEVE (Cui et al., 2009) (Fig. 6.5). The headgroup of this lipopeptide had alternating hydrophilic (E) and hydrophobic residues (V). Another interesting work is that on the formation of nanofibres from self-assembled branched PAs. The PAs had a palmitoyl chain coupled to the peptide sequence containing RGDS as a cell adhesion epitope (Guler et al., 2006). Other lipopeptides like $C_{10}H_{19}OA_4G_3S(PO_4)$RGD and

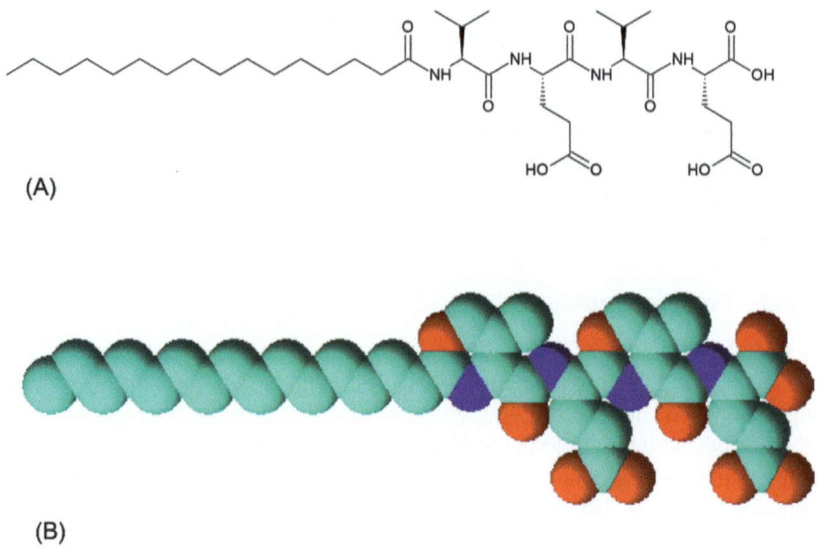

FIGURE 6.5 Peptide amphiphile $C_{16}H_{31}$OVEVE designed by Stupp and coworkers forming nanobelts. (A) Chemical structure, (B) Spacefill model showing the long hydrophobic alkyl tail and alternating hydrophilic (E) and hydrophobic residues (V) in the peptide headgroup (Cui et al., 2009).

$C_{16}H_{31}OC_4G_3S(PO_4)RGD$ have been shown to self-assemble into nanofibers at low pH and disassemble fully with increasing pH, thus making them pH-responsive (Hartgerink et al., 2002).

Greenfield et al. showed that the addition of salts in PAs resulted in nano-fibrillar gels having stronger rheological properties. The calcium-mediated ionic bridges in $CaCl_2-C_{16}-V_3A_3E_3$ lipopeptide resulted in gels with stronger intra and interfiber crosslinks, thereby enabling them to withstand higher strains than the HCl−PA lipopeptide gel (Greenfield et al., 2009). Deng et al. coupled a seven-residue peptide stretch from the β-amyloid ($A\beta_{11-17}$) to dodecanoic acid and found that the otherwise nonassembling $A\beta_{11-17}$ sequence, when synthesized as C_{12}-$A\beta_{11-17}$, was able to self-assemble into nanofibres and nanobelts at pH 3 and 10, respectively (Deng, Yu, Hou, & Wang, 2009). The lipopeptides have shown to form a variety of interesting superstructures. The peptide $C_{16}H_{31}-WA_4KA_4KA_4KA$ was found to self-assemble into spherical micelles with diameters around 10 nm that gradually transformed into worm-like micelles on prolonged incubation (Shimada, Lee, Bates, Hotta, & Tirrell, 2009).

Interestingly, fatty acids mixed with di- and multiamines have been investigated and shown to form distinct hydrogels (Basit, Pal, Sen, & Bhattacharya, 2008). This, however, is not an example of a PA but a coassembly of polar molecules with nonpolar ones. On a similar line, hydrogels were obtained from sodium salts of fatty acids mixing with oligolysines (Zhou et al., 2015).

6.3.3.2 *Peptide-only amphiphiles*

As the naturally occurring amino acid pool contains both hydrophilic and hydrophobic amino acids, the amphiphilic character can be imparted to the peptides without the need for an alkyl or a lipidic moiety. A number of such peptides have been investigated. One of the earliest works with such PAs was carried out by Walton and coworkers. They reported the self-assembly of $(SG)_n$ and $(AG)_n$ PAs into fibrillar β-sheet structures (Anderson, Chen, Rippon, & Walton, 1972; Bippon, Chen, & Walton, 1973). Zhang and coworkers studied several lipid-like peptides like A_6D, V_6D, V_6D_2, which formed the self-assembled structures like nanotubes, micelles, and vesicles (Vauthey, Santoso, Gong, Watson, & Zhang, 2002; Yang & Zhang, 2006).

In a similar study, several glycine-rich peptides behaved like "surfactant-like" molecules, thereby having the potential to stabilize membrane proteins (Santoso, Hwang, Hartman, & Zhang, 2002). The length of the hydrophobic tail has also been shown to modulate self-assembly in PAs. One such example is A_nK (where $n = 3$, 6, 9), which showed aggregates of different size and shape on increasing the length of the hydrophobic residues (Wang et al., 2009; Xu et al., 2008). Electron microscopy imaging showed the formation of sheets, fibrillar morphology (worm-like micelles), and nanorods formation for A3K, A6K, and A9K, respectively. Zhang and coworkers further investigated the self-complementary of the ionic peptides consisting of aspartic or glutamic acid residues (at N-terminus) and arginine or lysine residues (at C-terminus) to drive the self-assembly (Altman, Lee, Rich, & Zhang, 2000). Rich and coworkers studied the sequence Ac-$(AEAEAKAK)_2$-NH2, which resulted in β-sheet fibers of 10–20 nm in diameter (Zhang, Holmes, Lockshin, & Rich, 1993; Zhang, Lockshin, Cook, & Rich, 1994). Replacing the Glu with Arg and the Lys with Asp (RADA16 peptide) resulted in hydrogels that find application in tissue engineering (Holmes et al., 2000; Yokoi et al., 2005). Congo red birefringence proved that the fibrillar morphology is similar to that of the amyloid fibrils (Howie, Brewer, Howell, & Jones, 2008).

The peptide EFK16-II (FEFEFKFKFEFEFKFK) has been shown to form a β-sheet structure (Zhang & Altman, 1999). To investigate the effect of changing the size and sequence of the nonpolar residues, Mueller and coworkers studied peptide self-assembly by introducing residues like Val, Leu, Ile, Gly in the place of Phe in the sequence $(FEFEFKFK)_2$. The size and sequence of the nonpolar residues caused variations in the β-sheet secondary structure of the peptides that affected the self-assembly (Wang, Keasling, & Muller, 2005).

6.3.3.3 *Bolaamphiphiles*

A family of bolaamphiphile peptides was investigated by Matusi and coworkers, wherein the assembly resulted in tubular structures (Banerjee, Yu, & Matsui, 2003; Djalali, Chen, & Matsui, 2003). The histidine-rich sequences were able to immobilize Au and Cu nanocrystals on the nanotube surface. Another study by Banerjee et al. shows phenylalanine-based bolaamphiphile with an oligomethylene group to be a potent metallo-hydrogelator (Ray, Das, & Banerjee, 2007). The hydrogel can adsorb various toxic dyes from wastewater and also act as a carrier for vitamin B_{12}.

The literature available on the work of PAs is vast. Some of the excellent reports provide essential insights into the assembly (Bowerman & Nilsson, 2012; Cui, Webber, & Stupp, 2010; Hamley, 2011; Hartgerink et al., 2002; Mandal et al., 2014; Zhao et al., 2010). Some of the superstructures derived from PAs are listed in Table 6.2.

6.3.4 Turn containing peptides

Reverse turns happen to be an important structural motif in proteins. Tight turns are often involved in the sudden reversal of the peptide backbone. A β-hairpin is one of the most common outcomes of such a chain reversal. As a large majority of peptide self-assembled structures are composed of β-sheets, a β-hairpin could be an interesting monomeric unit to obtain

TABLE 6.2 Some of the peptide amphiphiles sequences and their self-assembly.

Sequence	Self-assembled structures	References
Ac−G_mD_n−OH	Nanotubes, nanovesicles	Santoso et al. (2002)
Ac−GAVILRR−NH_2	Nanodonuts, spherical micelles	Khoe, Yang, and Zhang (2008)
Ac−I_3K−NH_2	Nanotubes	Xu et al. (2010)
K_mL_n	Nanofibers	Breedveld, Nowak, Sato, Deming, and Pine (2004), Deming (2005)
NH_2−X_5−H_4R_8−$CONH_2$ (X = I, W, F)	Nanoparticles	Seow and Yang (2009)
$A_{12}H_5K_{10\ \text{or}\ 15}$	Core shell nanoparticles	Wiradharma, Khan, Tong, Wang, and Yang (2008)
$C_{16}H_{31}$−A_4K_4	Nanofibers	Yuwono and Hartgerink (2007)
$C_{16}H_{31}$−$V_3A_3E_3$	Nanofibrillar gels	Greenfield et al. (2009), Webber et al. (2010)
$C_{16}H_{31}$−LSQETFSDLWKLLPEN	Rod-like micelles	Missirlis, Khant, and Tirrell (2009)
Cholesterol−G_3R_6YGRKKRRQRRR	Nanoparticles	Liu et al. (2009)
$C_{12}H_{23}$−EVHHQKL	Nanofibrils	Deng et al. (2009)
Ac-QQRQQQQQEQQ-NH2	Twisted ribbon, nanotapes	Aggeli et al. (2001), Fishwick et al. (2003), Keen and McLeish (1997)
Ac-QQRFQWQFEQQ-NH2	Helical β-sheet tapes	Aggeli et al. (2001), Fishwick et al. (2003), Nyrkova, Semenov, Aggeli, and Boden (2000)
Palmitoyl-$GV_3A_3E_3$	Fibrils	Muraoka, Cui, and Stupp (2008)
VEVE	Nanobelt	Cui et al. (2009), Dong, Paramonov, Aulisa, Bakota, and Hartgerink (2007)

(Continued)

TABLE 6.2 (Continued)

Sequence	Self-assembled structures	References
VVEE	Cylindrical nanofibers	Cui et al. (2009)
R-GRKKRRQRRRPPQGSGG-FKFEFKFEFKFE	Nanoribbon	Lim, Lee, and Lee (2007)
C_{12}-GAGAGAGY	Nanofibers	Zhang et al. (2011)
C_{16}-KTTKS and C_{16}-ETTES (mixed)	Nanotapes	Hamley, Dehsorkhi, and Castelletto (2013)
H_2N-VKVKVKVK-V^DPPT-KVKVKVKV-NH_2	Fibrils	Ozbas, Kretsinger, et al. (2004), Ozbas, Rajagopal, et al. (2004), Schneider et al. (2002)
Fmoc-RGD	Fibrillar hydrogel	Zhou et al. (2009)
COOH-GE_3L_3-COC_5H_{10}-(urea oligomer), COOH-GE3L3-COC_5H_{10}-(urea oligomer), COOH-E3L3-COC_5H_{10}-(urea oligomer)	Nanofibrils	Claussen, Rabatic, and Stupp (2003)
C_{12}-EVHHQKL	Amyloid fibrillization	Deng et al. (2009)
C_{16}-$(G)_3(S)_2$PHSRN$(SG)_5$RGDSP	Nanofibrils	Shroff, Rexeisen, Arunagirinathan, and Kokkoli (2010)

β-sheet-rich superstructures. One of the most pioneering works on β-hairpin self-assembly has been carried out by Schneider and coworkers (Ozbas, Kretsinger, et al., 2004; Ozbas, Rajagopal, et al., 2004; Pochan et al., 2003; Schneider et al., 2002). A series of short peptides, named MAX peptides, were designed wherein the dipeptide ^{D}Pro-^{L}Pro was incorporated in the middle of the sequence. As ^{D}Pro-^{L}Pro is a well-known type-II′ β-turn-inducing motif, the designed peptides folded into β-hairpin structures and self-assembled to form hydrogels. The MAX1 peptide (H_2N-VKVKVKVK-V^DPPT-KVKVKVKVNH_2) folds into a β-hairpin conformation in an aqueous solution buffered at pH 9 and self-assembles to form hydrogels. Lowering the pH below the lysine side chain pKa causes interstrand electrostatic repulsion and thereby unfolding the hairpin conformation (Schneider et al., 2002). An investigation into the folding transition of two other analogs called the MAX2 (H_2N-VKVKVKVK-V^{D}PPT-KVKTKVKVNH$_2$) and MAX3 (H_2N-VKVKVKTK-V^{D}PPT-KVKTKVKVNH$_2$) showed MAX3 to undergo thermally reversible self-assembly leading to reversible hydrogels (Pochan et al., 2003). In another study, Chaudhary and coworkers designed a series of peptides having repeats of Aβ_{16-22} and Aβ_{16-22}(F20Y) connected through β-turn containing dipeptide motifs (Asn-Gly, ^{D}Pro-Gly, and Aib-^{D}Pro). All the six turn containing peptides formed transparent hydrogels (Datta, Kumar, Kumar, Nagaraj, & Chaudhary, 2019b).

6.3.5 Aromatic-moiety containing peptides and amino acids

Small aromatic peptides have been shown to form well-ordered nanostructures. The Phe-Phe aromatic cassette present in the Aβ_{16-22} is believed to play a critical role in its self-assembly (Görbitz, 2006). Görbitz and Gazit subsequently showed that the dipeptide Phe-Phe, by itself, can self-assemble into highly ordered nanotubes. Diphenylglycine, an analog wherein C_β is not there, and the phenyl rings are directly attached to the α-carbon, self-assembled into nanospheres in aqueous solution (Reches & Gazit, 2004). End capping with bulky aromatic groups like Fmoc (9-fluorenylmethoxycarbonyl) in peptides has been shown to promote self-assembly through aromatic $\pi - \pi$ stacking interactions. The earliest example of an aromatic self-assembling peptide was Fmoc-LD, which was reported to be a potent thermoreversible hydrogelator by Janmey and coworkers in 1995 (Vegners et al., 1995). Xu and coworkers were the first to report the self-assembly by a Fmoc-protected amino acid (Fmoc-Tyr-phosphate), which also showed enzyme responsiveness (Yang & Xu, 2004; Yang, Gu, Fu et al., 2004; Yang, Gu, Zhang et al., 2004). Ulijn and Gazit's groups carried out investigations on Fmoc-FF independently and were able to obtain hydrogels at physiological pH (Jayawarna et al., 2006; Mahler et al., 2006). In another study, Das and coworkers designed and developed a series of Fmoc-amino acid/peptide-functionalized cationic amphiphiles to test their gelation efficacy. Replacing the N-terminal Fmoc group with Boc (tert-butoxycarbonyl) moiety rendered most of the cationic dipeptides/amino acids nongelating. The study thus showed the presence of planar aromatic ring to be critical for the gelation efficiency of the cationic amphiphiles (Debnath, Shome, Das, & Das, 2010). Apart from Fmoc, various other aromatic moieties such as Boc, phenyl, naphthalene, azobenzene, and pyrene derivatives have been attached to the N-terminus of the peptides to drive self-assembly through aromatic interactions (Subbalakshmi, Basak, & Nagaraj, 2017; Zeng et al., 2014; Zhang et al., 2004) (Fig. 6.6). Peptides with an N-terminal naphthyl moiety have been shown to form structures like nanofibers and hydrogels that were found to have various applications as in inhibition of growth of glioblastoma cells, induce cell death, as potent anticancer agents, and inhibition of bacterial

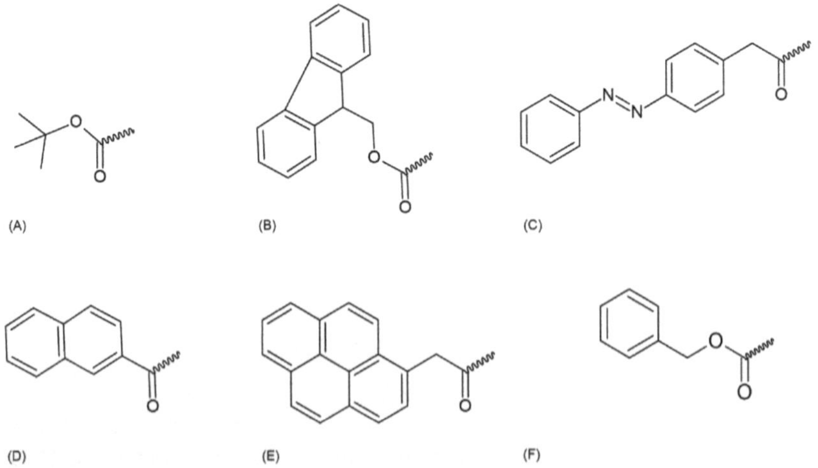

FIGURE 6.6 Structures of some common aromatic moieties that connect to the N-terminal end of peptidic sequences. (A) Boc (tert-butoxycarbonyl), (B) Fmoc (9-fluorenylmethoxycarbonyl), (C) azobenzene, (D) naphthalene, (E) pyrene, and (F) phenyl.

growth (Kuang & Xu, 2013; Kuang, Du, Zhou, & Xu, 2014; Kuang, Shi, et al., 2014; Yang, Liang, Guo, Guo, & Xu, 2007; Yang, Xu et al., 2007). Peptide derivatives having a perfluorinated phenyl moiety at the N-terminus have been shown to form hydrogels when compared to their nonfluorinated analogs (Hsu, Lin, Chang, Liu, & Lin, 2014). The assembly is believed to be directed by the quadrupole–quadrupole interactions between aromatic amino acid side chains and the perfluorinated N-terminal group. Another noteworthy example is the incorporation of an azobenzene moiety in a peptide sequence. Azobenzene is a photoresponsive molecule that can switch between cis and trans configurations by absorbing light of different excitation wavelengths. It is thus possible to obtain photoresponsive peptide hydrogels (having the potential to switch between gel − sol) by suitably incorporating azobenzene moiety in the peptide (Sahoo, Nalluri, Javid, Webb, & Ulijn, 2014). Various nucleotide bases like adenine, thymine, guanine, cytosine have been coupled with peptides to obtain hydrogels (Du et al., 2014; Li et al., 2012).

In addition to peptides, several small Fmoc-protected amino acids have also been investigated for their self-assembling propensity. Fmoc-Phe and Fmoc-Tyr, for example, were found to be potent hydrogelators (Sutton et al., 2009). The effect of covalent disulfide linkage along with aromatic interactions has also been explored. Fmoc-CF-OMe, for example, self-assembles to form a hydrogel in a reducing environment (Sadownik & Ulijn, 2010). Oxidation of the cysteine residues resulted in the collapse of the gel. An investigation on varying the C-terminus of the different Fmoc-phenylalanine derivatives revealed that COOH promotes gelation, COOMe promotes precipitation, and $CONH_2$ results in solutions generally (Ryan, Doran, Anderson, & Nilsson, 2011). An extensive review by Fleming and Ulijn (2014) on aromatic self-assembling amino acids and peptides explores various design possibilities.

6.4 Applications of peptide-based nanomaterials

The applications of the self-assembled peptide nanostructures are manifold. Owing to their versatility and biocompatibility, peptide-based superstructures have been used as drug delivery vehicles, tissue engineering matrices, scaffolds for catalysis, casting nanowires, and potential therapeutic options for aggregation-mediated disorders (Altunbas & Pochan, 2011; Aumüller & Fändrich, 2014; Goyal et al., 2021; Jerath, Goyal, Trivedi, Santhoshkumar, & Ramakrishnan, 2019; Jerath, Goyal, Trivedi, Santhoshkumar, & Ramakrishnan, 2020; Pandey & Ramakrishnan, 2020; Pandey, Morla, Kumar, & Ramakrishnan, 2020a, 2020b; Scheibel et al., 2003; Zhang, Li, Gao, Hou, & Xu, 2012).

Designed amyloid fibrils have been reported to possess intrinsic conductivity (Del Mercato et al., 2007). De novo designed peptides, short peptides from amyloidogenic peptides and proteins, and Fmoc-protected amyloidogenic peptides have been found to form hydrogels, which find applications in drug delivery and regenerative medicine (Das et al., 2016; Haines-Butterick et al., 2007). The ability of the hydrogel network to trap and release small molecules makes the peptide hydrogels promising candidates as drug delivery vehicles.

De novo designed peptides have found applications as ion channels (Ghadiri et al., 1994), antimicrobial agents (Fernandez-Lopez et al., 2001; Hazam, Jerath, Chaudhary, & Ramakrishnan, 2018; Hazam, Phukan, Akhil, Singh, & Ramakrishnan, 2021), cast for nanowires and coaxial wires (Carny et al., 2006; Reches & Gazit, 2003), and channels for fluid flow. Furthermore, peptide nanostructures have been employed as the scaffolds to provide physical support for cell culture and growth. This proved their potential to be used in tissue engineering (Chung & Park, 2009; Ellis-Behnke et al., 2006; Harrington et al., 2006; Holmes, 2002).

Fiber-forming peptides have been tagged with molecules to obtained self-assembled structures decorated with the tagged molecules. The tag could also be tethered to the preformed structures. Such derivatization has resulted in superstructures with biological, catalytic, optical, electrical, and electronic properties (Amdursky, Molotskii, Gazit, & Rosenman, 2009; Baldwin et al., 2006; Gazit, 2007b; Gras et al., 2008; Hamedi, Herland, Karlsson, & Inganäs, 2008; Herland, Thomsson, Mirzov, Scheblykin, & Inganäs, 2008; Pilkington, Roberts, Meade, & Gerrard, 2010; Tanaka, Herland, Lindgren, Tsutsui, & Andersson, 2008; Tu, Nilsson, & Forchheimer, 2013). Besides, the self-assembled peptide nanotubes have been utilized as supports for enzymes or, in some instances, to improve electrochemical parameters in an electrode (Yemini, Reches, Rishpon, & Gazit, 2005; Yu, Banerjee, Gao, Nuraje, & Matsui, 2005). This shows the application of peptide nanostructures as biosensors.

Insulin amyloid fibrils have been functionalized with conducting and luminescent polymers as electrochromic and electrochemical transistors, providing a single multifunctional device in biological sensing (Tu et al., 2013). Self-assembled FF nanowires have been modified with a conductive polymer and used as a biosensor and in cell culture for the growth of PC-12 and Hela cells (Sasso, Vedarethinam, Emnéus, Svendsen, & Castillo-Leóon, 2012). Research by Gras et al. showed amyloid fibers to be functionalized for biological cell adhesion via the RGD motif and thus can be used as biomaterials (Gras et al., 2008). In another study, insulin amyloid fibers in conjugation with glucose oxidase using glutaraldehyde have been utilized as a scaffold for enzyme immobilization (Pilkington et al., 2010). Protein amyloid fibrils from fusion protein made up of Cytochrome b_{562} with an SH3 dimer sequence have been functionalized to bind with metalloporphyrins (Baldwin et al., 2006). Moreover, conjugation of photochromic compounds with polypeptides and peptide nanostructures has been shown as potential photo modulating biosensors (Pieroni, Fissi, Angelini, & Lenci, 2001).

While mostly inorganic nanostructures have been known to have optical properties, recent studies on protein/peptide-based nanostructures have been functionalized like quantum dots. Self-assembly of tert-butoxycarbonyl-Phe-Phe-OH (Boc-FF) peptide nanospheres has been shown to exhibit optical characteristics of quantum dot (Amdursky et al., 2009). Self-assembled nanostructures and amyloid fibrils have also been conjugated with polymer complexes to be used as polymer light-emitting diodes (Schlieker, Bukau, & Mogk, 2002; Tanaka et al., 2008).

Another very promising material is the formation of gels like hydrogels, organogels, and aerogels from both amyloidogenic and nonamyloid peptides self-assembled structures (Datta & Chaudhary, 2021; Datta, Nagaraj, & Chaudhary, 2019; Kopeček & Yang, 2009; Nyström, Fernández-Ronco, Bolisetty, Mazzotti, & Mezzenga, 2016; Palui, Garai, Nanda, Nandi, & Banerjee, 2009). Hydrogel formation has been reported to form short Fmoc-dipeptides, some of which are suitable for three-dimensional cell culture (Diaferia et al., 2019; Jayawarna et al., 2009; Ma et al., 2010; Orbach et al., 2012; Smith et al., 2008). Krysmann et al. reported hydrogel formation from the amyloidogenic fragment of β-amyloid, KLVFF (Krysmann et al., 2008). An excellent review by Fichman et al. also shows the varied technological applications of short peptide-based gels (Fichman & Gazit, 2014). The presence of a network-like structure and their property to trap small molecules, several peptide-based hydrogels have widely been used as drug entrapment and release vehicles or as scaffolds in cell culture (Naskar, Palui, & Banerjee, 2009; Sutton et al., 2009). Apart from amyloidogenic protein sequences, designed PAs also give various superstructures (Cui et al., 2010; Zhao et al., 2010). An example includes the work by Stupp and coworkers where PAs have been modified with cell binding sequences to form hydrogels capable of the growth and differentiation of neuronal cells (Silva et al., 2004).

Besides tissue engineering and cell culture, functional peptide-based structures have also been shown to have potent antibacterial activities. Pioneering work by Schneider et al. on β-turn containing MAX1 sequence proved inherent antibacterial properties against both Gram-positive and Gram-negative bacteria (Rughani & Schneider, 2008; Salick, Kretsinger, Pochan, & Schneider, 2007). Peptides synthesized with strategic design have been used in bionanomaterials like three-dimensional cell culture, gene delivery, as well as membrane stabilization (Gelain, Bottai, Vescovi, & Zhang, 2006; Kokkoli et al., 2006; Yan et al., 2007; Zhao et al., 2006). Short designed self-assembling PAs behaved like surfactants and have been found to stabilize bovine rhodopsin in the presence of lipid molecules (Zhao et al., 2006). Several peptide-based hydrogels have been reported to have functions like antibacterial activity, removal of dyes from wastewater, as well as exhibit semiconducting and photoswitching behavior (Adhikari, Palui, & Banerjee, 2009; Nandi et al., 2017; Roy, Maiti, Panigrahi, Basak, & Banerjee, 2012). While the above are only a few examples, applications of peptide-based self-assembled structures are numerous.

The various properties and applications of the self-assembled structures obtained from proteins, amino acids, peptides, and peptidomimetics highlight the importance and possibilities of the bottom-up approach of fabricating smart, functional materials. In fact, the monomers need not restrict to the peptides, amino acids, or other biomolecules but could be any molecule appropriately designed to allow the required interactions for self-assembly. However, because of the inherent advantages like biocompatibility and biodegradability that they offer, the peptide and amino acids have an edge as the monomers over other molecules as far as biomedical applications are concerned.

6.5 Conclusion, challenges, and future directions

Over the last two decades, peptide-based nanomaterials have been extensively designed, developed, and examined. While many of these functional superstructures have emerged as promising materials with an exciting potential to be utilized in diverse nanotechnology related applications, still many hurdles need to be addressed and resolved. A few of these factors are listed next:

1. One of the main challenges lies in controlling the properties like dimensions, aspect ratio, and trigger-based assembly and tuning of these

superstructures. These issues have been addressed for a handful of peptide nanostructures but remain to be addressed for most. One of the reasons for lack of this control is poor understanding of the forces, pathways, and mechanisms that drive this assembly. A large number of peptide and peptide analogs have been reported to self-assemble but most of these studies are simply phenomenological reports.

2. The rational designing of the peptide sequences are challenging due to the presence of the combination of the multiple natural and unnatural amino acids. Even if considering only the 20 natural amino acids, designing a peptide sequence with two amino acids have the possibility of formation of 20^2 or 400 sequences. It is critical to find the link between the peptide sequence to its formed structure.
3. Controlling the self-assembly process in different conditions is important to obtain reproducible structures and avoid artifacts/defects. A change in the solvent, pH, temperature system can change the time point of nucleation initiation, thereby leading to varied growth and formation of different structures.
4. While many of the functional peptide materials have been promising biomaterials as in drug delivery vehicles when tested in vitro, they fail to retain their activity when tested under physiological conditions in vivo. The issue arises in the stability of the self-assembled structures, mechanical property, immunogenicity, cytotoxicity, cell-adhesion potential to name a few.
5. The cost of production for large-scale synthesis of peptides is crucial for their translation from the lab to the market. The challenges include the presence of impurities, failure of their removal, and low yield of the peptide.

Our chapter highlighted some of the pioneering works, current trends, and the recent advances in the field of peptide nanomaterials. The overall aim is to stimulate the interest in this intellectually challenging and exciting field of peptide-based molecular self-assembly. The possibilities in exploiting the rational design and synthesis of smart nanomaterials are indeed vast. It opens up a plethora of opportunities in the fabrication of peptide and peptidomimetic molecules for applying in the field of biology, chemistry, electronics, and material science.

References

Adhikari, B., Palui, G., & Banerjee, A. (2009). Self-assembling tripeptide based hydrogels and their use in removal of dyes from waste-water. *Soft Matter*, *5*(18), 3452–3460.

Adler-Abramovich, L., Aronov, D., Beker, P., Yevnin, M., Stempler, S., Buzhansky, L., ... Gazit, E. (2009). Self-assembled arrays of peptide nanotubes by vapour deposition. *Nature Nanotechnology*, *4*(12), 849.

Aggeli, A., Bell, M., Boden, N., N. Keen, J., C. B. McLeish, T., Nyrkova, I., ... Semenov, A. (1997). Engineering of peptide β-sheet nanotapes. *Journal of Materials Chemistry*, *7*(7), 1135–1145. Available from https://doi.org/10.1039/A701088E.

Aggeli, A., Nyrkova, I. A., Bell, M., Harding, R., Carrick, L., McLeish, T. C., ... Boden, N. (2001). Hierarchical self-assembly of chiral rod-like molecules as a model for peptide β-sheet tapes, ribbons, fibrils, and fibers. *Proceedings of the National Academy of Sciences*, *98*(21), 11857–11862.

Alivisatos, A. P., Johnsson, K. P., Peng, X., Wilson, T. E., Loweth, C. J., Bruchez, M. P., Jr, & Schultz, P. G. (1996). Organization of 'nanocrystal molecules' using DNA. *Nature*, *382* (6592), 609.

Altman, M., Lee, P., Rich, A., & Zhang, S. (2000). Conformational behavior of ionic self-complementary peptides. *Protein Science*, *9*(6), 1095–1105.

Altunbas, A., & Pochan, D. J. (2011). *Peptide-based and polypeptide-based hydrogels for drug delivery and tissue engineering. Peptide-Based Materials* (pp. 135–167). Springer.

Amdursky, N., Molotskii, M., Gazit, E., & Rosenman, G. (2009). Self-assembled bioinspired quantum dots: Optical properties. *Applied Physics Letters*, *94*(26), 261907. Available from https://doi.org/10.1063/1.3167354.

Anderson, J., Chen, H., Rippon, W., & Walton, A. (1972). Characterization of three sequential polydipeptides containing glycine. *Journal of Molecular Biology*, *67*(3), 459–468.

Aumüller, T., & Fändrich, M. (2014). Protein chemistry: Catalytic amyloid fibrils. *Nature Chemistry*, *6*(4), 273.

Balbach, J. J., Ishii, Y., Antzutkin, O. N., Leapman, R. D., Rizzo, N. W., Dyda, F., ... Tycko, R. (2000). Amyloid fibril formation by Aβ16–22, a seven-residue fragment of the Alzheimer's β-amyloid peptide, and structural characterization by solid state NMR. *Biochemistry*, *39*(45), 13748–13759.

Baldwin, A. J., Bader, R., Christodoulou, J., MacPhee, C. E., Dobson, C. M., & Barker, P. D. (2006). Cytochrome display on amyloid fibrils. *Journal of the American Chemical Society*, *128*(7), 2162–2163.

Ball, P. (1994). Polymers made to measure. *Nature*, *367*, 323–324.

Banerjee, I. A., Yu, L., & Matsui, H. (2003). Cu nanocrystal growth on peptide nanotubes by biomineralization: Size control of Cu nanocrystals by tuning peptide conformation. *Proceedings of the National Academy of Sciences*, *100*(25), 14678. Available from https://doi.org/10.1073/pnas.2433456100.

Basit, H., Pal, A., Sen, S., & Bhattacharya, S. (2008). Two-component hydrogels comprising fatty acids and amines: Structure, properties, and application as a template for the synthesis of metal nanoparticles. *Chemistry—A European Journal*, *14*(21), 6534–6545.

Bhak, G., Lee, S., Park, J. W., Cho, S., & Paik, S. R. (2010). Amyloid hydrogel derived from curly protein fibrils of α-synuclein. *Biomaterials*, *31*(23), 5986–5995.

Bippon, W., Chen, H., & Walton, A. (1973). Spectroscopic characterization of poly (Glu-Ala). *Journal of Molecular Biology*, *75*(2), 369–375.

Bong, D. T., Clark, T. D., Granja, J. R., & Ghadiri, M. R. (2001). Self-assembling organic nanotubes. *Angewandte Chemie International Edition*, *40*(6), 988–1011.

Bowerman, C. J., & Nilsson, B. L. (2012). Review self-assembly of amphipathic β-sheet peptides: Insights and applications. *Peptide Science*, *98*(3), 169–184. Available from https://doi.org/10.1002/bip.22058.

Brändén, C.-I., & Tooze, J. (1999). *Introduction to Protein Structure*. Taylor & Francis.

Breedveld, V., Nowak, A. P., Sato, J., Deming, T. J., & Pine, D. J. (2004). Rheology of block copolypeptide solutions: Hydrogels with tunable properties. *Macromolecules*, *37*(10), 3943–3953.

Briggs, B. D., & Knecht, M. R. (2012). Nanotechnology meets biology: Peptide-based methods for the fabrication of functional materials. *The Journal of Physical Chemistry Letters*, *3*(3), 405–418.

Buchanan, L. E., Dunkelberger, E. B., Tran, H. Q., Cheng, P.-N., Chiu, C.-C., Cao, P., ... Zanni, M. T. (2013). Mechanism of IAPP amyloid fibril formation involves an intermediate with a transient β-sheet. *Proceedings of the National Academy of Sciences*, *110*(48), 19285–19290.

Cao, H., Yuan, Q., Zhu, X., Zhao, Y.-P., & Liu, M. (2012). Hierarchical self-assembly of achiral amino acid derivatives into dendritic chiral nanotwists. *Langmuir*, *28*(43), 15410–15417.

Carny, O., Shalev, D. E., & Gazit, E. (2006). Fabrication of coaxial metal nanocables using a self-assembled peptide nanotube scaffold. *Nano Letters*, *6*(8), 1594–1597.

Castelletto, V., Hamley, I. W., Adamcik, J., Mezzenga, R., & Gummel, J. (2012). Modulating self-assembly of a nanotape-forming peptide amphiphile with an oppositely charged surfactant. *Soft Matter*, *8*(1), 217–226. Available from https://doi.org/10.1039/C1SM06677C.

Chapman, M. R., Robinson, L. S., Pinkner, J. S., Roth, R., Heuser, J., Hammar, M., ... Hultgren, S. J. (2002). Role of *Escherichia coli* curli operons in directing amyloid fiber formation. *Science*, *295*(5556), 851–855.

Chapman, R., Danial, M., Koh, M. L., Jolliffe, K. A., & Perrier, S. (2012). Design and properties of functional nanotubes from the self-assembly of cyclic peptide templates. *Chemical Society Reviews*, *41*(18), 6023–6041.

Chaudhary, N., & Nagaraj, R. (2011). Impact on the replacement of Phe by Trp in a short fragment of Aβ amyloid peptide on the formation of fibrils. *Journal of Peptide Science*, *17*(2), 115–123.

Chen, C. L., & Rosi, N. L. (2010). Peptide-based methods for the preparation of nanostructured inorganic materials. *Angewandte Chemie International Edition*, *49*(11), 1924–1942.

Chu-Kung, A. F., Bozzelli, K. N., Lockwood, N. A., Haseman, J. R., Mayo, K. H., & Tirrell, M. V. (2004). Promotion of peptide antimicrobial activity by fatty acid conjugation. *Bioconjugate Chemistry*, *15*(3), 530–535. Available from https://doi.org/10.1021/bc0341573.

Chung, H. J., & Park, T. G. (2009). Self-assembled and nanostructured hydrogels for drug delivery and tissue engineering. *Nano Today*, *4*(5), 429–437.

Claussen, R. C., Rabatic, B. M., & Stupp, S. I. (2003). Aqueous self-assembly of unsymmetric peptide bolaamphiphiles into nanofibers with hydrophilic cores and surfaces. *Journal of the American Chemical Society*, *125*(42), 12680–12681. Available from https://doi.org/10.1021/ja035882r.

Cohen, A. S., & Calkins, E. (1959). Electron microscopic observations on a fibrous component in amyloid of diverse origins. *Nature*, *183*(4669), 1202.

Cui, H., Muraoka, T., Cheetham, A. G., & Stupp, S. I. (2009). Self-assembly of giant peptide nanobelts. *Nano Letters*, *9*(3), 945–951.

Cui, H., Webber, M. J., & Stupp, S. I. (2010). Self-assembly of peptide amphiphiles: From molecules to nanostructures to biomaterials. *Peptide Science*, *94*(1), 1–18. Available from https://doi.org/10.1002/bip.21328.

Das, S., Zhou, K., Ghosh, D., Jha, N. N., Singh, P. K., Jacob, R. S., ... Maji, S. K. (2016). Implantable amyloid hydrogels for promoting stem cell differentiation to neurons. *Npg Asia Materials*, *8*, e304. Available from https://doi.org/10.1038/am.2016.116.

Dasgupta, A., Mondal, J. H., & Das, D. (2013). Peptide hydrogels. *RSC Advances*, *3*(24), 9117–9149.

Datta, D., & Chaudhary, N. (2021). *Peptide-based hydrogels for biomedical applications. Translational Biotechnology* (pp. 203–232). Elsevier.

Datta, D., Kumar, V., Kumar, S., Nagaraj, R., & Chaudhary, N. (2019a). Hydrogel formation by an aromatic analogue of a β-amyloid fragment, Aβ16–22: A scaffold for 3D cell culture. *ACS Omega*, *4*(1), 620–627.

Datta, D., Kumar, V., Kumar, S., Nagaraj, R., & Chaudhary, N. (2019b). Limpid hydrogels from β-turn motif-connected tandem repeats of Aβ 16–22. *Soft Matter*, *15*(24), 4827–4835.

Datta, D., Nagaraj, R., & Chaudhary, N. (2019). Water-alcohol bigels from fatty acylated dipeptides. *The Journal of Physical Chemistry B*, *124*(3), 577–588.

Debnath, S., Shome, A., Das, D., & Das, P. K. (2010). Hydrogelation through self-assembly of Fmoc-peptide functionalized cationic amphiphiles: Potent antibacterial agent. *The Journal of Physical Chemistry B*, *114*(13), 4407–4415. Available from https://doi.org/10.1021/jp909520w.

Del Mercato, L. L., Pompa, P. P., Maruccio, G., Della Torre, A., Sabella, S., Tamburro, A. M., ... Rinaldi, R. (2007). Charge transport and intrinsic fluorescence in amyloid-like fibrils. *Proceedings of the National Academy of Sciences*, *104*(46), 18019–18024.

Deming, T. J. (2005). Polypeptide hydrogels via a unique assembly mechanism. *Soft Matter*, *1*(1), 28–35. Available from https://doi.org/10.1039/B500307E.

Deng, M., Yu, D., Hou, Y., & Wang, Y. (2009). Self-assembly of peptide − amphiphile C12 − Aβ (11 − 17) into nanofibrils. *The Journal of Physical Chemistry B*, *113*(25), 8539–8544.

Diaferia, C., Ghosh, M., Sibillano, T., Gallo, E., Stornaiuolo, M., Giannini, C., ... Accardo, A. (2019). Fmoc-FF and hexapeptide-based multicomponent hydrogels as scaffold materials. *Soft Matter*, *15*(3), 487–496. Available from https://doi.org/10.1039/C8SM02366B.

Djalali, R., Chen, Y.-f, & Matsui, H. (2003). Au nanocrystal growth on nanotubes controlled by conformations and charges of sequenced peptide templates. *Journal of the American Chemical Society*, *125*(19), 5873–5879. Available from https://doi.org/10.1021/ja0299598.

Dobson, C. M. (2004). Principles of protein folding, misfolding and aggregation. In: *Paper Presented at the Seminars in Cell & Developmental Biology*.

Dong, H., Paramonov, S. E., Aulisa, L., Bakota, E. L., & Hartgerink, J. D. (2007). Self-assembly of multidomain peptides: Balancing molecular frustration controls conformation and nanostructure. *Journal of the American Chemical Society*, *129*(41), 12468–12472.

Du, X., Zhou, J., Guvench, O., Sangiorgi, F. O., Li, X., Zhou, N., & Xu, B. (2014). Supramolecular assemblies of a conjugate of nucleobase, amino acids, and saccharide act as agonists for proliferation of embryonic stem cells and development of zygotes. *Bioconjugate Chemistry*, *25*(6), 1031–1035. Available from https://doi.org/10.1021/bc500187m.

Eanes, E., & Glenner, G. (1968). X-ray diffraction studies on amyloid filaments. *Journal of Histochemistry & Cytochemistry*, *16*(11), 673–677.

Elliot, M. A., Karoonuthaisiri, N., Huang, J., Bibb, M. J., Cohen, S. N., Kao, C. M., & Buttner, M. J. (2003). The chaplins: A family of hydrophobic cell-surface proteins involved in aerial mycelium formation in Streptomyces coelicolor. *Genes & Development*, *17*(14), 1727–1740.

Ellis-Behnke, R. G., Liang, Y.-X., You, S.-W., Tay, D. K., Zhang, S., So, K.-F., & Schneider, G. E. (2006). Nano neuro knitting: Peptide nanofiber scaffold for brain repair and axon regeneration with functional return of vision. *Proceedings of the National Academy of Sciences*, *103*(13), 5054–5059.

Feltwell, J. (1990). *The Story of Silk*. Sutton.

Fernandez-Lopez, S., Kim, H.-S., Choi, E. C., Delgado, M., Granja, J. R., Khasanov, A., ... Ghadiri, M. R. (2001). Antibacterial agents based on the cyclic d,l-α-peptide architecture. *Nature, 412*(6845), 452–455. Available from https://doi.org/10.1038/35086601.

Fichman, G., & Gazit, E. (2014). Self-assembly of short peptides to form hydrogels: Design of building blocks, physical properties and technological applications. *Acta Biomaterialia, 10* (4), 1671–1682.

Fishwick, C. W., Beevers, A. J., Carrick, L. M., Whitehouse, C. D., Aggeli, A., & Boden, N. (2003). Structures of helical β-tapes and twisted ribbons: The role of side-chain interactions on twist and bend behavior. *Nano Letters, 3*(11), 1475–1479.

Fleming, S., & Ulijn, R. V. (2014). Design of nanostructures based on aromatic peptide amphiphiles. *Chemical Society Reviews, 43*(23), 8150–8177. Available from https://doi.org/10.1039/C4CS00247D.

Fletcher, N. L., Lockett, C. V., & Dexter, A. F. (2011). A pH-responsive coiled-coil peptide hydrogel. *Soft Matter, 7*(21), 10210–10218. Available from https://doi.org/10.1039/C1SM06261A.

Fowler, D. M., Koulov, A. V., Alory-Jost, C., Marks, M. S., Balch, W. E., & Kelly, J. W. (2005). Functional amyloid formation within mammalian tissue. *PLoS Biology, 4*(1), e6.

Fraenkel-Conrat, H., Singer, B., & Williams, R. (1957). Infectivity of viral nucleic acid. *Biochimica et Biophysica Acta, 25*, 87–96.

Friedreich, N., & Kekulé, A. (1859). Zur amyloidfrage. *Virchows Archiv, 16*(1), 50–65.

Gao, J., Wang, H., Wang, L., Wang, J., Kong, D., & Yang, Z. (2009). Enzyme promotes the hydrogelation from a hydrophobic small molecule. *Journal of the American Chemical Society, 131*(32), 11286–11287.

Gazit, E. (2007a). Plenty of room for biology at the bottom. In: *Published by Imperial College Press and distributed by World Scientific Publishing Co.*

Gazit, E. (2007b). Self-assembled peptide nanostructures: The design of molecular building blocks and their technological utilization. *Chemical Society Reviews, 36*(8), 1263–1269.

Gazit, E. (2008). *Self-assembly of short peptides for nanotechnological applications. Nanobiotechnology* (pp. 385–395). Springer.

Gelain, F., Bottai, D., Vescovi, A., & Zhang, S. (2006). Designer self-assembling peptide nanofiber scaffolds for adult mouse neural stem cell 3-dimensional cultures. *PloS One, 1*(1), e119.

Ghadiri, M. R., Granja, J. R., & Buehler, L. K. (1994). Artificial transmembrane ion channels from self-assembling peptide nanotubes. *Nature, 369*(6478), 301.

Ghadiri, M. R., Granja, J. R., Milligan, R. A., McRee, D. E., & Khazanovich, N. (1993). Self-assembling organic nanotubes based on a cyclic peptide architecture. *Nature, 366*(6453), 324.

Ghosh, S., Adler-Abramovich, L., Gazit, E., & Verma, S. (2013). Spacer driven morphological twist in Phe-Phe dipeptide conjugates. *Tetrahedron, 69*(8), 2004–2009.

Glabe, C. G. (2006). Common mechanisms of amyloid oligomer pathogenesis in degenerative disease. *Neurobiology of Aging, 27*(4), 570–575.

Görbitz, C. H. (2006). The structure of nanotubes formed by diphenylalanine, the core recognition motif of Alzheimer's β-amyloid polypeptide. *Chemical Communications, 22*, 2332–2334. Available from https://doi.org/10.1039/B603080G.

Gortner, R., & Hoffman, W. (1921). An interesting colloidal gel. *Journal of the American Chemical Society, 43*, 2199–2202.

Goux, W. J., Kopplin, L., Nguyen, A. D., Leak, K., Rutkofsky, M., Shanmuganandam, V. D., ... Kirschner, D. A. (2004). The formation of straight and twisted filaments from short tau peptides. *Journal of Biological Chemistry*, *279*(26), 26868–26875.

Goyal, R., Jerath, G., Akhil, R., Chandrasekharan, A., Puppala, E. R., Ponneganti, S., ... Ramakrishnan, V. (2021). Geometry encoded functional programming of tumor homing peptides for targeted drug delivery. *Journal of Controlled Release*, *333*, 16–27.

Gras, S. L., Tickler, A. K., Squires, A. M., Devlin, G. L., Horton, M. A., Dobson, C. M., & MacPhee, C. E. (2008). Functionalised amyloid fibrils for roles in cell adhesion. *Biomaterials*, *29*(11), 1553–1562.

Greenfield, M. A., Hoffman, J. R., Olvera de la Cruz, M., & Stupp, S. I. (2009). Tunable mechanics of peptide nanofiber gels. *Langmuir*, *26*(5), 3641–3647.

Guerrero-Ferreira, R., Taylor, N. M., Mona, D., Ringler, P., Lauer, M. E., Riek, R., ... Stahlberg, H. (2018). Cryo-EM structure of alpha-synuclein fibrils. *Elife*, *7*, e36402.

Guler, M. O., Hsu, L., Soukasene, S., Harrington, D. A., Hulvat, J. F., & Stupp, S. I. (2006). Presentation of RGDS epitopes on self-assembled nanofibers of branched peptide amphiphiles. *Biomacromolecules*, *7*(6), 1855–1863. Available from https://doi.org/10.1021/bm060161g.

Haines-Butterick, L., Rajagopal, K., Branco, M., Salick, D., Rughani, R., Pilarz, M., ... Schneider, J. P. (2007). Controlling hydrogelation kinetics by peptide design for three-dimensional encapsulation and injectable delivery of cells. *Proceedings of the National Academy of Sciences*, *104*(19), 7791–7796.

Hamedi, M., Herland, A., Karlsson, R. H., & Inganäs, O. (2008). Electrochemical devices made from conducting nanowire networks self-assembled from amyloid fibrils and alkoxysulfonate PEDOT. *Nano Letters*, *8*(6), 1736–1740.

Hamley, I., Dehsorkhi, A., & Castelletto, V. (2013). Coassembly in binary mixtures of peptide amphiphiles containing oppositely charged residues. *Langmuir*, *29*(16), 5050–5059.

Hamley, I. W. (2011). Self-assembly of amphiphilic peptides. *Soft Matter*, *7*(9), 4122–4138. Available from https://doi.org/10.1039/C0SM01218A.

Hardy, J., & Selkoe, D. J. (2002). The amyloid hypothesis of Alzheimer's disease: Progress and problems on the road to therapeutics. *Science*, *297*(5580), 353–356.

Harrington, D. A., Cheng, E. Y., Guler, M. O., Lee, L. K., Donovan, J. L., Claussen, R. C., & Stupp, S. I. (2006). Branched peptide-amphiphiles as self-assembling coatings for tissue engineering scaffolds. *Journal of Biomedical Materials Research Part A: An Official Journal of The Society for Biomaterials, The Japanese Society for Biomaterials, and The Australian Society for Biomaterials and the Korean Society for Biomaterials*, *78*(1), 157–167.

Hartgerink, J. D., Beniash, E., & Stupp, S. I. (2001). Self-assembly and mineralization of peptide-amphiphile nanofibers. *Science*, *294*(5547), 1684–1688.

Hartgerink, J. D., Beniash, E., & Stupp, S. I. (2002). Peptide-amphiphile nanofibers: A versatile scaffold for the preparation of self-assembling materials. *Proceedings of the National Academy of Sciences*, *99*(8), 5133. Available from https://doi.org/10.1073/pnas.072699999.

Hartley, D. M., Walsh, D. M., Chianping, P. Y., Diehl, T., Vasquez, S., Vassilev, P. M., ... Selkoe, D. J. (1999). Protofibrillar intermediates of amyloid β-protein induce acute electrophysiological changes and progressive neurotoxicity in cortical neurons. *Journal of Neuroscience*, *19*(20), 8876–8884.

Hazam, P. K., Jerath, G., Chaudhary, N., & Ramakrishnan, V. (2018). Peptido-mimetic approach in the design of syndiotactic antimicrobial peptides. *International Journal of Peptide Research and Therapeutics*, *24*(2), 299–307.

Hazam, P. K., Phukan, C., Akhil, R., Singh, A., & Ramakrishnan, V. (2021). Antimicrobial effects of syndiotactic polypeptides. *Scientific Reports*, *11*(1), 1–10.

Heerklotz, H., & Seelig, J. (2001). Detergent-like action of the antibiotic peptide surfactin on lipid membranes. *Biophysical Journal*, *81*(3), 1547–1554. Available from https://doi.org/10.1016/S0006-3495(01)75808-0.

Heise, H., Hoyer, W., Becker, S., Andronesi, O. C., Riedel, D., & Baldus, M. (2005). Molecular-level secondary structure, polymorphism, and dynamics of full-length α-synuclein fibrils studied by solid-state NMR. *Proceedings of the National Academy of Sciences*, *102*(44), 15871–15876.

Herland, A., Thomsson, D., Mirzov, O., Scheblykin, I. G., & Inganäs, O. (2008). Decoration of amyloid fibrils with luminescent conjugated polymers. *Journal of Materials Chemistry*, *18* (1), 126–132.

Hervás, R., Li, L., Majumdar, A., del Carmen Fernández-Ramírez, M., Unruh, J. R., Slaughter, B. D., … Nagai, Y. (2016). Molecular basis of Orb2 amyloidogenesis and blockade of memory consolidation. *PLoS Biology*, *14*(1), e1002361.

Hirst, A. R., Escuder, B., Miravet, J. F., & Smith, D. K. (2008). High-tech applications of self-assembling supramolecular nanostructured gel-phase materials: From regenerative medicine to electronic devices. *Angewandte Chemie International Edition*, *47*(42), 8002–8018.

Holmes, T. C. (2002). Novel peptide-based biomaterial scaffolds for tissue engineering. *Trends in Biotechnology*, *20*(1), 16–21.

Holmes, T. C., de Lacalle, S., Su, X., Liu, G., Rich, A., & Zhang, S. (2000). Extensive neurite outgrowth and active synapse formation on self-assembling peptide scaffolds. *Proceedings of the National Academy of Sciences*, *97*(12), 6728–6733.

Horne, W. S., Ashkenasy, N., & Ghadiri, M. R. (2005). Modulating charge transfer through cyclic d, l-α-peptide self-assembly. *Chemistry—A European Journal*, *11*(4), 1137–1144.

Howie, A. J., Brewer, D. B., Howell, D., & Jones, A. P. (2008). Physical basis of colors seen in Congo red-stained amyloid in polarized light. *Laboratory Investigation*, *88*(3), 232.

Hsu, S.-M., Lin, Y.-C., Chang, J.-W., Liu, Y.-H., & Lin, H.-C. (2014). Intramolecular interactions of a phenyl/perfluorophenyl pair in the formation of supramolecular nanofibers and hydrogels. *Angewandte Chemie International Edition*, *53*(7), 1921–1927. Available from https://doi.org/10.1002/anie.201307500.

Huang, Y., Qiu, Z., Xu, Y., Shi, J., Lin, H., & Zhang, Y. (2011). Supramolecular hydrogels based on short peptides linked with conformational switch. *Organic & Biomolecular Chemistry*, *9*(7), 2149–2155. Available from https://doi.org/10.1039/C0OB01057J.

Inouye, H., Gleason, K. A., Zhang, D., Decatur, S. M., & Kirschner, D. A. (2010). Differential effects of phe19 and phe20 on fibril formation by amyloidogenic peptide Aβ16–22 (Ac-KLVFFAE-NH2). *Proteins: Structure, Function, and Bioinformatics*, *78*(10), 2306–2321.

Irwansyah, I., Li, Y.-Q., Shi, W., Qi, D., Leow, W. R., Tang, M. B. Y., … Chen, X. (2015). Gram-positive antimicrobial activity of amino acid-based hydrogels. *Advanced Materials*, *27* (4), 648–654. Available from https://doi.org/10.1002/adma.201403339.

Jarrett, J. T., & Lansbury, P. T., Jr (1993). Seeding “one-dimensional crystallization” of amyloid: a pathogenic mechanism in Alzheimer’s disease and scrapie? *Cell*, *73*(6), 1055–1058.

Jayawarna, V., Ali, M., Jowitt, T. A., Miller, A. F., Saiani, A., Gough, J. E., & Ulijn, R. V. (2006). Nanostructured hydrogels for three-dimensional cell culture through self-assembly of fluorenylmethoxycarbonyl–dipeptides. *Advanced Materials*, *18*(5), 611–614. Available from https://doi.org/10.1002/adma.200501522.

Jayawarna, V., Richardson, S. M., Hirst, A. R., Hodson, N. W., Saiani, A., Gough, J. E., & Ulijn, R. V. (2009). Introducing chemical functionality in Fmoc-peptide gels for cell culture.

Acta Biomaterialia, *5*(3), 934–943. Available from https://doi.org/10.1016/j.actbio.2009.01.006.

Jerath, G., Goyal, R., Trivedi, V., Santhoshkumar, T., & Ramakrishnan, V. (2019). Syndiotactic peptides for targeted delivery. *Acta Biomaterialia*, *87*, 130–139.

Jerath, G., Goyal, R., Trivedi, V., Santhoshkumar, T. R., & Ramakrishnan, V. (2020). Conformationally constrained peptides for drug delivery. *Journal of Peptide Science*, *26* (4–5), e3244.

Keen, J., & McLeish, T. B. (1997). Engineering of peptide β-sheet nanotapes. *Journal of Materials Chemistry*, *7*(7), 1135–1145.

Khoe, U., Yang, Y., & Zhang, S. (2008). Self-assembly of nanodonut structure from a cone-shaped designer lipid-like peptide surfactant. *Langmuir*, *25*(7), 4111–4114.

Kholkin, A., Amdursky, N., Bdikin, I., Gazit, E., & Rosenman, G. (2010). Strong piezoelectricity in bioinspired peptide nanotubes. *ACS Nano*, *4*(2), 610–614.

King, C.-Y., Tittmann, P., Gross, H., Gebert, R., Aebi, M., & Wüthrich, K. (1997). Prion-inducing domain 2–114 of yeast Sup35 protein transforms in vitro into amyloid-like filaments. *Proceedings of the National Academy of Sciences*, *94*(13), 6618–6622.

Kokkoli, E., Mardilovich, A., Wedekind, A., Rexeisen, E. L., Garg, A., & Craig, J. A. (2006). Self-assembly and applications of biomimetic and bioactive peptide-amphiphiles. *Soft Matter*, *2*(12), 1015–1024.

Koley, P., & Pramanik, A. (2011). Nanostructures from single amino acid-based molecules: Stability, fibrillation, encapsulation, and fabrication of silver nanoparticles. *Advanced Functional Materials*, *21*(21), 4126–4136. Available from https://doi.org/10.1002/adfm.201101465.

Kopeček, J., & Yang, J. (2009). Peptide-directed self-assembly of hydrogels. *Acta Biomaterialia*, *5*(3), 805–816.

Kretsinger, J. K., Haines, L. A., Ozbas, B., Pochan, D. J., & Schneider, J. P. (2005). Cytocompatibility of self-assembled β-hairpin peptide hydrogel surfaces. *Biomaterials*, *26* (25), 5177–5186. Available from https://doi.org/10.1016/j.biomaterials.2005.01.029.

Krone, M. G., Hua, L., Soto, P., Zhou, R., Berne, B., & Shea, J.-E. (2008). Role of water in mediating the assembly of Alzheimer amyloid-β Aβ16 − 22 protofilaments. *Journal of the American Chemical Society*, *130*(33), 11066–11072.

Krysmann, M. J., Castelletto, V., Kelarakis, A., Hamley, I. W., Hule, R. A., & Pochan, D. J. (2008). Self-assembly and hydrogelation of an amyloid peptide fragment. *Biochemistry*, *47* (16), 4597–4605.

Kuang, Y., Du, X., Zhou, J., & Xu, B. (2014). Supramolecular nanofibrils inhibit cancer progression in vitro and in vivo. *Advanced Healthcare Materials*, *3*(8), 1217–1221. Available from https://doi.org/10.1002/adhm.201300645.

Kuang, Y., Shi, J., Li, J., Yuan, D., Alberti, K. A., Xu, Q., & Xu, B. (2014). Pericellular hydrogel/nanonets iInhibit cancer cells. *Angewandte Chemie International Edition*, *53*(31), 8104–8107. Available from https://doi.org/10.1002/anie.201402216.

Kuang, Y., & Xu, B. (2013). Disruption of the dynamics of microtubules and selective inhibition of glioblastoma cells by nanofibers of small hydrophobic molecules. *Angewandte Chemie International Edition*, *52*(27), 6944–6948. Available from https://doi.org/10.1002/anie.201302658.

Lambert, M. P., Barlow, A., Chromy, B. A., Edwards, C., Freed, R., Liosatos, M., ... Viola, K. L. (1998). Diffusible, nonfibrillar ligands derived from Aβ1–42 are potent central nervous system neurotoxins. *Proceedings of the National Academy of Sciences*, *95*(11), 6448–6453.

Lehn, J.-M. (1993). Supramolecular Chemistry. *Science*, *260*(5115), 1762–1763. Available from https://doi:10.1126/science.8511582.

Letchford, K., & Burt, H. (2007). A review of the formation and classification of amphiphilic block copolymer nanoparticulate structures: micelles, nanospheres, nanocapsules and polymersomes. *European Journal of Pharmaceutics and Biopharmaceutics*, *65*(3), 259–269.

Li, X., Du, X., Li, J., Gao, Y., Pan, Y., Shi, J., … Xu, B. (2012). Introducing d-amino acid or simple glycoside into small peptides to enable supramolecular hydrogelators to resist proteolysis. *Langmuir*, *28*(37), 13512–13517. Available from https://doi.org/10.1021/la302583a.

Liang, G., Yang, Z., Zhang, R., Li, L., Fan, Y., Kuang, Y., … Xu, B. (2009). Supramolecular hydrogel of a d-amino acid dipeptide for controlled drug release in vivo. *Langmuir*, *25*(15), 8419–8422. Available from https://doi.org/10.1021/la804271d.

Liang, Y., Guo, P., Pingali, S. V., Pabit, S., Thiyagarajan, P., Berland, K. M., & Lynn, D. G. (2008). Light harvesting antenna on an amyloid scaffold. *Chemical Communications*, *48*, 6522–6524.

Lim, Y.-b, Kwon, O.-J., Lee, E., Kim, P.-H., Yun, C.-O., & Lee, M. (2008). A cyclic RGD-coated peptide nanoribbon as a selective intracellular nanocarrier. *Organic & Biomolecular Chemistry*, *6*(11), 1944–1948. Available from https://doi.org/10.1039/B802470G.

Lim, Y. b, Lee, E., & Lee, M. (2007). Cell-penetrating-peptide-coated nanoribbons for intracellular nanocarriers. *Angewandte Chemie International Edition*, *46*(19), 3475–3478.

Liu, C., Zhao, M., Jiang, L., Cheng, P. N., Park, J., Sawaya, M. R., … Eisenberg, D. (2012). Out-of-register beta-sheets suggest a pathway to toxic amyloid aggregates. *Proceedings of the National Academy of Sciences of the United States of America*, *109*(51), 20913–20918, 1218792109 [pii]10.1073/pnas.1218792109.

Liu, L., Xu, K., Wang, H., Tan, P. J., Fan, W., Venkatraman, S. S., … Yang, Y.-Y. (2009). Self-assembled cationic peptide nanoparticles as an efficient antimicrobial agent. *Nature Nanotechnology*, *4*(7), 457.

Lu, K., Jacob, J., Thiyagarajan, P., Conticello, V. P., & Lynn, D. G. (2003). Exploiting amyloid fibril lamination for nanotube self-assembly. *Journal of the American Chemical Society*, *125* (21), 6391–6393.

Ma, M., Kuang, Y., Gao, Y., Zhang, Y., Gao, P., & Xu, B. (2010). Aromatic – aromatic interactions induce the self-assembly of pentapeptidic derivatives in water to form nanofibers and supramolecular hydrogels. *Journal of the American Chemical Society*, *132*(8), 2719–2728.

Mahler, A., Reches, M., Rechter, M., Cohen, S., & Gazit, E. (2006). Rigid, self-assembled hydrogel composed of a modified aromatic dipeptide. *Advanced Materials*, *18*(11), 1365–1370. Available from https://doi.org/10.1002/adma.200501765.

Mandal, D., Shirazi, A. N., & Parang, K. (2014). Self-assembly of peptides to nanostructures. *Organic & Biomolecular Chemistry*, *12*(22), 3544–3561.

Mandal, D., Tiwari, R. K., Shirazi, A. N., Oh, D., Ye, G., Banerjee, A., … Parang, K. (2013). Self-assembled surfactant cyclic peptide nanostructures as stabilizing agents. *Soft Matter*, *9* (39), 9465–9475.

Mart, R. J., Osborne, R. D., Stevens, M. M., & Ulijn, R. V. (2006). Peptide-based stimuli-responsive biomaterials. *Soft Matter*, *2*(10), 822–835.

Maury, C. (2009). The emerging concept of functional amyloid. *Journal of Internal Medicine*, *265*(3), 329–334.

Mirkin, C. A., Letsinger, R. L., Mucic, R. C., & Storhoff, J. J. (1996). A DNA-based method for rationally assembling nanoparticles into macroscopic materials. *Nature*, *382*(6592), 607.

Missirlis, D., Khant, H., & Tirrell, M. (2009). Mechanisms of peptide amphiphile internalization by SJSA-1 cells in vitro. *Biochemistry*, *48*(15), 3304–3314.

Muraoka, T., Cui, H., & Stupp, S. I. (2008). Quadruple helix formation of a photoresponsive peptide amphiphile and its light-triggered dissociation into single fibers. *Journal of the American Chemical Society*, *130*(10), 2946–2947.

Nandi, N., Gayen, K., Ghosh, S., Bhunia, D., Kirkham, S., Sen, S. K., ... Banerjee, A. (2017). Amphiphilic peptide-based supramolecular, noncytotoxic, stimuli-responsive hydrogels with antibacterial activity. *Biomacromolecules*, *18*(11), 3621–3629.

Naskar, J., Palui, G., & Banerjee, A. (2009). Tetrapeptide-based hydrogels: Ffor encapsulation and slow release of an anticancer drug at physiological pH. *The Journal of Physical Chemistry B*, *113*(35), 11787–11792.

Nelson, R., Sawaya, M. R., Balbirnie, M., Madsen, A. Ø., Riekel, C., Grothe, R., & Eisenberg, D. (2005). Structure of the cross-β spine of amyloid-like fibrils. *Nature*, *435*(7043), 773.

Niece, K. L., Hartgerink, J. D., Donners, J. J., & Stupp, S. I. (2003). Self-assembly combining two bioactive peptide-amphiphile molecules into nanofibers by electrostatic attraction. *Journal of the American Chemical Society*, *125*(24), 7146–7147.

Nyrkova, I., Semenov, A., Aggeli, A., & Boden, N. (2000). Fibril stability in solutions of twisted-sheet peptides: A new kind of micellization in chiral systems. *The European Physical Journal B—Condensed Matter and Complex Systems*, *17*(3), 481–497.

Nyström, G., Fernández-Ronco, M. P., Bolisetty, S., Mazzotti, M., & Mezzenga, R. (2016). Amyloid templated gold aerogels. *Advanced Materials*, *28*(3), 472–478.

Orbach, R., Mironi-Harpaz, I., Adler-Abramovich, L., Mossou, E., Mitchell, E. P., Forsyth, V. T., ... Seliktar, D. (2012). The rheological and structural properties of Fmoc-peptide-based hydrogels: The effect of aromatic molecular architecture on self-assembly and physical characteristics. *Langmuir*, *28*(4), 2015–2022. Available from https://doi.org/10.1021/la204426q.

Ozbas, B., Kretsinger, J., Rajagopal, K., Schneider, J. P., & Pochan, D. J. (2004). Salt-triggered peptide folding and consequent self-assembly into hydrogels with tunable modulus. *Macromolecules*, *37*(19), 7331–7337.

Ozbas, B., Rajagopal, K., Schneider, J. P., & Pochan, D. J. (2004). Semiflexible chain networks formed via self-assembly of β-hairpin molecules. *Physical Review Letters*, *93*(26), 268106.

Pachahara, S. K., Chaudhary, N., Subbalakshmi, C., & Nagaraj, R. (2012). Hexafluoroisopropanol induces self-assembly of β-amyloid peptides into highly ordered nanostructures. *Journal of Peptide Science*, *18*(4), 233–241.

Pachahara, S. K., & Nagaraj, R. (2015). Probing the role of aromatic residues in the self-assembly of Aβ (16–22) in fluorinated alcohols and their aqueous mixtures. *Biochemistry and Biophysics Reports*, *2*, 1–13.

Palui, G., Garai, A., Nanda, J., Nandi, A. K., & Banerjee, A. (2009). Organogels from different self-assembling new dendritic peptides: Morphology, reheology, and structural investigations. *The Journal of Physical Chemistry B*, *114*(3), 1249–1256.

Pandey, G., Morla, S., Kumar, S., & Ramakrishnan, V. (2020a). Modulating Aβ fibrillogenesis with 'Trojan'peptides. *Neuropeptides*, *81*, 102030.

Pandey, G., Morla, S., Kumar, S., & Ramakrishnan, V. (2020b). Modulation of tau protein aggregation using 'Trojan' sequences. *Biochimica et Biophysica Acta (BBA)—General Subjects*, *1864*(7), 129569.

Pandey, G., & Ramakrishnan, V. (2020). Invasive and non-invasive therapies for Alzheimer's disease and other amyloidosis. *Biophysical Reviews*, 1–12.

Pandey, G., Saikia, J., Sasidharan, S., Joshi, D. C., Thota, S., Nemade, H. B., ... Ramakrishnan, V. (2017). Modulation of peptide based nano-assemblies with electric and magnetic fields. *Scientific Reports*, *7*(1), 1–9.

Pauling, L. (1960). *The Nature of the Chemical Bond* (Vol. 260). Ithaca, NY: Cornell University Press.

Petkova, A. T., Ishii, Y., Balbach, J. J., Antzutkin, O. N., Leapman, R. D., Delaglio, F., & Tycko, R. (2002). A structural model for Alzheimer's β-amyloid fibrils based on experimental constraints from solid state NMR. *Proceedings of the National Academy of Sciences, 99* (26), 16742–16747.

Pieroni, O., Fissi, A., Angelini, N., & Lenci, F. (2001). Photoresponsive polypeptides. *Accounts of Chemical Research, 34*(1), 9–17.

Pilkington, S. M., Roberts, S. J., Meade, S. J., & Gerrard, J. A. (2010). Amyloid fibrils as a nanoscaffold for enzyme immobilization. *Biotechnology Progress, 26*(1), 93–100.

Pochan, D. J., Schneider, J. P., Kretsinger, J., Ozbas, B., Rajagopal, K., & Haines, L. (2003). Thermally reversible hydrogels via intramolecular folding and consequent self-assembly of a de novo designed peptide. *Journal of the American Chemical Society, 125*(39), 11802–11803.

Rajagopal, K., Lamm, M. S., Haines-Butterick, L. A., Pochan, D. J., & Schneider, J. P. (2009). Tuning the pH responsiveness of β-hairpin peptide folding, self-assembly, and hydrogel material formation. *Biomacromolecules, 10*(9), 2619–2625. Available from https://doi.org/10.1021/bm900544e.

Ray, S., Das, A. K., & Banerjee, A. (2007). pH-responsive, bolaamphiphile-based smart metallo-hydrogels as potential dye-adsorbing agents, water purifier, and vitamin B_{12} carrier. *Chemistry of Materials, 19*(7), 1633–1639.

Reches, M., & Gazit, E. (2003). Casting metal nanowires within discrete self-assembled peptide nanotubes. *Science, 300*(5619), 625–627.

Reches, M., & Gazit, E. (2004). Formation of closed-cage nanostructures by self-assembly of aromatic dipeptides. *Nano Letters, 4*(4), 581–585.

Reches, M., & Gazit, E. (2006). Controlled patterning of aligned self-assembled peptide nanotubes. *Nature Nanotechnology, 1*(3), 195.

Reches, M., Porat, Y., & Gazit, E. (2002). Amyloid fibril formation by pentapeptide and tetrapeptide fragments of human calcitonin. *Journal of Biological Chemistry, 277*(38), 35475–35480.

Roy, S., Maiti, D. K., Panigrahi, S., Basak, D., & Banerjee, A. (2012). A new hydrogel from an amino acid-based perylene bisimide and its semiconducting, photo-switching behaviour. *RSC Advances, 2*(29), 11053–11060.

Rughani, R. V., & Schneider, J. P. (2008). Molecular design of β-hairpin peptides for material construction. *MRS Bulletin, 33*(5), 530–535.

Ryan, D. M., Anderson, S. B., Senguen, F. T., Youngman, R. E., & Nilsson, B. L. (2010). Self-assembly and hydrogelation promoted by F 5-phenylalanine. *Soft Matter, 6*(3), 475–479.

Ryan, D. M., Doran, T. M., Anderson, S. B., & Nilsson, B. L. (2011). Effect of C-terminal modification on the self-assembly and hydrogelation of fluorinated Fmoc-Phe derivatives. *Langmuir, 27*(7), 4029–4039. Available from https://doi.org/10.1021/la1048375.

Ryu, J., & Park, C. B. (2008). High-temperature self-assembly of peptides into vertically well-aligned nanowires by aniline vapor. *Advanced Materials, 20*(19), 3754–3758.

Sadownik, J. W., & Ulijn, R. V. (2010). Locking an oxidation-sensitive dynamic peptide system in the gel state. *Chemical Communications, 46*(20), 3481–3483. Available from https://doi.org/10.1039/C001982H.

Sahoo, J. K., Nalluri, S. K. M., Javid, N., Webb, H., & Ulijn, R. V. (2014). Biocatalytic amide condensation and gelation controlled by light. *Chemical Communications, 50*(41), 5462–5464. Available from https://doi.org/10.1039/C4CC01431F.

Salick, D. A., Kretsinger, J. K., Pochan, D. J., & Schneider, J. P. (2007). Inherent antibacterial activity of a peptide-based β-hairpin hydrogel. *Journal of the American Chemical Society, 129*(47), 14793–14799.

Santoso, S., Hwang, W., Hartman, H., & Zhang, S. (2002). Self-assembly of surfactant-like peptides with variable glycine tails to form nanotubes and nanovesicles. *Nano Letters, 2*(7), 687–691. Available from https://doi.org/10.1021/nl025563i.

Sarikaya, M., Tamerler, C., Jen, A. K.-Y., Schulten, K., & Baneyx, F. (2003). Molecular biomimetics: Nanotechnology through biology. *Nature Materials, 2*(9), 577.

Sasso, L., Vedarethinam, I., Emnéus, J., Svendsen, W. E., & Castillo-Leóon, J. (2012). Self-assembled diphenylalanine nanowires for cellular studies and sensor applications. *Journal of Nanoscience and Nanotechnology, 12*(4), 3077–3083.

Scheibel, T., Parthasarathy, R., Sawicki, G., Lin, X.-M., Jaeger, H., & Lindquist, S. L. (2003). Conducting nanowires built by controlled self-assembly of amyloid fibers and selective metal deposition. *Proceedings of the National Academy of Sciences, 100*(8), 4527–4532.

Schlieker, C., Bukau, B., & Mogk, A. (2002). Prevention and reversion of protein aggregation by molecular chaperones in the *E. coli* cytosol: Implications for their applicability in biotechnology. *Journal of Biotechnology, 96*(1), 13–21.

Schneider, J. P., Pochan, D. J., Ozbas, B., Rajagopal, K., Pakstis, L., & Kretsinger, J. (2002). Responsive hydrogels from the intramolecular folding and self-assembly of a designed peptide. *Journal of the American Chemical Society, 124*(50), 15030–15037.

Schnur, J. M. (1993). Lipid tubules: A paradigm for molecularly engineered structures. *Science, 262*(5140), 1669. Available from https://doi.org/10.1126/science.262.5140.1669.

Schuster, B., Györvary, E., Pum, D., & Sleytr, U. B. (2005). *Nanotechnology with S-layer proteins. Protein Nanotechnology* (pp. 101–123). Springer.

Seow, W. Y., & Yang, Y. Y. (2009). A class of cationic triblock amphiphilic oligopeptides as efficient gene-delivery vectors. *Advanced Materials, 21*(1), 86–90.

Shimada, T., Lee, S., Bates, F. S., Hotta, A., & Tirrell, M. (2009). Wormlike micelle formation in peptide-lipid conjugates driven by secondary structure transformation of the headgroups. *The Journal of Physical Chemistry B, 113*(42), 13711–13714. Available from https://doi.org/10.1021/jp901727q.

Shroff, K., Rexeisen, E. L., Arunagirinathan, M. A., & Kokkoli, E. (2010). Fibronectin-mimetic peptide-amphiphile nanofiber gels support increased cell adhesion and promote ECM production. *Soft Matter, 6*(20), 5064–5072. Available from https://doi.org/10.1039/C0SM00321B.

Sigler, P. B., Xu, Z., Rye, H. S., Burston, S. G., Fenton, W. A., & Horwich, A. L. (1998). Structure and function in GroEL-mediated protein folding. *Annual Reviews of Biochemistry, 67*, 581–608.

Silva, G. A., Czeisler, C., Niece, K. L., Beniash, E., Harrington, D. A., Kessler, J. A., & Stupp, S. I. (2004). Selective differentiation of neural progenitor cells by high-epitope density nanofibers. *Science, 303*(5662), 1352–1355.

Smith, A. M., Williams, R. J., Tang, C., Coppo, P., Collins, R. F., Turner, M. L., … Ulijn, R. V. (2008). Fmoc-diphenylalanine self assembles to a hydrogel via a novel architecture based on π–π interlocked β-sheets. *Advanced Materials, 20*(1), 37–41.

Smith, J. F., Knowles, T. P., Dobson, C. M., MacPhee, C. E., & Welland, M. E. (2006). Characterization of the nanoscale properties of individual amyloid fibrils. *Proceedings of the National Academy of Sciences, 103*(43), 15806–15811.

Spector, M. S., Easwaran, K. R., Jyothi, G., Selinger, J. V., Singh, A., & Schnur, J. M. (1996). Chiral molecular self-assembly of phospholipid tubules: A circular dichroism study. *Proceedings of the National Academy of Sciences*, *93*(23), 12943–12946.

Subbalakshmi, C., Basak, P., & Nagaraj, R. (2017). Self-assembly of t-butyloxycarbonyl protected dipeptide methyl esters composed of leucine, isoleucine, and valine into highly organized structures from alcohol and aqueous alcohol mixtures. *Peptide Science*, *108*(6), e23033.

Sutton, S., Campbell, N. L., Cooper, A. I., Kirkland, M., Frith, W. J., & Adams, D. J. (2009). Controlled release from modified amino acid hydrogels governed by molecular size or network dynamics. *Langmuir*, *25*(17), 10285–10291. Available from https://doi.org/10.1021/la9011058.

Tanaka, H., Herland, A., Lindgren, L. J., Tsutsui, T., & Andersson, M. R. (2008). Enhanced current efficiency from bio-organic light-emitting diodes using decorated amyloid fibrils with conjugated polymer. *Nano Letters*, *8*(9), 2858–2861.

Tarabout, C., Roux, S., Gobeaux, F., Fay, N., Pouget, E., Meriadec, C., ... Besselievre, F. (2011). Control of peptide nanotube diameter by chemical modifications of an aromatic residue involved in a single close contact. *Proceedings of the National Academy of Sciences*, *108*(19), 7679–7684.

Tenidis, K., Waldner, M., Bernhagen, J., Fischle, W., Bergmann, M., Weber, M., ... Kapurniotu, A. (2000). Identification of a penta-and hexapeptide of islet amyloid polypeptide (IAPP) with amyloidogenic and cytotoxic properties. *Journal of Molecular Biology*, *295*(4), 1055–1071.

Tjernberg, L., Hosia, W., Bark, N., Thyberg, J., & Johansson, J. (2002). Charge attraction and β propensity are necessary for amyloid fibril formation from tetrapeptides. *Journal of Biological Chemistry*, *277*(45), 43243–43246.

Toledano, S., Williams, R. J., Jayawarna, V., & Ulijn, R. V. (2006). Enzyme-triggered self-assembly of peptide hydrogels via reversed hydrolysis. *Journal of the American Chemical Society*, *128*(4), 1070–1071.

Török, M., Milton, S., Kayed, R., Wu, P., McIntire, T., Glabe, C. G., & Langen, R. (2002). Structural and dynamic features of Alzheimer's Aβ peptide in amyloid fibrils studied by site-directed spin labeling. *Journal of Biological Chemistry*, *277*(43), 40810–40815.

Tu, D., Nilsson, D., & Forchheimer, R. (2013). Electrochromic electrochemical transistors gated with polyelectrolyte-decorated amyloid fibrils. *Journal of Display Technology*, *9*(9), 755–759.

Ulijn, R. V., & Woolfson, D. N. (2010). Peptide and protein based materials in 2010: From design and structure to function and application. *Chemical Society Reviews*, *39*(9), 3349–3350.

Valéry, C., Paternostre, M., Robert, B., Gulik-Krzywicki, T., Narayanan, T., Dedieu, J.-C., ... Artzner, F. (2003). Biomimetic organization: Octapeptide self-assembly into nanotubes of viral capsid-like dimension. *Proceedings of the National Academy of Sciences*, *100*(18), 10258. Available from https://doi.org/10.1073/pnas.1730609100.

Vauthey, S., Santoso, S., Gong, H., Watson, N., & Zhang, S. (2002). Molecular self-assembly of surfactant-like peptides to form nanotubes and nanovesicles. *Proceedings of the National Academy of Sciences*, *99*(8), 5355. Available from https://doi.org/10.1073/pnas.072089599.

Vegners, R., Shestakova, I., Kalvinsh, I., Ezzell, R. M., & Janmey, P. A. (1995). Use of a gel-forming dipeptide derivative as a carrier for antigen presentation. *Journal of Peptide Science*, *1*(6), 371–378. Available from https://doi.org/10.1002/psc.310010604.

Wagoner, V. A., Cheon, M., Chang, I., & Hall, C. K. (2014). Impact of sequence on the molecular assembly of short amyloid peptides. *Proteins: Structure, Function, and Bioinformatics*, *82*(7), 1469–1483.

Wang, J., Han, S., Meng, G., Xu, H., Xia, D., Zhao, X., ... Lu, J. R. (2009). Dynamic self-assembly of surfactant-like peptides A 6 K and A 9 K. *Soft Matter*, *5*(20), 3870–3878.

Wang, J., Liu, K., Xing, R., & Yan, X. (2016). Peptide self-assembly: Thermodynamics and kinetics. *Chemical Society Reviews*, *45*(20), 5589–5604.

Wang, K., Keasling, J. D., & Muller, S. J. (2005). Effects of the sequence and size of non-polar residues on self-assembly of amphiphilic peptides. *International Journal of Biological Macromolecules*, *36*(4), 232–240. Available from https://doi.org/10.1016/j.ijbiomac.2005.06.006.

Wang, M., Du, L., Wu, X., Xiong, S., & Chu, P. K. (2011). Charged diphenylalanine nanotubes and controlled hierarchical self-assembly. *ACS Nano*, *5*(6), 4448–4454.

Webber, M. J., Tongers, J., Renault, M.-A., Roncalli, J. G., Losordo, D. W., & Stupp, S. I. (2010). Development of bioactive peptide amphiphiles for therapeutic cell delivery. *Acta Biomaterialia*, *6*(1), 3–11.

Whitesides, G. M., Mathias, J. P., & Seto, C. T. (1991). Molecular self-assembly and nanochemistry: A chemical strategy for the synthesis of nanostructures. *Science*, *254*(5036), 1312–1319.

Winfree, E., Liu, F., Wenzler, L. A., & Seeman, N. C. (1998). Design and self-assembly of two-dimensional DNA crystals. *Nature*, *394*(6693), 539–544. Available from https://doi.org/10.1038/28998.

Winkler, S., Szela, S., Avtges, P., Valluzzi, R., Kirschner, D. A., & Kaplan, D. (1999). Designing recombinant spider silk proteins to control assembly. *International Journal of Biological Macromolecules*, *24*(2–3), 265–270.

Wiradharma, N., Khan, M., Tong, Y. W., Wang, S., & Yang, Y. Y. (2008). Self-assembled cationic peptide nanoparticles capable of inducing efficient gene expression in vitro. *Advanced Functional Materials*, *18*(6), 943–951.

Xie, Y., Wang, X., Huang, R., Qi, W., Wang, Y., Su, R., & He, Z. (2015). Electrostatic and aromatic interaction-directed supramolecular self-assembly of a designed Fmoc-tripeptide into helical nanoribbons. *Langmuir*, *31*(9), 2885–2894.

Xu, H., Wang, J., Han, S., Wang, J., Yu, D., Zhang, H., ... Lu, J. R. (2008). Hydrophobic-region-induced transitions in self-assembled peptide nanostructures. *Langmuir*, *25*(7), 4115–4123.

Xu, H., Wang, Y., Ge, X., Han, S., Wang, S., Zhou, P., ... Lu, J. R. (2010). Twisted nanotubes formed from ultrashort amphiphilic peptide I3K and their templating for the fabrication of silica nanotubes. *Chemistry of Materials*, *22*(18), 5165–5173.

Yan, X., He, Q., Wang, K., Duan, L., Cui, Y., & Li, J. (2007). Transition of cationic dipeptide nanotubes into vesicles and oligonucleotide delivery. *Angewandte Chemie International Edition*, *46*(14), 2431–2434.

Yan, X., Li, J., & Möhwald, H. (2011). Self-assembly of hexagonal peptide microtubes and their optical waveguiding. *Advanced Materials*, *23*(25), 2796–2801.

Yang, S. J., & Zhang, S. (2006). Self-assembling behavior of designer lipid-like peptides. *Supramolecular Chemistry*, *18*(5), 389–396. Available from https://doi.org/10.1080/10615800600658586.

Yang, Z., Gu, H., Fu, D., Gao, P., Lam, J. K., & Xu, B. (2004). Enzymatic formation of supramolecular hydrogels. *Advanced Materials*, *16*(16), 1440–1444. Available from https://doi.org/10.1002/adma.200400340.

Yang, Z., Gu, H., Zhang, Y., Wang, L., & Xu, B. (2004). Small molecule hydrogels based on a class of antiinflammatory agents. *Chemical Communications*, *2*, 208–209.

Yang, Z., Liang, G., Guo, Z., Guo, Z., & Xu, B. (2007). Intracellular hydrogelation of small molecules inhibits bacterial growth. *Angewandte Chemie International Edition*, *46*(43), 8216–8219. Available from https://doi.org/10.1002/anie.200701697.

Yang, Z., Liang, G., Ma, M., Gao, Y., & Xu, B. (2007). In vitro and in vivo enzymatic formation of supramolecular hydrogels based on self-assembled nanofibers of a β-amino acid derivative. *Small*, *3*(4), 558–562. Available from https://doi.org/10.1002/smll.200700015.

Yang, Z., & Xu, B. (2004). A simple visual assay based on small molecule hydrogels for detecting inhibitors of enzymes. *Chemical Communications*, *21*, 2424–2425. Available from https://doi.org/10.1039/B408897B.

Yang, Z. M., Xu, K. M., Guo, Z. F., Guo, Z. H., & Xu, B. (2007). Intracellular enzymatic formation of nanofibers results in hydrogelation and regulated cell death. *Advanced Materials*, *19* (20), 3152–3156. Available from https://doi.org/10.1002/adma.200701971.

Yemini, M., Reches, M., Rishpon, J., & Gazit, E. (2005). Novel electrochemical biosensing platform using self-assembled peptide nanotubes. *Nano Letters*, *5*(1), 183–186.

Yokoi, H., Kinoshita, T., & Zhang, S. (2005). Dynamic reassembly of peptide RADA16 nanofiber scaffold. *Proceedings of the National Academy of Sciences*, *102*(24), 8414–8419.

Yu, L., Banerjee, I. A., Gao, X., Nuraje, N., & Matsui, H. (2005). Fabrication and application of enzyme-incorporated peptide nanotubes. *Bioconjugate Chemistry*, *16*(6), 1484–1487.

Yuwono, V. M., & Hartgerink, J. D. (2007). Peptide amphiphile nanofibers template and catalyze silica nanotube formation. *Langmuir*, *23*(9), 5033–5038.

Zeng, G., Liu, L., Xia, D., Li, Q., Xin, Z., Wang, J., ... Dong, M. (2014). Transition of chemically modified diphenylalanine peptide assemblies revealed by atomic force microscopy. *RSC Advances*, *4*(15), 7516–7520. Available from https://doi.org/10.1039/C3RA46718J.

Zhang, J., Hao, R., Huang, L., Yao, J., Chen, X., & Shao, Z. (2011). Self-assembly of a peptide amphiphile based on hydrolysed *Bombyx mori* silk fibroin. *Chemical Communications*, *47* (37), 10296–10298.

Zhang, L., Li, N., Gao, F., Hou, L., & Xu, Z. (2012). Insulin amyloid fibrils: An excellent platform for controlled synthesis of ultrathin superlong platinum nanowires with high electrocatalytic activity. *Journal of the American Chemical Society*, *134*(28), 11326–11329.

Zhang, S., & Altman, M. (1999). Peptide self-assembly in functional polymer science and engineering. *Reactive and Functional Polymers*, *41*(1), 91–102. Available from https://doi.org/10.1016/S1381-5148(99)00031-0.

Zhang, S., Holmes, T., Lockshin, C., & Rich, A. (1993). Spontaneous assembly of a self-complementary oligopeptide to form a stable macroscopic membrane. *Proceedings of the National Academy of Sciences*, *90*(8), 3334. Available from https://doi.org/10.1073/pnas.90.8.3334.

Zhang, S., Holmes, T. C., DiPersio, C. M., Hynes, R. O., Su, X., & Rich, A. (1995). Self-complementary oligopeptide matrices support mammalian cell attachment. *Biomaterials*, *16* (18), 1385–1393.

Zhang, S., Lockshin, C., Cook, R., & Rich, A. (1994). Unusually stable β-sheet formation in an ionic self-complementary oligopeptide. *Biopolymers: Original Research on Biomolecules*, *34*(5), 663–672.

Zhang, Y., Gu, H., Yang, Z., & Xu, B. (2003). Supramolecular hydrogels respond to ligand – receptor interaction. *Journal of the American Chemical Society*, *125*(45), 13680–13681.

Zhang, Y., Yang, Z., Yuan, F., Gu, H., Gao, P., & Xu, B. (2004). Molecular recognition remolds the self-assembly of hydrogelators and increases the elasticity of the hydrogel by 106-fold. *Journal of the American Chemical Society*, *126*(46), 15028–15029. Available from https://doi.org/10.1021/ja044401g.

Zhao, X., Nagai, Y., Reeves, P. J., Kiley, P., Khorana, H. G., & Zhang, S. (2006). Designer short peptide surfactants stabilize G protein-coupled receptor bovine rhodopsin. *Proceedings of the National Academy of Sciences*, *103*(47), 17707–17712.

Zhao, X., Pan, F., Xu, H., Yaseen, M., Shan, H., Hauser, C. A. E., ... Lu, J. R. (2010). Molecular self-assembly and applications of designer peptide amphiphiles. *Chemical Society Reviews*, *39*(9), 3480–3498. Available from https://doi.org/10.1039/B915923C.

Zhao, X., & Zhang, S. (2006). Molecular designer self-assembling peptides. *Chemical Society Reviews*, *35*(11), 1105–1110.

Zhou, K., Yang, S., Zhao, G., Ning, Y., & Xu, C. (2015). Self-assembly of the sodium salts of fatty acids into limpid hydrogels through non-covalent interactions with peptides. *RSC Advances*, *5*(76), 61719–61724.

Zhou, M., Smith, A. M., Das, A. K., Hodson, N. W., Collins, R. F., Ulijn, R. V., & Gough, J. E. (2009). Self-assembled peptide-based hydrogels as scaffolds for anchorage-dependent cells. *Biomaterials*, *30*(13), 2523–2530.

Zhu, P., Yan, X., Su, Y., Yang, Y., & Li, J. (2010). Solvent-induced structural transition of self-assembled dipeptide: From organogels to microcrystals. *Chemistry—A European Journal*, *16*(10), 3176–3183.

Chapter 7

Peptide nanocatalysts

Jahnu Saikia and Vibin Ramakrishnan
Department of Biosciences and Bioengineering, Indian Institute of Technology Guwahati, Guwahati, Assam, India

7.1 Introduction

The rate of a chemical reaction is a function of multiple parameters, of which frequency of collision of the reactants is the most predominant one (Goldberger & Watson, 2004). Product formation occurs when the collision between the reacting molecules is sufficient to overcome certain threshold energy called the activation energy (E_a) (Fig. 7.1). However, the collision involves reactants with random orientations, and therefore E_a is formally the average energy barrier that must be overcome for product formation. The lower the E_a threshold of a reaction, the greater the proportion of successful collisions, and the faster the reaction rate. The rate of a chemical reaction is given by Arrhenius equation,

$$k = Ae^{-Ea/RT}$$

where k is rate constant, A is frequency factor, e is a mathematical quantity, Ea is the activation energy, T is temperature, and R is gas constant.

Catalysis is the process of increasing the reaction rate by adding a substance called a catalyst. A catalyst provides a new route to the chemical reaction with lower E_a without getting effectively consumed at the end of the reaction (Figure 2.5). This is achieved by (1) orienting the reacting particles to promote successful collision and (2) by binding the reactant to form an intermediate complex with the lower activation energy. The intermediate complex is also referred to as the activation complex.

The characteristics of a catalytic reaction are the turnover frequency (TOF) and the turnover number (TON). The TOF is the number of rotations of the catalytic cycle per unit time (Boudart, 1995). It is calculated as the number of molecules converted per active site with respect to time (Schüth, Ward, & Buriak, 2018), and is given by,

$$TOF = \frac{NAv}{S} \cdot \frac{dNi}{dt}$$

De Novo Peptide Design. DOI: https://doi.org/10.1016/B978-0-323-99917-5.00006-8

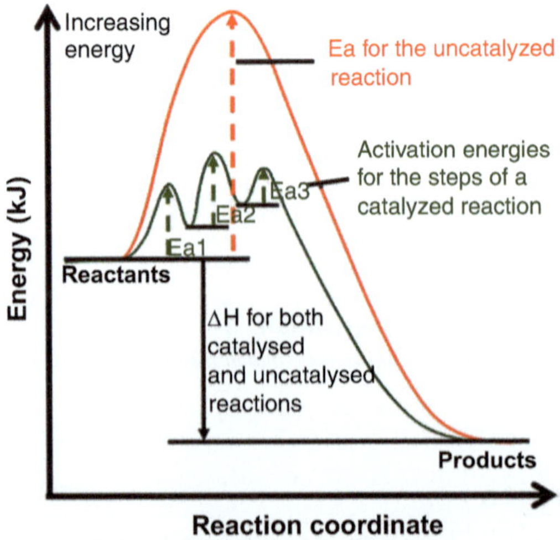

FIGURE 7.1 Comparison of the activation energies of catalytic (E_a 1–3, olive) and noncatalytic (Ea, red) reactions. The energy versus reaction coordinate diagram showing a two-step enzyme-catalyzed reaction, having two catalytic intermediates and three transition states.

where *dNi/dt* is the differential concentration change of *i* with time, *NAv* is Avogadro's number, and *S* is the number of active sites.

The TON is a measure of catalyst stability and is highly sensitive to the reaction time considered. It is calculated as the number of catalytic cycles a site can perform until it is completely deactivated (Schüth et al., 2018). The TON is expressed as a function of the TOF,

$$TON = \int_0^\infty TOF(t)dt$$

Catalysts accelerate reactions by several orders of magnitude, enabling them to be carried out under the most favorable thermodynamic regime and at much lower temperature and pressure.

7.2 Types of catalysts

Depending on their reaction phase, catalysts can be broadly classified into two types: homogeneous and heterogeneous catalysts.

7.2.1 Homogeneous catalysts

Homogeneous catalysts and the substrate are present in one phase, most often the liquid phase. They contain well-defined, single-site centers that can be prepared using organic synthesis and offers high activity with remarkable selectivity (Corma & Garcia, 2008). Single or multiple-site centers in homogeneous catalysis generally imply the presence of one or more metal centers. The metal atom usually has an easily replaceable group, also called an open-site, involved in substrate binding, which is also the site for bond formation or cleavage. In addition, they extend the scope for fine-tuning and optimization through functionalization or subtle structure variations. Important examples include Brønsted and Lewis acids, metals ions, metal and organometallic complexes, and biocatalysts such as enzymes. Here, every single catalytic entity can act as a single active site that corroborates the underlying enhanced activity and selectivity compared to the heterogeneous catalysts. Homogeneous catalysts, however, are not suitable to perform continuous flow reactions due to the high cost of separation and recovery. Besides, they have limited thermal stability up to <100°C (Haibach, Kundu, Brookhart, & Goldman, 2012). The recognition of such inherent limitations in homogeneous catalysis led to the development of heterogenize-homogeneous catalysts with an aim to combine the superior activity and selectivity of homogenous catalysts with the ability to recycle as seen in heterogeneous catalysts.

7.2.2 Heterogeneous catalysts

In contrast to homogeneous catalysts, a heterogeneous catalyst and the substrate are in different phases, typically solid − liquid or solid − gas. They offer ease of separation from the reaction mixture and, therefore, can be recycled and reused until loss of catalytic activity. This makes them the ideal catalyst to perform continuous flow reactions amid robust conditions. Numerous reports also suggest that such catalysts can be reactivated to perform like fresh materials (Tiltscher, Wolf, & Schelchshorn, 1981). Heterogeneous catalysis lacks the degree of knowledge of the transformation at the molecular level and demands further research to reach activity and selectivity values similar to homogeneous ones.

According to the heterogeneous catalyst theory, the entire surface of the solid catalyst is not responsible for catalyzing the reaction. Only specified locations called the "active site" on the catalyst surface will effectively take part in the reaction (Muhler, 1997). The active site may be generated due to unsaturated functional groups resulting from the surface irregularities or atoms with inherent chemical properties that enable the interaction with the adsorbed reactant at the atomic or molecular scale. In the case of porous catalysts, such as a gel, the steps 1, 2, 6, and 7 (Fig. 7.2) involving the diffusion and mass transfer determine the reaction rate irrespective of the number of the active sites in the catalyst.

However, in recent times, the frontiers between homogeneous and heterogeneous catalysis have been surpassed and becoming less defined. The development of various analytical techniques has allowed researchers to gain insight and characterize the catalytic sites of complex materials based on

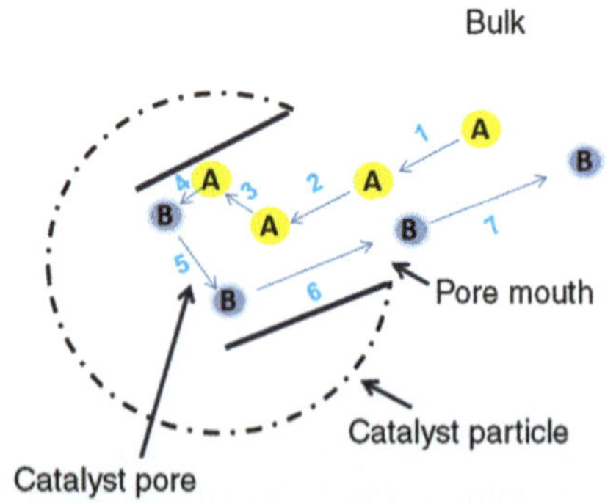

FIGURE 7.2 Steps involved in a typical heterogeneous catalytic reaction. (1) Transportation of reactant (A). (2) Diffusion of the reactant (A) to the internal catalytic surface. (3) Adsorption of reactant (A) on the catalytic surface. (4) Surface reaction to produce (B). (5) Desorption of the product (B) from the catalytic surface. (6) Diffusion of the product (B) to the pore mouth. (7) Transfer of the product (B) to the bulk fluid. Steps 1, 2, 6, and 7 are the diffusion steps that determine the mass transfer, and steps 3, 4, and 5 are the adsorption and reaction steps in the reaction (Campbell, 1988).

their composition, structure, and distribution. For example, it is now possible to synthesize heterogeneous catalysts with a single catalytic site that resembles a homogeneous catalyst (Joubert et al., 2006). The introduction of nanoassemblies in catalysis is aimed to make all the active sites on a solid support accessible for reactants, thus increasing the rate and selectivity of the catalysts.

7.2.3 Nanocatalysts

Nanomaterials often present unique properties that are not seen in bulk materials. Due to their large surface area, tailor-made nanostructures have displayed the potential to meet the interface of homo and heterogeneous catalysis. Higher surface area implies increased exposed active sites, thus have a larger contact area with reactants. Nanocatalysts are reported to be catalytically more active than conventional heterogeneous catalysts (Singh & Tandon, 2014). In addition, they possess strong mechanical strengths, high resistance to temperature, and organic solvents. The Brownian motion of nanoparticles, the confining effect of nanopores, and the self-assembling behaviors of discrete nanostructures make them an excellent candidate for efficient catalysts. Characteristic properties of nanomaterials that control their catalytic efficiency depend on their size, morphology, and chemical composition (Singh & Tandon, 2014; Somwanshi, Somvanshi, & Kharat, 2020). For chiral catalysis and synthesis of natural products, nanocatalysts play an important role in increasing the yield TON, facilitating the purification of products, and recyclization of the catalyst. Fig. 7.3 compares the

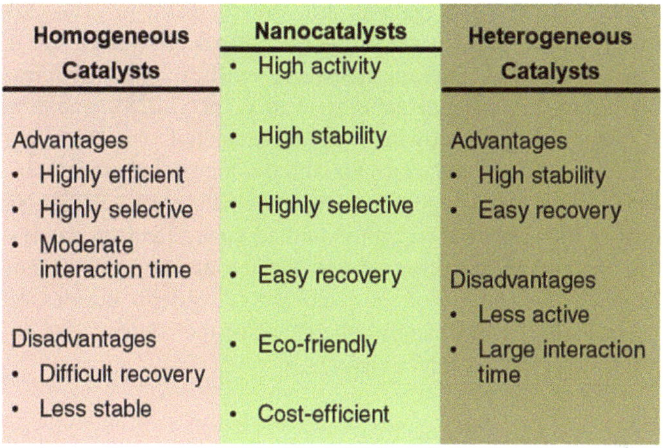

FIGURE 7.3 Characteristics of homogeneous, heterogeneous, and nanocatalysts.

characteristics of nanocatalysts that behave as an interface between homo and heterogeneous catalysts. Some of the factors affecting nanocatalysis are discussed next.

7.2.3.1 *Effect of size*

Nanocatalysts, as compared to their bulk counterparts, offer a much higher surface-to-volume ratio. When the size decreases to a certain nanoregime (typically a couple or even subnanometer), significant changes in the electronic states and coordination environment of the surface atoms of the catalyst molecule is observed. This, in turn exposes more catalytically active sites for substrate binding, thus leading to an increased TOF (Kalidindi & Jagirdar, 2012). The relation of catalytic activity and nanoparticle size is influenced markedly by the proportion of the surface atoms having high coordination numbers (Umpierre, de Jesús, & Dupont, 2011). Importantly, however, decreasing size does not always lead to improved catalytic characteristics as the proportion of the surface atoms might change drastically (Umpierre et al., 2011). According to the Boudart classification, catalytic reactions leading to variation in the TON with metal dispersion are called structure-sensitive, and those that do not depend on the dispersion are called structure-insensitive (Pan, Shen, Yao, Bentalib, & Peng, 2018). This is based on the idea that it is possible to change the type of site geometry or the nature of the exposed facets by changing the particle size.

7.2.3.2 *Effect of shape*

Given that different surfaces of a nanoparticle exhibit different morphology and electronic state, they could display different activities and selectivity for the same catalytic reaction (Zhang, Jin, Xiong, Lim, & Xia, 2013). The shape dependence of catalytic performance intrinsically results from the differences in geometry of the participating entities. Nanoparticles can be divided into zero-dimensional spheres (Pan et al., 2018), one-dimensional nanotubes, nanorods or nanowires, two-dimensional nanobelts, nanosheets (Li, Liu, & Liu, 2012), and complex three-dimensional structures such as nanoflowers, nanostars, polygonal nanoframes (Lin et al., 2012). Narayanan and El-Sayed (2004) have thoroughly studied such effect on catalysis. They have found that the distribution of the surface atoms in the nanostructure strongly impacts the activity of the nanocatalyst system. Besides, Lee et al. have demonstrated that by changing the shape of the nanocatalyst, it is possible to alter its activity and selectivity.

7.2.3.3 *Effect of composition*

The third important factor influencing the effect of nanocatalysts is their composition. The incorporation of a molecule such as an electron donor or an acceptor to the catalyst's surface significantly influences the activation energy of the reaction pathway apart from modulating its selectivity (Somorjai, Frei, & Park, 2009). The surface composition also contributes to the durability and recoverability of the nanocatalyst. For example, incorporating ferro- or paramagnetic metal ion allows the catalyst separation using a magnet. Another elegant way of recovering the catalyst from the reaction mixture is to render its surface pH-sensitive.

7.3 Enantioselectivity in catalytic reactions

Enantioselective catalysis involves the selective and reproducible generation of one enantiomer of a chiral product from achiral precursors. Last few decades have seen an ever-increasing demand for enantiopure molecules in various industrial sectors that includes pharmaceutical, agrochemical, food industry, and cosmetics. Enantioselectivity in asymmetric reactions involves the preferential selection of one enantiomer over the other chiral product. Without exception, such reactions are mediated by chiral and nonracemic catalysts. The enantioselective C–H, C–C, C–O, and C–N bond formation have been extensively studied using asymmetric homogeneous catalysts (Mlynarski & Paradowska, 2008). However, they do not comply with the atom economy at the industrial level, primarily due to the inability to recover the catalyst. On the other hand, stereochemical selectivity can be introduced in a heterogeneous catalyst to ensure that a chiral transition state is achieved (McMorn & Hutchings, 2004). The major disadvantage of heterogeneous catalysts, however, is the presence of an invariable range of active sites that makes it difficult to control the overall enantioselectivity. Asymmetric catalysis is categorized into metal-catalysis, organocatalysis, and biocatalysis

(Xiang & Tan, 2020). Biocatalysis involves the use of isolated enzymes as well as enzymes present inside living cells to perform high stereoselectivity and chemical transformations of organic compounds. While an impressive number of such catalysts with great efficiency have been reported to date, most of them are susceptible to improvement in terms of their product selectivity, stability, TOF and TON, cost and availability, ease of preparation, and recyclability.

7.4 Enzyme catalysis

Enzymes are enantiomerically pure, natural catalysts, responsible for conserving the chirality in living systems (Liu, Zhang, & Wang, 2015). They catalyze the synthesis of a vast number of biomolecules and act as a model for energy-efficient, environmentally benign chemical agents. Enzymes differ from ordinary synthetic chemical catalysts as they can catalyze reactions at physiological temperature and pH and with enviable rates and selectivity. Ever since their discovery a century ago, enzymes have fascinated chemists and biologists alike. Because of their specificity and affinity, there is an ever-increasing demand for enzyme mimics or artificial enzymes that can overcome the shortcomings of a natural enzyme (Fig. 7.4).

The form and function are inseparable in biology, and this key concept equally applies to enzyme design. Although a quantitative accounting of enzyme efficiency still eludes us, extensive biochemical, mechanistic, and computational analyses, coupled with an explosion in sequence and structural data, have provided a solid qualitative understanding of how these catalysts function (Hilvert, 2013). Designing an enzyme combines a rational vision of the net outcome and its translation into a sequence of amino acids. It involves the creation of proteins or peptides with novel activity and purpose with an aim to advance understanding of protein functions. With the

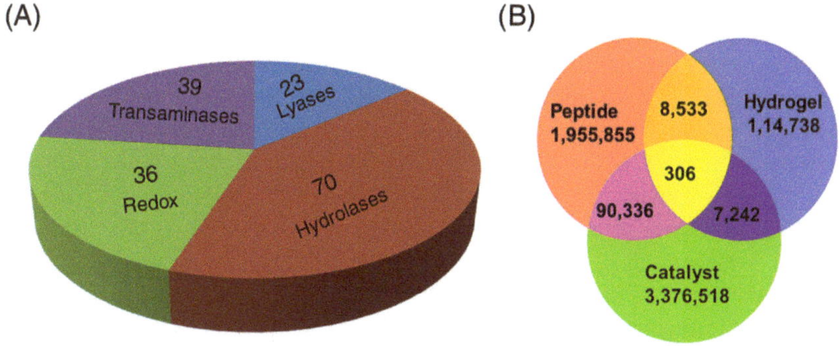

FIGURE 7.4 (A) Patents granted on some of the industrially important enzyme mimics between 2014 and 2020. (B) Current status of peptide-based catalyst hits in Scifinder as of April 30, 2020.

advancements in various computational tools, the field of protein design has grown enormously in the past two decades. It is easy to make large libraries of proteins with diverse sequences, but the actual challenge lies in getting predetermined properties such as a three-dimensional fold and catalytic functions. The efficiency and selectivity of catalytic proteins enzymes have inspired a generation of scientists to develop various approaches to design de novo catalytic protein and peptides. The process provides a challenging test of our understanding of enzyme function, while laying the groundwork for the design of more efficient and robust catalysts. Various computational tools capable of screening large virtual protein libraries have contributed to understanding the basis of protein folding and design (Jerath, Hazam, & Ramakrishnan, 2014; Jerath, Hazam, Shekhar, & Ramakrishnan, 2016; Kumar, Ramakrishnan, Ranbhor, Patel, & Durani, 2009; Mehra, Jerath, Ramakrishnan, & Trivedi, 2015; Metri et al., 2014; Ranbhor et al., 2018; Shaw et al., 2010).

Before diving deep into peptide-based artificial enzymes, it is highly relevant to acknowledge those factors that are responsible for such exquisite catalytic success.

The mechanism of enzyme catalysis is best described by the transition state theory, first reported by Eyring in 1935. According to this theory, enzyme catalysis starts with the formation of the enzyme − substrate complex (ES) that undergoes a series of chemical transformations to give the activated complex (ES*). The substrate is then converted to the product, thus forming the enzyme-product (EP) complex before the final product is being released. Pauling in 1948 proposed that enzyme catalysis occurs due to its ability to stabilize the transition state structure for the reaction relative to that of the substrate's ground state (Schowen, 1978). Fig. 7.1 illustrates the free energy diagram for an unimolecular enzyme-catalyzed reaction.

7.4.1 Enzyme active site

An enzyme active site is a cleft formed by an array of amino acids that binds to the substrate in an intermediate conformation and determines the efficiency of enzymatic catalysis. Often, the proximity of the amino acids in the active site is due to the precise three-dimensional folding of the primary amino acid sequence. The structural flexibility of the active site offers high catalytic efficiency to an enzyme. The active site, in general, is characterized by the presence of polar amino acids such as histidine, serine, aspartate, lysine, glutamate, and cysteine, which directly interact with the substrate to form an intermediate complex. Researchers over the years have found that only 11 out of the 20 naturally occurring amino acids possess catalytic properties. Table 7.1 summarizes the most common amino acids found in the active site and the characteristic properties of their catalytic groups.

TABLE 7.1 The function of catalytic groups in amino acids.

Amino acid	Catalytic group	p*K*a	Function
Histidine 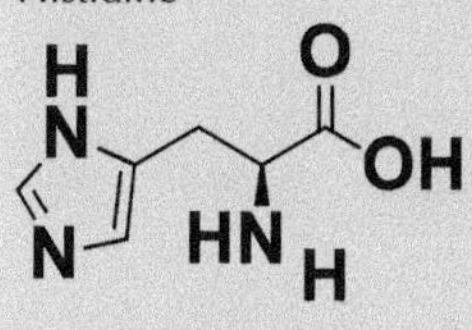	Imidazole	6 (approx.)	General acid, general base, nucleophile
Proline 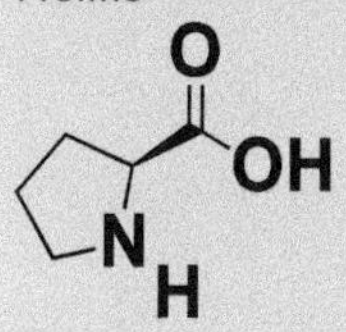	Imine	–	Nucleophile, especially to C=O
Cysteine 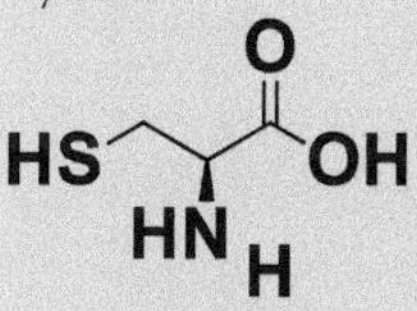	Thiol	8.3	Strong nucleophile
Arginine 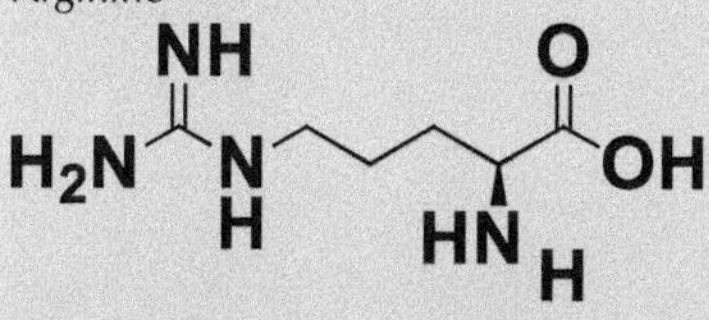	Guanidinium	12.5	Stabilizes double H-bonding interactions with CO_2^- and PO_2^-
Lysine 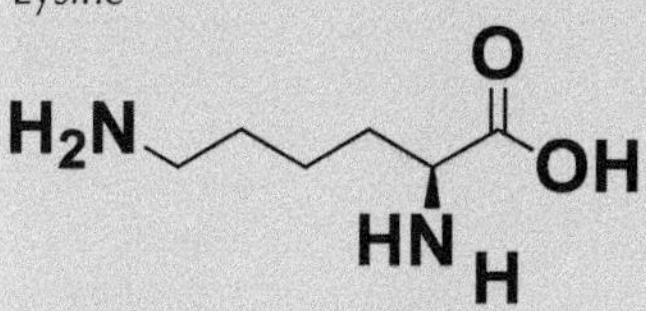	Amine	10.8	Nucleophile, especially for C=O
Glutamate 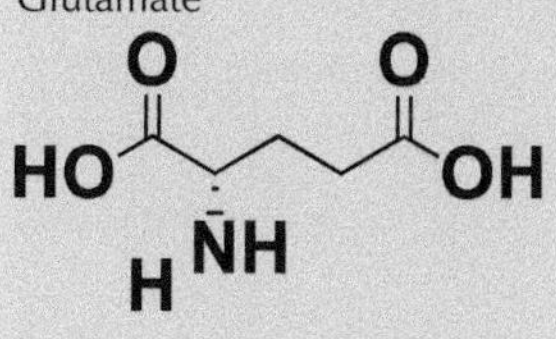	Carboxylate	4.3	General base or a nucleophile

(*Continued*)

TABLE 7.1 (Continued)

Amino acid	Catalytic group	p*K*a	Function
Aspartate 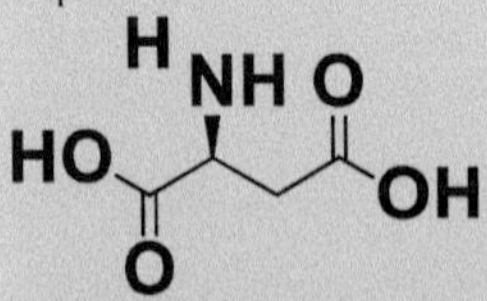	Carboxylate	3.9	General acid near neutral pH
Glutamine 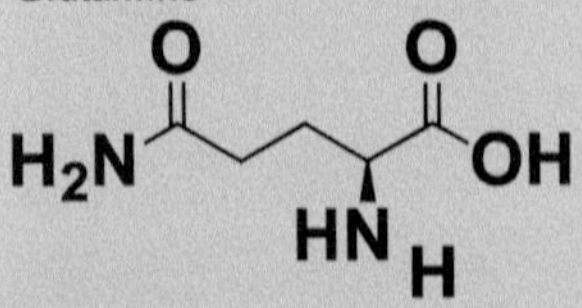	Amide	–	Unreactive unless they are primed by interaction with another residue
Asparagine 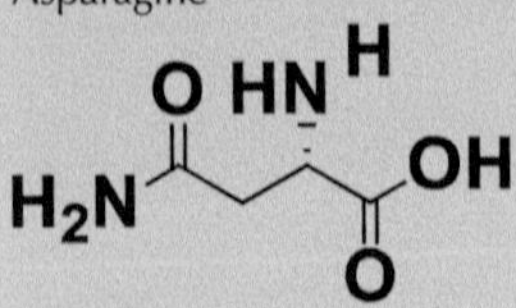	Amide	–	Unreactive unless they are primed by interaction with another residue
Serine 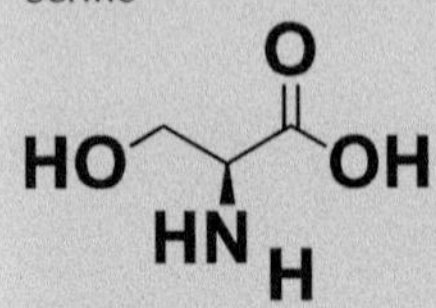	Hydroxyl	13 (approx.)	Nucleophile when the proton is removed from OH by a strong general base
Threonine 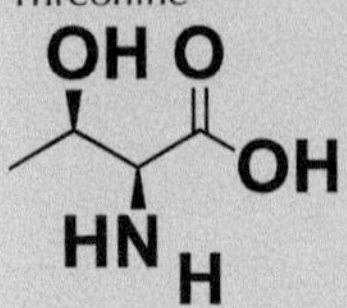	Hydroxyl	13 (approx.)	Nucleophile when the proton is removed from OH by a strong general base
Tyrosine 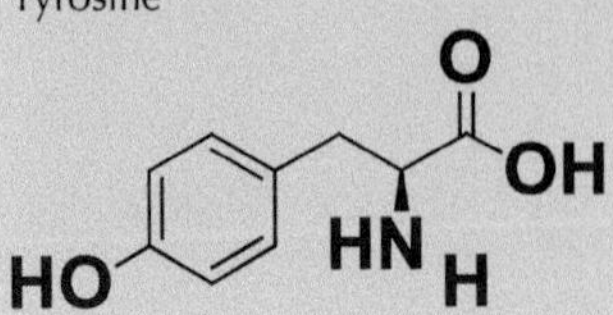	Hydroxyl	11	Nucleophile, especially for phosphorus

Amino acid residues like aspartate, glutamate, and histidine, in particular, serve as proton donors or acceptors during the chemical transformation. The active site conformation is further supported by the secondary structure (α-helix or β-sheet structure) of the polypeptide chain. In addition, some enzymes contain a nonproteinaceous component called the prosthetic group that is tightly bound to the enzyme and facilitates its function by acting as an electron acceptor. One such example is the heme-iron prosthetic group in catalase. Cofactors are another type of nonprotein molecule in certain enzymes that directly affects the enzyme function. They are different from a prosthetic group as they are loosely bound to the enzyme and may undergo modification during the enzymatic reaction. Cofactors can be either metallic or organic in nature. Several factors affect the substrate binding to the enzyme active site. A few of them are discussed in the following section.

7.4.1.1 Proximity and orientation effect

The binding of the enzyme and substrate results in the ES complex that acts as a single molecule having a faster rate of reaction due to the low entropy of the reactants. The proximity and the orientation of the ES complex align the reactive groups into optimal orientation for the reaction to be successful. This is known as the proximity and orientation effect and plays a significant role in intramolecular catalysis. It is reported that the transition state has relatively better interaction with enzymes as compared to the substrate itself (Lienhard, 1973). This induces the structural arrangement, which strains substrate bond into a position closer to the conformation of the transition state. This in turn decreases the energy difference between transition state and substrate, which means lowering of the activation energy, and resultingly enhancement in reaction kinetics.

7.4.1.2 Proton donors or acceptors

Enzymes are large molecules and may possess both acidic and basic groups in their active sites. Such proton donor or acceptor groups may help to stabilize the charge accumulated during the transition state. This results in the effect of activating electrophile and nucleophile groups or stabilizing leaving groups. Histidine is predominantly found in the active site of many acid–base catalysis reactions because it has a pK_a around 7, which allows it to act as both an acid and a base. In addition, the presence of charged residues such as Glu, Lys, Arg, and Asp promotes ionic interactions, thus stabilizing the charge transition state of the catalyst. Sometimes, the involvement of metal ions in electrostatic catalysis, as in the case of carboxypeptidase, decreases the pKa of water, making it an effective nucleophile (Makinen, Kuo, Dymowski, & Jaffer, 1979).

Most highly evolved enzymes depend on constellations of catalytic residues to promote demanding chemical transformations. Incorporating arrays of precisely positioned functional groups into protein binding pockets and tuning their reactivity, therefore, represent important challenges for protein design. Proton transfer reactions, the key to the function of many extraordinarily efficient enzymes, are ideal for exploring strategies for properly positioning acid and base groups.

The inherent properties of an enzyme, such as its enantioselectivity, have been widely exploited in organic synthesis for the synthesis of chiral products from achiral precursors. However, the industrial application of enzymes is restricted largely due to their instability in harsh conditions such as elevated temperature, pH, presence of organic solvents, and oxidative factors (Chapman, Ismail, & Dinu, 2018). This has led to a new frontier in catalysis research called enzyme mimetic.

7.5 Enzyme mimetics

Enzyme mimics are artificial enzymes that have heralded much interest as a viable alternative to natural enzymes in terms of their functions. Besides, they offer additional advantages such as tunable structures, high stability, and lower cost of synthesis (Kuah, Toh, Yee, Ma, & Gao, 2016). Enzyme mimics can be designed using two approaches; functional and structural. Functional mimics generally incorporate metal complexes with a similar activity of the natural enzyme, while structural mimics translate the enzyme's active site structure using functional moieties (Labádi, Benkő, Markó, & Szilágyi, 2009). To date, different types of enzyme mimics have been reported, ranging from protein redesign, nanocatalysts, metallopeptides, and supramolecular structures (Fig. 7.5) (Kuah et al., 2016).

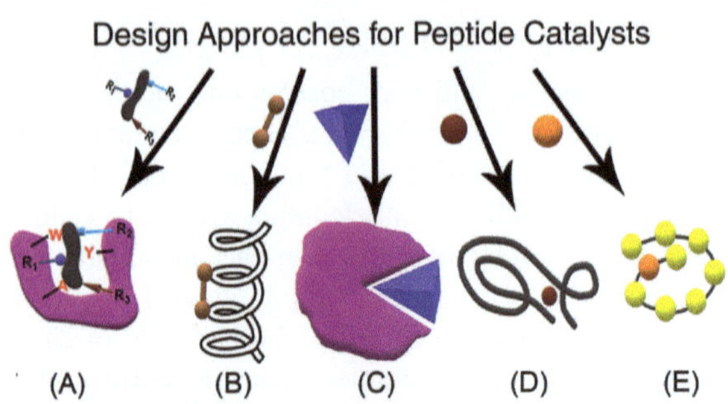

FIGURE 7.5 Various strategies employed in the development of artificial enzyme catalysts. It includes (A) quantitative assessment of catalytic function using theozyme, (B) developing metallopeptides, (C) incorporation of nonnative cofactors, (D) Substitution of natural metal ion, (E) incorporation of unnatural amino acids in the protein and peptide scaffolds.

7.5.1 Single amino acid mutations can impart high catalytic activity onto a protein

Redesigning of natural protein scaffolds represents the most expeditious yet less ambitious route to new catalytic activity. The combination of rational and random mutation in the existing active site can drastically alter its function (Shao & Arnold, 1996). For example, a single mutation in the active site of the pyridoxal phosphate-dependent racemase resulted in an aldolase. Researchers rely on several techniques even at the genetic level to generate the desired properties, such as improve stability, broaden substrate specificity, increase tolerance to nonnatural conditions, augment stereoselectivity, and enhance trace activities (Chen, 2001; Korendovych, 2018; Shao & Arnold, 1996). Also, the substitution of active site residues is likely to yield profound changes in terms of the enzyme's kinetic parameters as well as their selectivity.

The genetic level mutation is not restricted to the active site only. Instead, distant mutation can also subtly re-orient the binding pockets substantially. A striking demonstration by Fa et al. highlights the power of single amino acid mutation in re-engineering DNA-polymerase (Fa, Radeghieri, Henry, & Romesberg, 2004). The designed enzyme showed promising results in the encoded synthesis of unnatural biopolymers. Considering the large prospects of such mutations, it is important to have a fair amount of knowledge to predict which amino acid substitution is more likely to cause meaningful modifications.

Carter et al. have earlier reported that the replacement of Asp-32 with Ala in the paradigmatic Asp-His-Ser catalytic triad of the serine protease produced a variant that is 10^{-4} times as active as the wild-type (WT) (Carter & Wells, 1988). Their study has further highlighted the presence of hydrogen bonding between Asp-32 and His-68 residues. Hokenson et al., in a separate study, reported nucleophile eliminating mutations in β-lactamases by replacing Ser70 with Ala. This yielded a $k_{cat} \sim 10^6$-fold higher opening of the β-lactam ring of penicillin antibiotics than the background (Peracchi, 2001). The advancements in this field of research have led to the design of enzymes that display catalytic promiscuity for reactions that are not seen in nature (Leveson-Gower, Mayer, & Roelfes, 2019). Johnsson et al., in their early attempts, have reported that the introduction of a catalytically active amino acid residue, lysine, into a nascent hydrophobic cavity can promote the catalytic reaction of oxaloacetate to pyruvate at 3–4 orders of magnitude faster than the control reaction (Johnsson, Allemann, Widmer, & Benner, 1993).

In a similar study, Baltzer et al. introduced a histidine residue onto the surface of a helix-loop-helix containing polypeptide and found that the peptide could catalyze the hydrolysis of *p*-nitrophenyl esters in water as well as water-trifluoroethanol mixtures at three orders of magnitude above the control (Broo, Brive, Ahlberg, & Baltzer, 1997). They have extended their study by changing the percentage of histidine residues or introduced glutamine, arginine, and lysine mutations in the sequence. While the rate of the *p*-nitrophenyl ester hydrolysis reaction has reduced significantly, the designed catalysts showed an enantiomeric preference for D-norleucine *p*-nitrophenyl ester being hydrolyzed faster (Broo, Brive, Sott, & Baltzer, 1998).

7.5.2 Incorporation of unnatural amino acids as a minimalistic approach for enzyme design

UAAs consist of a diverse range of side-chain modifications in amino acids. They have been widely studied to generate structures and functions that are absent in natural amino acids (Gaurav & Santhoshkumar, 2020; Goyal, Jerath, Akhil, et al., 2021; Goyal, Jerath, Chandrasekharan, et al., 2021; Goyal, Jerath, Chandrasekharan, Kumar, & Ramakrishnan, 2020; Hazam, Akhil, Jerath, Saikia, & Ramakrishnan, 2019; Hazam, Jerath, Chaudhary, & Ramakrishnan, 2018; Hazam, Jerath, Kumar, Chaudhary, & Ramakrishnan, 2017; Jerath, Goyal, Trivedi, Santhoshkumar, & Ramakrishnan, 2019; Kumar et al., 2009; Prakash, Ranbhor, & Ramakrishnan, 2020; Ramakrishnan, Ranbhor, Kumar, & Durani, 2006; Ranbhor, Ramakrishnan, Kumar, & Durani, 2006). Incorporation of UAAs beyond the 20 proteinogenic residues to mimic post-translational modifications may result in enzymes with significantly enhanced catalytic activity (Saikia et al., 2021). The genetically coded amino acids with the rare exception of selenocystein85 and pyrrolysine86 drastically limit the chemistry with their functional groups like hydroxyl, carboxyl, thiol, and aromatic side chains. Among the various strategies, residue-specific and site-specific chemical modifications of amino acid side chains are becoming the preferred approach. For example, the substitution of the phenylalanine or tryptophan residues with a naphthyl or biphenyl aromatic groups increases the side-chain hydrophobicity apart from affecting the π-stacking interaction with the respective protein-ligand (Nannemann, Birmingham, Scism, & Bachmann, 2011). Green et al. chemically designed N_δ-methyl histidine ligand in an ascorbate peroxidase (APX2) that overcomes the inherent reliance on the Asp-His hydrogen bonding interaction. This resulted in significantly higher TON compared to the native enzyme without compromising the catalytic efficiency (Green, Hayashi, Mittl, & Hilvert, 2016).

7.5.3 De novo design of peptide-based enzyme mimics

The late 1960s and early 1970s of the last century mark the beginning of a new era of protein catalysts. Small fragments of natural proteins such as adrenocorticotropic hormone and RNase A had been shown to retain the substantial biological activity of the parent molecule (Boissonnas, Waller, & Jaquenoud, 1956). Over the years, the revolutionary invention of the solid phase peptide synthesis method by Merrifield (1963) (Merrifield & Woolley, 1956)

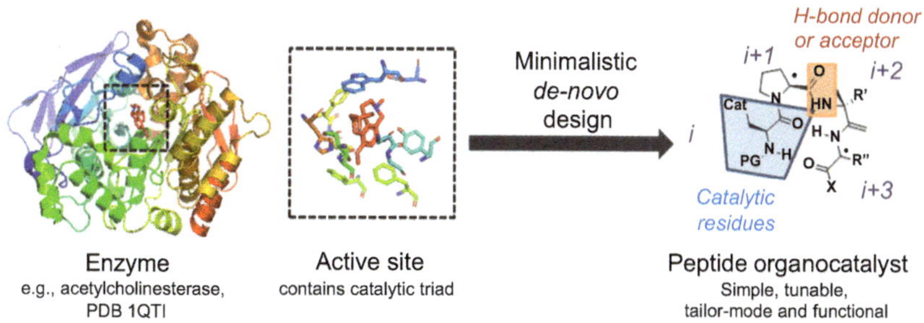

FIGURE 7.6 Summary of peptide catalysts. The active site containing the catalytic triad is carefully studied with an aim to design its structural or functional mimics using synthetic peptides.

and Chou and Fasman's finding (Chou & Fasman, 1978) of the distribution of 20 natural amino acids in α-helix, β-sheet, and β-turn have allowed to successfully construct artificial polypeptides. Fig. 7.6 summarizes the strategy of developing a biomimetic peptide catalyst.

Researchers over the years have made great progress in the design of polypeptides possessing stable structures even at the amino acid level (Bhardwaj et al., 2016; Rufo et al., 2014). One promising strategy to redesign or mimic an enzyme is to take advantage of existing features of the original active site, such as the substrate-binding pockets and the most abundant functional groups. This reduces the effort to mimic the activity, at least in a mechanistic sense, to the original function of the protein. Another strategy is to insert amino acids with novel functional capabilities into the designed sequence. This may significantly expand the range of accessible reactions (Shaw, 2012). The incorporation of redox-active or metal-binding residues in the active site will further extend this approach (Zozulia, Dolan, & Korendovych, 2018).

However, the construction of an artificial enzyme is a complex process and requires the formation of an active site in a tertiary fold.

7.5.4 Peptide-based minimalist approach of enzyme catalysis

The ability to generate molecules that are capable of mimicking the active site of an enzyme represents a promising strategy in enzyme design. Synthetic peptides are able to generate exact copies of a protein fragment apart from allowing diverge chemical modifications, including the incorporation of nonproteinogenic amino acids and modification of peptide backbone. Such modification in synthetic catalytic peptides increases their proteolytic stability and diversity in terms of substrate selectivity.

The field of peptide-based catalysis had a humble beginning with the simplest form of designs containing homopolymers of amino acids such as poly-His (Merrifield & Woolley, 1958) and dipeptides, poly(Leu-Lys) (Barbier & Brack, 1992) that could hydrolyze suitable esters and oligoribonucleotides but with very low activity. Such findings have paved the way for the true de novo design of peptide catalysts. Chakravarty et al. in 1973 reported weak glycosidase activity in an α-helical decapeptide EFAAEEAASF that was designed to mimic the active site of lysozyme (Chipman & Sharon, 1969). They concluded that the enzymatic activity is triggered by the increased pKa of the side-chain carboxyl group of the active-site Glu (Glu6) that was sandwiched between the aromatic rings of Phe2 and Phe10. In another study, Gutte et al. designed a 34-residue polypeptide containing histidine and two cysteine residues that form an antiparallel ββα-fold. The peptide showed preferential binding to cytidine phosphates and had substantial ribonuclease activity (2.5% of the specific activity of natural RNase A). The peptide complex was stable even at 0.1 M formic acid. This discovery showed that the design of the de novo catalytic peptide was in reach and can be directed to emulate natural enzymes.

In general, peptide mimetics involve three different approaches based on the sequence, structural, and functional information about the enzyme's active site. The ability to screen large combinatorial libraries, biased to favor specific folds, has significantly enhanced the discovery of peptide catalysts for a broad range of chemical transformations such as aldol and redox reactions (Fig. 7.7) (Davie, Mennen, Xu, & Miller, 2007). The main features of the minimalist approach for peptide catalysts are outlined in the following sections.

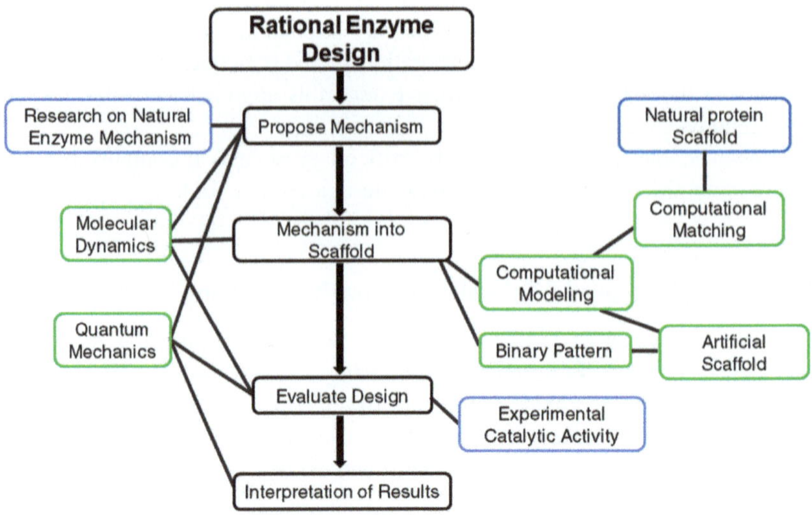

FIGURE 7.7 General approach followed for rational enzyme design. The main steps are in black squares. Computational tools (*blue boxes*) and wet-lab experiments (*green boxes*) aid along the process (*thin arrow*).

7.6 Structural design of the artificial peptide-based enzyme

X-ray crystallography and NMR studies revealed that the most well-defined secondary structures in a protein structure are α-helices and β-sheets. These structures are held together by the intramolecular hydrogen bonds, for example, between the hydrogen bond donor NH and the acceptor COOH in the protein backbone, that provide a rigid and stable framework for proteins (Branden & Tooze, 1999). In the case of an enzyme, the catalytic groups are attached to this framework through their side chains. The catalytic activity largely depends on the structural stability of the enzyme and the microenvironment of the active site. Therefore the focus on peptide-based enzyme mimics has mostly been on designing helical or sheet peptides. The positioning of the catalytic groups must be carefully examined during such designs. For example, an HPPHPPP sequence, where H represents a hydrophobic group and P represents a polar group, forms an α-helical coiled-coil while an alternating H and P residues in the sequence (HPHP) results in a sheet conformation (Zozulia et al., 2018). One strategy is to rationally replace a P with a catalytic residue, thus retaining the hydrophobic skeleton of the peptide. Another strategy involves exploiting other intermolecular interactions, such as the $\pi - \pi$ stacking between the aromatic groups that would facilitate the formation of β-sheet structures (Fleming & Ulijn, 2014). However, it is not always necessary to mimic the secondary structure of the natural enzyme to achieve its functionality as long as it maintains the core features of the catalytic active site. In one such example, Zastrow et al. have inserted the active site of carbonic anhydrase (CA) and nitrite reductase into helical-stranded coiled-coil scaffolds and achieved considerable catalytic activity (Zastrow, Peacock, Stuckey, & Pecoraro, 2012). In another study, Qi et al. have found a structure-activity correlation among the designed peptides. While the helix-forming assemblies showed high activity for phosphoester hydrolysis, the β-sheet assemblies did not catalyze the reaction at all. They attributed this "partiality" to the distinct dipole moment present in the helical peptides, resulting in a greater affinity to the phosphate group.

7.6.1 Optimization of the catalytic microenvironment

The catalytic microenvironment is the spatial region surrounding the catalytic groups that comprise both the catalytic and noncatalytic residue side chains. In the case of peptide-based enzyme mimics, these amino acid side chains often form a stereospecific structure responsible for substrate recognition and orientation. The pKa(s) of the catalytic groups are affected by the neighboring residue side chains. In addition, they also determine the polarity and pH of the microenvironment, which in turn affects the TON and substrate binding. Studies by Yang et al. demonstrated the effect of the X residue on the catalytic activity of a β-sheet-forming peptide (Fmoc-FFXAH-

NH_2) (Wang & Qi, 2020). By replacing the X with other residues, they could alter the hydrophobicity and charge of the peptide, which in turn resulted in different catalytic efficiencies. Their findings also highlight the role of self-assembly on the catalytic microenvironment. During the self-assembly of a hydrophobic peptide, a hydrophobic core is formed where the substrate concentration is usually higher. This results in enhanced reaction kinetics.

7.7 Self-assembling peptide catalysts

During the primordial times on earth, there were no enzymes as we know them today. Instead, it is believed that short peptide units were the "primitive enzymes" that helped in the production of more complex molecules from simple precursors. Based on evolutionary biology, it is reasonable to simulate the functions of natural enzymes using self-assembling peptides because, just like in an enzyme, the catalytic groups in a peptide unit come from the amino acid side chain. In addition, peptide assemblies formed using various noncovalent interactions generate a hydrophobic microenvironment that facilitates the accumulation of reactants at their active site. Researchers over the years have developed short self-assembling peptides, tripeptides to heptapeptides, that can efficiently catalyze simple chemical reactions under physiological conditions with or without cofactors. Supramolecular self-assembly of short peptides offers a minimalistic approach to mimic the functional properties of a multifaceted biocatalyst such as the active site or a hydrophobic pocket (Makam & Gazit, 2018). They have been utilized to assemble catalytic functional groups in a nanoenvironment that is adequate for efficient organic transformations. Micelles formed by amphiphilic peptides (Cleij, Drenth, & Nolte, 1991), peptide-nanoparticle complexes (Zaramella, Scrimin, & Prins, 2012), and rigid-rod β-barrels (Das, Sakai, & Matile, 2002) are the most significant examples.

The efficiency of their catalysis may be attributed to the combination of two factors; (1) affinity to substrate binding and (2) proximity effects of the catalytic moieties (Singh, Tena-Solsona, Miravet, & Escuder, 2015). Recent studies also suggest catalytic efficiency in peptide-based nanotube and nanofiber structures (Díaz, Kühbeck, & Koopmans, 2011). Amyloid-like materials are extremely robust over a range of temperatures and pressures, highlighting their potential use in the development of catalysis beyond the tolerance of most protein enzymes.(Luong et al., 2016) In the first report of supramolecular assembled peptide catalyst, Stupp et al. showed well-defined peptide-based catalytic nanostructures in aqueous media (Guler & Stupp, 2007). The initial discovery of de novo designed amyloid catalysts led different groups to expand the sequence space of the peptides and the reactions catalyzed. The ability to perform organic reactions in aqueous media motivated the designing of amphiphilic peptide molecules that formed hydrogels. Escuder and Miravet's groups have extensively studied the catalytic activity of L-proline-based supramolecular gelators in C − C bond forming reactions, such as aldol addition reactions (Rodriguez-Llansola, Escuder, & Miravet, 2009; Rodríguez-Llansola, Miravet, & Escuder, 2009; Rodríguez-Llansola, Miravet, & Escuder, 2010). In addition, Lynn et al. have demonstrated enantioselective chemical reactions using simple peptide assemblies (Omosun et al., 2017). Fluorenylmethyloxycarbonyl containing Fmoc-tripeptide with histidine as a catalytically active residue (Fmoc-FFH-$CONH_2$) has been reported, which self-assembles at pH 7.5 to form nanotubes capable of catalyzing the hydrolysis of *p*-nitrophenyl acetate (PNPA) in aqueous conditions (Huang et al., 2013). Our group has studied the role of proline position and aromaticity on the epitaxial growth and hydrogelation properties of a series of tripeptides at physiological pH (Saikia et al., 2021).

While there is an ever-increasing number of report on peptide-based supramolecular catalysts, the stereoselectivity and efficiency of such systems are yet to match with that of an enzyme. Several parameters, such as temperature, solvent, and chirality, may affect the catalysis and must be considered in detail, with an aim to control the reaction performance by tuning the structure at the molecular or supramolecular level.

7.8 Peptide mimics with enhanced catalytic efficiency

It is desirable for a catalyst to be highly specific and catalyze one or a certain type of reaction. This property is important for industrial applications, especially in the pharmaceutical industries that demand the production of single enantiomers for reduced health hazards. In order to preserve the high specificity of a natural enzyme, the peptide-based enzyme mimic must have greater interactions between the exposed hydrophobic surface of the assembly and hydrophobic moieties of the substrate (Berdugo, Miravet, & Escuder, 2013). Zhu et al. developed a strategy for constructing an artificial hydrolase with customized selectivity by coating the surface of catalytic peptide nanofibers with the imprinted polymer layer (Zhu, Wang, Qi, Su, & He, 2019). By replacing the polymerization process with supramolecular assembly, Wang et al. obtained a hydrolase mimic through the co-assembly of Fmoc-tripeptides (Wang et al., 2016). In the case of proline-based asymmetric aldol or Mannich reactions, the selectivity arises from the steric effects during different intermediate states of the catalytic pathways (Fig. 7.2B) (Rodriguez-Llansola et al., 2009; Singh & Escuder, 2017).

7.9 Peptide-based artificial metalloenzyme

The active site of a metalloenzyme comprises of a metal complex where a transition metal monomer or dimer is coordinated by a few amino acids (Vallee & Williams, 1968). Biomimicry of such active sites using designed peptides and metal ions is envisaged as a powerful strategy to create artificial catalysts. The complementary effect of a highly reactive metal ion and the peptide's structural and functional versatility can be translated into reactive and selective catalysts. Also, exploring this field provides information regarding the fundamental aspects of molecular recognition in catalysis, protein design, and protein folding (DeGrado, Summa, Pavone, Nastri, & Lombardi, 1999). The combination of peptides and metal catalysts has been used in several asymmetric transformations such as hydrosilylation (Greenfield, Agarkov, & Gilbertson, 2003), allylic alkylation (Breit & Laungani, 2003), conjugate additions to enones (Degrado, Mizutani, & Hoveyda, 2001), and Strecker reaction (Josephsohn, Kuntz, Snapper, & Hoveyda, 2001).

Artificial metalloenzymes (ArMs) are assembled using four design strategies (Fig. 7.8) (Schwizer et al., 2018). The most common design of ArMs involves a scaffold peptide containing uniquely reactive nucleophilic residues (e.g., histidine) and a corresponding electrophilic reaction partner (e.g., Cu^{2+}, Fe^{2+}, Fe^{3+}). A reductionist approach has shown that ultra-short peptides (3–7 residues) can form a well-defined primary coordination sphere around the metal ion (Mori, Abet, & Inoue, 1995). Kaiser's helichrome catalyst is among the earlier examples of synthetic metallopeptide catalysts designed by covalently linking

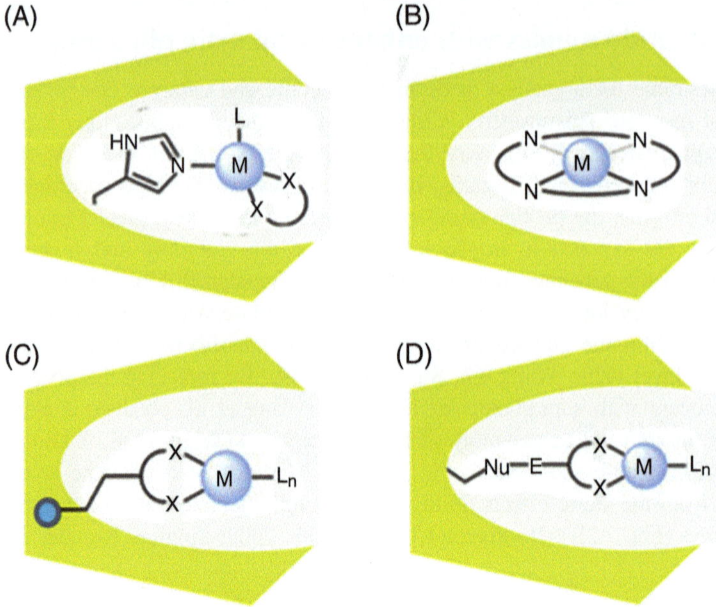

FIGURE 7.8 Four general approaches to artificial metalloenzyme assembly. (A) Nucleophilic residue positioned within a cavity may interact with the transition metal ion. (B) Metal substitution approach, where another metal substitutes the metalloenzyme's native metal cofactor, thus conferring novel catalytic activity to the complex. (C) Supramolecular coordination through a high-affinity inhibitor embedded inside. (D) Covalent immobilization due to reaction between the complementary functional groups present in the ligand and the host peptide (Davis & Ward, 2019). *Concept from reference Davis, H. J., & Ward, T. R. (2019). Artificial metalloenzymes: Challenges and opportunities.* ACS Central Science, *5(7), 1120–1136.*

four 15-residue peptides to tetrahydroxysuccinamide esters through their N-terminus (Sasaki & Kaiser, 1989). The design resulted in helix formation with a hydrophobic substrate-binding pocket. In another study, Dervan et al. reported a metallopeptide catalyst by rationally modifying a conserved fold with synthetic metal-binding moiety (Sluka, Horvath, Bruist, Simon, & Dervan, 1987). The group also reported a Gly-Gly-His tripeptide based on the consensus sequence present in the copper-binding domain of serum albumin (Mack, Iverson, & Dervan, 1988). Sambasivan and Ball have convincingly demonstrated the potential application of the metallopeptides KADAALDAK for enantioselective catalysis upon binding to Rhodium(II) (Sambasivan & Ball, 2010; Sambasivan & Ball, 2012; Sambasivan & Ball, 2013).

7.10 Exploring unnatural amino acids for de novo peptide-based metalloenzymes

Few research groups have explored UAAs, including D-amino acids, to enable designs and functions beyond those accessible using natural amino acids (Artner et al., 2012). Gilbertson et al. demonstrated that it is even possible to design a nonhelical secondary structure that specifically binds to the metal ion and performs catalysis. Using different turn-inducing motifs such as Pro-d-X, where d-X is a D-amino acid, they have constructed peptides containing Yyy-Pro-d-X-Zzz, where Yyy and Zzz are phosphoserine residues (Agarkov, Greenfield, Ohishi, Collibee, & Gilbertson, 2004; Gilbertson & Pawlick, 1996). The peptides showed excellent metal binding affinity and moderate catalytic efficiency. Ramakrishnan et al. mimicked the histidine brace found in lytic polysaccharide monooxygenases (LPMO) by incorporating a D-His in the C-terminal of a tripeptide sequence. The peptide binds to Cu(II) at pH 7 to form flake-like heterogeneous structures. Functionally, the designed catalysts can oxidize benzyl alcohol to benzaldehyde (82% in 3 h) with 100% selectivity at 40°C (unpublished). Pecoraro mimicked the active site of the CA using three-helix bundles of TRI peptides with well-defined metal binding sites for Hg(II) and Zn(II) that are facilitated by three histidines and three cysteine residues (Zastrow et al., 2012).

While such developments are exciting, there is still room for improvement, particularly in the realm of chemical reactivity, where metallopeptides as catalysts are dwarfed by the sheer number of industrially important chemical reactions available to small-molecule catalysts.

Case studies

7.11 Aldolase mimic

Aldol reaction is one of the most persuasive carbon – carbon bond forming reactions. The addition of a nucleophilic ketone donor to an electrophilic aldehyde acceptor in the presence of a chiral catalyst gives access to enantiomeric rich aldol products (List, Hoang, & Martin, 2004).

Aldolases are a member of the glycolytic enzyme family that is able to catalyze the aldol reaction in a reversible manner. Aldolase has found application in organic chemistry due to its ability to form a new C – C bond from two carbonyl compounds, together with the formation of a β-hydroxyl carbonyl skeleton. This occurs with great regio- and stereoselectivity under physiological conditions.(Dean, Greenberg, & Wong, 2007) Based on their mode of catalysis, natural aldolases are divided into two types: Type I and Type II aldolases. Over the years, researchers have concentrated on mimicking them using different strategies.(Machajewski & Wong, 2000) Type I mimics is amino acid-based catalysts, which activate the donor via enamine formation and the acceptor through a hydrogen bond with an acid functionality (Agami, Puchot, & Sevestre, 1986). Type II aldolase mimics consist of bimetallic catalysts containing a Lewis acidic metal for aldehyde activation and a Brønsted base for enolate generation to form an active complex (Yamada, Yoshikawa, Sasai, & Shibasaki, 1997).

7.11.1 Metal ion-free aldolase mimic (Type I aldolases)

In natural Type I aldolase, the catalytic function usually relates to tyrosine and lysine residues in the active site. Tyrosine, in particular, acts as an acid-base catalyst and is involved in substrate protonation. Lysine, on the other hand, stabilized the charge of the resulting conjugate base of tyrosine and the strained configuration of the C-terminus (Fig. 7.9A). However, during the design of peptide mimics, L-proline is commonly used as a catalytic group due to its proven catalytic efficiency for asymmetric aldol coupling reactions (Eder, Sauer, & Wiechert, 1971; Font, Sayalero, Bastero, Jimeno, & Pericas, 2008). Fig. 7.9B shows the proposed catalytic function of L-proline involving the enamine mechanism. Escuder and team have extensively studied the proline-based self-assembling peptide aldolases (Rodriguez-Llansola et al., 2009; Rodríguez-Llansola et al., 2010; Rodríguez-Llansola, Escuder, & Miravet, 2009; Rodríguez-Llansola, Escuder, Hamley, Hayes, & Miravet, 2012; Rodríguez-Llansola, Miravet, et al., 2009). Using a series of Pro-Val containing amphiphilic and bolaamphiphilic molecules, they have investigated the effect of various solvents, temperatures, and composition on the structure and activity of the catalytic assemblies. They have observed that, compared with the solution state, the base catalytic ability of the proline residue increases upon self-assembly. They have attributed this effect to the steric effect of the heterogeneous organogels that inhibit or reduce the formation of byproducts. Their study also highlights the importance of supramolecular assembly in improving the catalytic activity and selectivity of the Type I aldolase mimics (Rodriguez-Llansola et al., 2009).

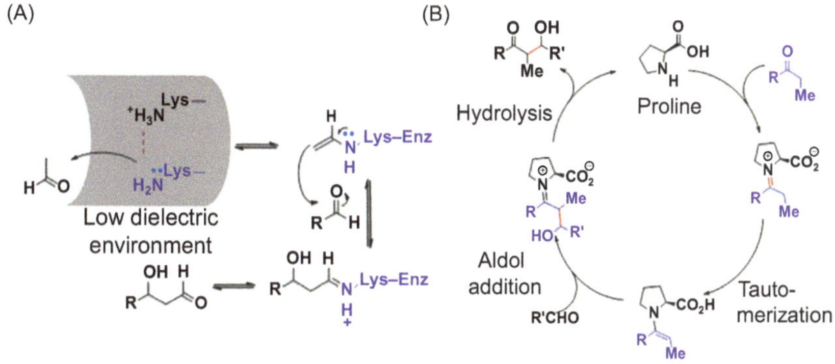

FIGURE 7.9 The proposed catalytic mechanism of aldolase I. (A) A reactive, nucleophilic lysine residue attacks the carbonyl of the aldehyde to form a carbinolamine that subsequently collapses to a Schiff base. The Schiff base tautomerizes to an enamine group and can then attack a second aldehyde or ketone. After hydrolysis and release of the enzyme, the β-hydroxy-ketone product is formed. The reactivity of the Schiff base-forming lysine is enhanced by low dielectric environment. (B) Catalytic mechanism of proline-based enamine catalysis involving tautomerization to nucleophilic enamine, aldol addition, and finally hydrolysis of enamine that releases the aldol adduct. In appropriate conditions, the chirality on the proline can be transferred to the aldol adduct.

In addition, proline residues incorporated into a self-assembling peptide overcome the limitations possessed by free proline, such as its inability to catalyze reactions in an aqueous medium. Our group has studied the role of peptide sequence and its stereochemistry on a series of proline-containing tri-peptides that functionally mimic Type I aldolase mimics (Saikia et al., 2021). The hydrogel-forming peptides were tested for their ability to catalyze the aldol reaction between cyclohexanone and 4, nitrobenzaldehyde. Our observations suggest that the position of the proline residue and its stereochemistry directly mediates the catalytic efficiency and the product stereoselectivity. In another study, Lee et al. reported a novel hydrophilic di-peptide containing Pro and Lys residues and a hydrophobic 1,4,5,8-naphthalenetetracarboxylic acid diimide side chain that formed catalytic nanotube structures (Lee & Parquette, 2015). In the presence of water, the water-insoluble substrates (p-nitrobenzaldehyde and cyclohexanone) were encapsulated in the nanotubes that provided a microenvironment for aldol coupling. In a similar study, Alves et al. designed a lipopeptide containing Pro-Arg-Try tripeptide that formed catalytic micelles upon self-assembly (Soares et al., 2017). The designed catalyst had enhanced enantioselectivity (85%ee) and yield even in the presence of water.

Interestingly, in natural protein, proline is responsible for the formation of helix or turn configuration in the secondary structure, rather than acting as a catalytic group. Even so, considering that effective proline-based catalysts are capable of replacing aldolases in some simple aldol reactions, they can be considered as an efficient Type I aldolase mimics.

7.12 Hydrolase mimic

Hydrolases catalyze bond cleavages by reaction with water (Busto, Gotor-Fernández, & Gotor, 2010). They are useful to the organic chemist as they have broad substrate specificity with high stereoselectivity. In addition, they are independent of any cofactors and can catalyze reactions in an aqueous medium with high efficiency.

Histidine is predominantly found in the active site of many natural hydrolases (Busto et al., 2010). The role of histidine in enzyme catalysis mainly depends on the features of the imidazole group. Under physiological conditions, imidazole is likely half-protonated, making it available to act as a general acid, a general base, or a nucleophile. Thus histidine tends to form hydrogen bonds, combines a donor and an acceptor, and can take part in either nucleophilic or basic catalysis. In most of the peptide-based hydrolase mimics, histidine is introduced in the peptide sequence, and therefore the proton always comes from water. This generates a nucleophilic hydroxide that reacts with the ester substrate. In the last two decades, histidine-containing peptide-based hydrolase mimics of varying lengths have been reported with different secondary structures (Guler & Stupp, 2007; Huang et al., 2013; Singh, Conte, Ulijn, Miravet, & Escuder, 2015; Zhang et al., 2014). In addition, the supramolecular mimic contains closely aligned histidine residues such that enhanced basicity of the imidazole side chain is achieved due to deprotonation caused by neighboring histidine residues. This consequently enhances the deprotonation of water, resulting in higher hydrolytic activity as compared to the free peptide monomers (Wang et al., 2016).

In another design strategy, a more natural catalytic triad is followed that comprises of an acid, a base, and a nucleophile. A number of reports suggests synthesis of artificial hydrolases through the co-assembly of peptide sequences such as Ser-His-Asp, Cys-His-Asp, and Ser-His-His (Burton, Thomson, Dawson, Brady, & Woolfson, 2016; Gulseren, Khalily, Tekinay, & Guler, 2016; Wang et al., 2016; Wong, Masunaga, Chuah, Sudesh, & Numata, 2016). Wang et al. have demonstrated that the co-assembly of Fmoc-FFH, Fmoc-FFS, and Fmoc-FFD at a 40:1:1 ratio has enhanced hydrolytic activity for CoA − HSD (3α-hydroxysteroid dehydrogenase) as compared to the individual peptides (Wang et al., 2016). Apart from histidine, enzymes such as glycoside hydrolases comprise of aspartic acid and glutamic acid in their active sites (Moracci, Capalbo, Ciaramella, & Rossi, 1996) (Craik, Roczniak, Largman, & Rutter, 1987). One of them acts as a general acid and assists the departure of the leaving group while the other acts as a basic group and attacks the substrate directly or assists the attack as a general base. The complexity of the active site of glycoside hydrolase has limited the number of reports of its active site mimics. Wang et al. reported a series of seven residue β-sheet-forming peptides that contain more than one glutamic acid or aspartic acid residue (He, Zhang, Zhang, et al., 2017). The peptides were found to be highly active for the hydrolysis of cellobiose, suggesting that peptides with ordered molecular confirmation and regular distribution of intramolecular hydrogen bonds have higher catalytic activities. Similarly, He et al. reported that Glu-rich Fmoc-heptapeptide nanofiber exhibited hydrolytic activity toward cellobiose and cellopentaose. They attributed this to the hydrophobic microenvironment that facilitated proton transfer, an essential step in catalytic hydrolysis (He, Zhang, Liu, Fang, & Wang, 2017).

7.13 Oxidase mimic

Redox reactions are an integral part of a number of physiological processes such as photosynthesis, nerve function, muscle contraction, and N_2 fixation. A typical redox reaction involves an oxidoreductase that catalyzes the transfer of electrons from an electron donor (reductant) to an electron acceptor (oxidant), thus changing the oxidation state. Transition metals such as copper [Cu(II) and Cu(I)] and iron [Fe(III) and Fe(II)] with variable valance states are effective electron donors. Based on the knowledge about the interactions between metals and natural polypeptides, researchers have developed artificial oxidoreductase mimics with redox activity. It is achieved through metal ion coordination through noncovalent anchoring of metal and incorporating metal cofactors. Korendovych et al. designed an oxidase mimic using assembling peptides in the presence of Cu (II) that are able to oxidize dimethoxyphenol (DMP) in the presence of molecular oxygen (Makhlynets, Gosavi, & Korendovych, 2016). By screening a series of β-strand forming peptides, they also concluded that the catalytic activity was highly sequence-dependent. The most efficient catalytic peptide (Ac-IHIHIQI-$CONH_2$) forms a coordination sphere for Cu(II), while the nonfibril forming peptide ($NH_2 - IHIHIQI - COOH$) did not show catalysis. Moreover, the replacement of His in position 2 or 4 in the sequence with Ala resulted in significantly reduced activity. In addition to copper and iron, few researchers have also employed rare transition metals to design oxidase mimics through direct coordination. Lee et al. developed a peroxidase mimic using histidyl bola amphiphiles that are coordinated to Mn(II) ions (Kim & Lee, 2015). In the presence of H_2O_2, the catalyst showed a high conversion of the organic substrate. In the case of an ArM system, based on the properties of the binding sites and the metal ions, the coordination between the protein and the metal ion varies drastically.

7.13.1 Lytic polysaccharide monooxygenase

LPMOs are metalloenzymes that activate molecular oxygen and cleave the $C - H$ bond in polysaccharides (Ciano, Davies, Tolman, & Walton, 2018) and has found major industrial implications in recent years (Arora, Bharval, Sarswati, Sen, & Yennamalli, 2019; Horn, Vaaje-Kolstad, Westereng, &

Eijsink, 2012). LPMOs utilize copper as their functional, active site metal for oxidizing recalcitrant polysaccharides in nature. The LPMO domains usually comprise 200–250 amino acids that drastically restrict their commercial application due to the excessive cost associated with their synthesis and stability at a range of temperature and pH. Interestingly, various spectroscopic and computational investigations have provided insights into the three-dimensional copper-containing active center featuring the "histidine brace" (Ciano et al., 2018; Kjaergaard et al., 2014; Span & Marletta, 2015; Walton & Davies, 2016). The histidine brace is conserved and comprised of a single divalent copper ion, which is chelated by two nitrogen atoms of histidine at position 36 (one atom from the backbone and one from the side chain) and a nitrogen atom from a second histidine at position 135, forming an overall T-shaped 3 N configuration (Ciano et al., 2018; Vaaje-Kolstad, Forsberg, Loose, Bissaro, & Eijsink, 2017). The main chain amino group of the N-terminal histidine (His1) and imidazole side chain contributes two of its nitrogen while the second conserved histidine contributes the third nitrogen.

7.14 Conclusions and future prospects

Replicating the enzyme functions using synthetic materials is not new; however, using a biologically derived component such as peptides to construct enzyme mimics has only begun recently. Over the years, the progress in our understanding of how enzyme functions has allowed us to induce nascent catalytic activities using strategically placed single amino acid mutation. This also suggests that the application of even crude design concepts can generate significant catalytic activity, thus paving the way for subsequent, direct evolution. The developments in the field of peptide chemistry and its fusion with biology have allowed researchers to overcome several major challenges associated with protein design. A critical advantage of using short oligopeptides is their small size compared to an enzyme, thus enabling easy molecular design. Also, a short sequence of 8–10 residues offers high-throughput screening methods to completely probe all the possible peptide sequences. Self-assembling short peptides and their ability to form supramolecular assemblies have unique advantages from the point of practical applicability.

Nevertheless, the growing interest in peptide-based enzyme mimics also presents severe bottlenecks, and therefore research must be focused on a number of challenges. First, the design of a de novo peptide sequence that balances the basic principles of self-assembly and enzymology. By now, it is evident that the size, shape, and microenvironment of the peptide assembly are strictly related to the catalytic efficiency of the enzyme mimic. Second, the incorporation of UAAs such as D-amino acids in the peptide sequence may overcome the steric hindrance present in a syndiotactic peptide design. This may yield structures beyond the scope of poly L-amino acids and with reduced sequence lengths. For example, the formation of right- or left-handed helices may generate a novel configuration for stereoselective catalysis. Third, there is a lack of general principles for the design and reconstruction of peptide-based enzyme mimics. Most of the reported peptide catalysts exploit a limited number of essential reactive groups to form the active sites. The ability to reproduce a natural microenvironment with adequate proximity and orientation is challenging in artificial mimics. This might be the reason why the TOF and the catalytic efficiency of the supramolecular catalysts is rarely comparable to the natural enzyme even though they are robust and recyclable. Fourth, in the case of metalloenzymes, the coordination energy between the metal ion and functional groups ($\sim$40–80 kJ/mol) is significantly higher than the driving force for the monomer self-assembly. The co-assembly of the metal ions results in competition between the metal bonding and the peptide self-assembly, which in turn affects the formation of well-ordered catalytic structures. Therefore it is important to investigate the metal type, coordination number, peptide sequence, and metal:peptide stoichiometry in order to design effective enzyme mimics.

References

Agami, C., Puchot, C., & Sevestre, H. (1986). Is the mechanism of the proline-catalyzed enantioselective aldol reaction related to biochemical processes? *Tetrahedron Letters*, *27*(13), 1501–1504.

Agarkov, A., Greenfield, S. J., Ohishi, T., Collibee, S. E., & Gilbertson, S. R. (2004). Catalysis with phosphine-containing amino acids in various "turn" motifs. *The Journal of Organic Chemistry*, *69*(23), 8077–8085.

Arora, R., Bharval, P., Sarswati, S., Sen, T. Z., & Yennamalli, R. M. (2019). Structural dynamics of lytic polysaccharide monoxygenases reveals a highly flexible substrate binding region. *Journal of Molecular Graphics and Modelling*, *88*, 1–10.

Artner, L. M., Merkel, L., Bohlke, N., Beceren-Braun, F., Weise, C., Dernedde, J., ... Hackenberger, C. P. (2012). Site-selective modification of proteins for the synthesis of structurally defined multivalent scaffolds. *Chemical Communications*, *48*(4), 522–524.

Barbier, B., & Brack, A. (1992). Conformation-controlled hydrolysis of polyribonucleotides by sequential basic polypeptides. *Journal of the American Chemical Society*, *114*(9), 3511–3515.

Berdugo, C., Miravet, J. F., & Escuder, B. (2013). Substrate selective catalytic molecular hydrogels: The role of the hydrophobic effect. *Chemical Communications*, *49*(90), 10608–10610.

Bhardwaj, G., Mulligan, V. K., Bahl, C. D., Gilmore, J. M., Harvey, P. J., Cheneval, O., ... Eletsky, A. (2016). Accurate de novo design of hyperstable constrained peptides. *Nature*, *538*(7625), 329–335.

Boissonnas, R., Waller, J.-P., & Jaquenoud, P.-A. (1956). Synthesis of a polypeptide with ACTH-like structure. *Experientia*, *12*(11), 446–448.

Boudart, M. (1995). Turnover rates in heterogeneous catalysis. *Chemical Reviews*, *95*(3), 661–666.

Branden, C., & Tooze, J. (1999). *Introduction to protein structure*. Garland Science.

Breit, B., & Laungani, A. C. (2003). Copper-catalyzed enantioselective conjugate addition of dialkylzinc reagents to enones with new peptidyl phosphane ligands. *Tetrahedron: Asymmetry*, *14*(24), 3823–3826.

Broo, K. S., Brive, L., Ahlberg, P., & Baltzer, L. (1997). Catalysis of hydrolysis and transesterification reactions of p-nitrophenyl esters by a designed helix – loop – helix dimer. *Journal of the American Chemical Society*, *119*(47), 11362–11372.

Broo, K. S., Brive, L., Sott, R. S., & Baltzer, L. (1998). Site-selective control of the reactivity of surface-exposed histidine residues in designed four-helix-bundle catalysts. *Folding and Design*, *3*(4), 303–312.

Burton, A. J., Thomson, A. R., Dawson, W. M., Brady, R. L., & Woolfson, D. N. (2016). Installing hydrolytic activity into a completely de novo protein framework. *Nature Chemistry*, *8*(9), 837–844.

Busto, E., Gotor-Fernández, V., & Gotor, V. (2010). Hydrolases: Catalytically promiscuous enzymes for non-conventional reactions in organic synthesis. *Chemical Society Reviews*, *39* (11), 4504–4523.

Campbell, I. M. (1988). *Catalysis at surfaces*. Springer Science & Business Media.

Carter, P., & Wells, J. A. (1988). Dissecting the catalytic triad of a serine protease. *Nature*, *332* (6164), 564–568.

Chapman, J., Ismail, A. E., & Dinu, C. Z. (2018). Industrial applications of enzymes: Recent advances, techniques, and outlooks. *Catalysts*, *8*(6), 238.

Chen, R. (2001). Enzyme engineering: rational redesign vs directed evolution. *Trends in Biotechnology*, *19*(1), 13–14.

Chipman, D. M., & Sharon, N. (1969). Mechanism of lysozyme action. *Science*, *165*(3892), 454–465.

Chou, P. Y., & Fasman, G. D. (1978). Empirical predictions of protein conformation. *Annual Review of Biochemistry*, *47*(1), 251–276.

Ciano, L., Davies, G. J., Tolman, W. B., & Walton, P. H. (2018). Bracing copper for the catalytic oxidation of C–H bonds. *Nature Catalysis*, *1*(8), 571–577.

Cleij, M. C., Drenth, W., & Nolte, R. J. (1991). Mechanism of enantioselective ester cleavage by histidine containing dipeptides at a micellar interface. *The Journal of Organic Chemistry*, *56*(12), 3883–3891.

Corma, A., & Garcia, H. (2008). Crossing the borders between homogeneous and heterogeneous catalysis: developing recoverable and reusable catalytic systems. *Topics in Catalysis*, *48*(1), 8–31.

Craik, C. S., Roczniak, S., Largman, C., & Rutter, W. J. (1987). The catalytic role of the active site aspartic acid in serine proteases. *Science*, *237*(4817), 909–913.

Das, G., Sakai, N., & Matile, S. (2002). Toward catalytic rigid-rod β-barrels: A hexamer with multiple histidines. *Chirality*, *14*(1), 18–24.

Davie, E. A. C., Mennen, S. M., Xu, Y., & Miller, S. J. (2007). Asymmetric catalysis mediated by synthetic peptides. *Chemical Reviews*, *107*(12), 5759–5812.

Davis, H. J., & Ward, T. R. (2019). Artificial metalloenzymes: Challenges and opportunities. *ACS Central Science*, *5*(7), 1120–1136.

Dean, S. M., Greenberg, W. A., & Wong, C. H. (2007). Recent advances in aldolase-catalyzed asymmetric synthesis. *Advanced Synthesis & Catalysis*, *349*(8-9), 1308–1320.

Degrado, S. J., Mizutani, H., & Hoveyda, A. H. (2001). Modular peptide-based phosphine ligands in asymmetric catalysis: efficient and enantioselective Cu-catalyzed conjugate additions to five-, six-, and seven-membered cyclic enones. *Journal of the American Chemical Society*, *123*(4), 755–756.

DeGrado, W. F., Summa, C. M., Pavone, V., Nastri, F., & Lombardi, A. (1999). De novo design and structural characterization of proteins and metalloproteins. *Annual Review of Biochemistry*, *68*(1), 779–819.

Díaz, D. D., Kühbeck, D., & Koopmans, R. J. (2011). Stimuli-responsive gels as reaction vessels and reusable catalysts. *Chemical Society Reviews*, *40*(1), 427–448.

Eder, U., Sauer, G., & Wiechert, R. (1971). New type of asymmetric cyclization to optically active steroid CD partial structures. *Angewandte Chemie International Edition in English*, *10*(7), 496–497.

Fa, M., Radeghieri, A., Henry, A. A., & Romesberg, F. E. (2004). Expanding the substrate repertoire of a DNA polymerase by directed evolution. *Journal of the American Chemical Society*, *126*(6), 1748–1754.

Fleming, S., & Ulijn, R. V. (2014). Design of nanostructures based on aromatic peptide amphiphiles. *Chemical Society Reviews*, *43*(23), 8150–8177.

Font, D., Sayalero, S., Bastero, A., Jimeno, C., & Pericas, M. A. (2008). Toward an artificial aldolase. *Organic Letters*, *10*(2), 337–340.

Gaurav, J., & Santhoshkumar, T. R. (2020). Conformationally constrained peptides for drug delivery. *Peptide Science*, *26*(6).

Gilbertson, S. R., & Pawlick, R. V. (1996). Synthesis of thiophosphoryl derivatives of proline: Building blocks for phosphanyl-substituted peptides with β-turns. *Angewandte Chemie International Edition in English*, *35*(8), 902–904.

Goldberger, M. L., & Watson, K. M. (2004). *Collision theory*. Courier Corporation.

Goyal, R., Jerath, G., Akhil, R., Chandrasekharan, A., Puppala, E. R., Ponneganti, S., … Ramakrishnan, V. (2021). Geometry encoded functional programming of tumor homing peptides for targeted drug delivery. *Journal of Controlled Release*, *333*, 16–27.

Goyal, R., Jerath, G., Chandrasekharan, A., Christian, Y., Kumar, T., & Ramakrishnan, V. (2021). Molecular hybridization combining tumor homing and penetrating peptide domains for cellular targeting. *Drug Delivery and Translational Research*, 1–8.

Goyal, R., Jerath, G., Chandrasekharan, A., Kumar, T. S., & Ramakrishnan, V. (2020). Peptide-based delivery vectors with pre-defined geometrical locks. *RSC Medicinal Chemistry*, *11* (11), 1303–1313.

Green, A. P., Hayashi, T., Mittl, P. R., & Hilvert, D. (2016). A chemically programmed proximal ligand enhances the catalytic properties of a heme enzyme. *Journal of the American Chemical Society*, *138*(35), 11344–11352.

Greenfield, S. J., Agarkov, A., & Gilbertson, S. R. (2003). High asymmetric induction with β-turn-derived palladium phosphine complexes. *Organic Letters*, *5*(17), 3069–3072.

Guler, M. O., & Stupp, S. I. (2007). A self-assembled nanofiber catalyst for ester hydrolysis. *Journal of the American Chemical Society*, *129*(40), 12082–12083.

Gulseren, G., Khalily, M. A., Tekinay, A. B., & Guler, M. O. (2016). Catalytic supramolecular self-assembled peptide nanostructures for ester hydrolysis. *Journal of Materials Chemistry B*, *4*(26), 4605–4611.

Haibach, M. C., Kundu, S., Brookhart, M., & Goldman, A. S. (2012). Alkane metathesis by tandem alkane-dehydrogenation–olefin-metathesis catalysis and related chemistry. *Accounts of Chemical Research*, *45*(6), 947–958.

Hazam, P. K., Akhil, R., Jerath, G., Saikia, J., & Ramakrishnan, V. (2019). Topological effects on the designability and bactericidal potency of antimicrobial peptides. *Biophysical Chemistry*, *248*, 1–8.

Hazam, P. K., Jerath, G., Chaudhary, N., & Ramakrishnan, V. (2018). Peptido-mimetic approach in the design of syndiotactic antimicrobial peptides. *International Journal of Peptide Research and Therapeutics*, *24*(2), 299–307.

Hazam, P. K., Jerath, G., Kumar, A., Chaudhary, N., & Ramakrishnan, V. (2017). Effect of tacticity-derived topological constraints in bactericidal peptides. *Biochimica et Biophysica Acta (BBA) – Biomembranes*, *1859*(8), 1388–1395.

He, X., Zhang, F., Liu, J., Fang, G., & Wang, S. (2017). Homogenous graphene oxide-peptide nanofiber hybrid hydrogel as biomimetic polysaccharide hydrolase. *Nanoscale*, *9*(45), 18066–18074.

He, X., Zhang, F., Zhang, L., Zhang, Q., Fang, G., Liu, J., … Zhang, S. (2017). Probing the structure–activity relationship of a novel artificial cellobiose hydrolase. *Journal of Materials Chemistry B*, *5*(26), 5225–5233.

Hilvert, D. (2013). Design of protein catalysts. *Annual Review of Biochemistry*, *82*, 447–470.

Horn, S. J., Vaaje-Kolstad, G., Westereng, B., & Eijsink, V. (2012). Novel enzymes for the degradation of cellulose. *Biotechnology for Biofuels*, *5*(1), 45.

Huang, Z., Guan, S., Wang, Y., Shi, G., Cao, L., Gao, Y., … Liu, J. (2013). Self-assembly of amphiphilic peptides into bio-functionalized nanotubes: A novel hydrolase model. *Journal of Materials Chemistry B*, *1*(17), 2297–2304.

Jerath, G., Goyal, R., Trivedi, V., Santhoshkumar, T., & Ramakrishnan, V. (2019). Syndiotactic peptides for targeted delivery. *Acta Biomaterialia*, *87*, 130–139.

Jerath, G., Hazam, P. K., & Ramakrishnan, V. (2014). bPE toolkit: Toolkit for computational protein engineering. *Systems and Synthetic Biology*, *8*(4), 337–341.

Jerath, G., Hazam, P. K., Shekhar, S., & Ramakrishnan, V. (2016). Mapping the geometric evolution of protein folding motor. *PloS One*, *11*(10), e0163993.

Johnsson, K., Allemann, R. K., Widmer, H., & Benner, S. A. (1993). Synthesis, structure and activity of artificial, rationally designed catalytic polypeptides. *Nature*, *365*(6446), 530–532.

Josephsohn, N. S., Kuntz, K. W., Snapper, M. L., & Hoveyda, A. H. (2001). Mechanism of enantioselective Ti-catalyzed Strecker reaction: Peptide-based metal complexes as bifunctional catalysts. *Journal of the American Chemical Society*, *123*(47), 11594–11599.

Joubert, J., Delbecq, F., Sautet, P., Le Roux, E., Taoufik, M., Thieuleux, C., … Basset, J.-M. (2006). Molecular understanding of alumina supported single-site catalysts by a combination of experiment and theory. *Journal of the American Chemical Society*, *128*(28), 9157–9169.

Kalidindi, S. B., & Jagirdar, B. R. (2012). Nanocatalysis and prospects of green chemistry. *ChemSusChem*, *5*(1), 65–75.

Kim, M.-C., & Lee, S.-Y. (2015). Peroxidase-like oxidative activity of a manganese-coordinated histidyl bolaamphiphile self-assembly. *Nanoscale*, *7*(40), 17063–17070.

Kjaergaard, C. H., Qayyum, M. F., Wong, S. D., Xu, F., Hemsworth, G. R., Walton, D. J., … Johansen, K. S. (2014). Spectroscopic and computational insight into the activation of O_2 by the mononuclear Cu center in polysaccharide monooxygenases. *Proceedings of the National Academy of Sciences*, *111*(24), 8797–8802.

Korendovych, I. V. (2018). Rational and semirational protein design. *Protein Engineering*, 15–23.

Kuah, E., Toh, S., Yee, J., Ma, Q., & Gao, Z. (2016). Enzyme mimics: Advances and applications. *Chemistry–A European Journal*, *22*(25), 8404–8430.

Kumar, A., Ramakrishnan, V., Ranbhor, R., Patel, K., & Durani, S. (2009). Homochiral stereochemistry: The missing link of structure to energetics in protein folding. *The Journal of Physical Chemistry B*, *113*(51), 16435–16442.

Labádi, I., Benkő, M., Markó, K., & Szilágyi, I. (2009). Mimicking a superoxide dismutase (SOD) enzyme by copper (II) and zinc (II)-complexes. *Reaction Kinetics and Catalysis Letters*, *96*(2), 327–333.

Lee, K. S., & Parquette, J. R. (2015). A self-assembled nanotube for the direct aldol reaction in water. *Chemical Communications*, *51*(86), 15653–15656.

Leveson-Gower, R. B., Mayer, C., & Roelfes, G. (2019). The importance of catalytic promiscuity for enzyme design and evolution. *Nature Reviews Chemistry*, *3*(12), 687–705.

Li, J., Liu, C.-y, & Liu, Y. (2012). Au/graphene hydrogel: Synthesis, characterization and its use for catalytic reduction of 4-nitrophenol. *Journal of Materials Chemistry*, *22*(17), 8426–8430.

Lienhard, G. E. (1973). Enzymatic catalysis and transition-state theory. *Science*, *180*(4082), 149–154.

Lin, X., Ji, G., Liu, Y., Huang, Q., Yang, Z., & Du, Y. (2012). Formation mechanism and magnetic properties of hollow Fe_3O_4 nanospheres synthesized without any surfactant. *CrystEngComm*, *14*(24), 8658–8663.

List, B., Hoang, L., & Martin, H. J. (2004). New mechanistic studies on the proline-catalyzed aldol reaction. *Proceedings of the National Academy of Sciences*, *101*(16), 5839–5842.

Liu, M., Zhang, L., & Wang, T. (2015). Supramolecular chirality in self-assembled systems. *Chemical Reviews*, *115*(15), 7304–7397.

Luong, T. Q., Erwin, N., Neumann, M., Schmidt, A., Loos, C., Schmidt, V., ... Winter, R. (2016). Hydrostatic pressure increases the catalytic activity of amyloid fibril enzymes. *Angewandte Chemie International Edition*, *55*(40), 12412–12416.

Machajewski, T. D., & Wong, C. H. (2000). The catalytic asymmetric aldol reaction. *Angewandte Chemie International Edition*, *39*(8), 1352–1375.

Mack, D. P., Iverson, B. L., & Dervan, P. B. (1988). Design and chemical synthesis of a sequence-specific DNA-cleaving protein. *Journal of the American Chemical Society*, *110*(22), 7572–7574.

Makam, P., & Gazit, E. (2018). Minimalistic peptide supramolecular co-assembly: expanding the conformational space for nanotechnology. *Chemical Society Reviews*, *47*(10), 3406–3420.

Makhlynets, O. V., Gosavi, P. M., & Korendovych, I. V. (2016). Short self-assembling peptides are able to bind to copper and activate oxygen. *Angewandte Chemie International Edition*, *55*(31), 9017–9020.

Makinen, M. W., Kuo, L. C., Dymowski, J. J., & Jaffer, S. (1979). Catalytic role of the metal ion of carboxypeptidase A in ester hydrolysis. *Journal of Biological Chemistry*, *254*(2), 356–366.

McMorn, P., & Hutchings, G. J. (2004). Heterogeneous enantioselective catalysts: strategies for the immobilisation of homogeneous catalysts. *Chemical Society Reviews*, *33*(2), 108–122.

Mehra, A., Jerath, G., Ramakrishnan, V., & Trivedi, V. (2015). Characterization of ICAM-1 biophore to design cytoadherence blocking peptides. *Journal of Molecular Graphics and Modelling*, *57*, 27–35.

Merrifield, R., & Woolley, D. (1956). The synthesis of L-seryl-L-histidyl-L-leucyl-L-valyl-L-glutamic acid, a peptide with strepogenin activity. *Journal of the American Chemical Society*, *78*(18), 4646–4649.

Merrifield, R., & Woolley, D. (1958). The synthesis of peptides with strepogenin activity1. *Journal of the American Chemical Society*, *80*(24), 6635–6639.

Metri, R., Jerath, G., Kailas, G., Gacche, N., Pal, A., & Ramakrishnan, V. (2014). Structure-based barcoding of proteins. *Protein Science*, *23*(1), 117–120.

Mlynarski, J., & Paradowska, J. (2008). Catalytic asymmetric aldol reactions in aqueous media. *Chemical Society Reviews*, *37*(8), 1502–1511.

Moracci, M., Capalbo, L., Ciaramella, M., & Rossi, M. (1996). Identification of two glutamic acid residues essential for catalysis in the β-glycosidase from the thermoacidophilic archaeon *Sulfolobus solfataricus*. *Protein Engineering, Design and Selection*, *9*(12), 1191–1195.

Mori, A., Abet, H., & Inoue, S. (1995). Amino-acids, peptides and their derivatives: Powerful chiral ligands for metal-catalyzed asymmetric syntheses. *Applied Organometallic Chemistry*, *9*(3), 189–197.

Muhler, M. (1997). Berichte der Bunsengesellschaft für physikalische Chemie. In J.M. Thomas & W.J. Thomas (Eds.), *Principles and practice of heterogeneous catalysis*, VCH, Weinheim, 1997, ISBN 3-527-29239-X, Preis: 88, DM: Wiley Online Library.

Nannemann, D. P., Birmingham, W. R., Scism, R. A., & Bachmann, B. O. (2011). Assessing directed evolution methods for the generation of biosynthetic enzymes with potential in drug biosynthesis. *Future Medicinal Chemistry*, *3*(7), 809–819.

Narayanan, R., & El-Sayed, M. A. (2004). Shape-dependent catalytic activity of platinum nanoparticles in colloidal solution. *Nano Letters*, *4*(7), 1343–1348.

Omosun, T. O., Hsieh, M.-C., Childers, W. S., Das, D., Mehta, A. K., Anthony, N. R., . . . Lynn, D. G. (2017). Catalytic diversity in self-propagating peptide assemblies. *Nature Chemistry*, *9*(8), 805–809.

Pan, Y., Shen, X., Yao, L., Bentalib, A., & Peng, Z. (2018). Active sites in heterogeneous catalytic reaction on metal and metal oxide: Theory and practice. *Catalysts*, *8*(10), 478.

Peracchi, A. (2001). Enzyme catalysis: removing chemically 'essential' residues by site-directed mutagenesis. *Trends in Biochemical Sciences*, *26*(8), 497–503.

Prakash, V., Ranbhor, R., & Ramakrishnan, V. (2020). De novo designed heterochiral blue fluorescent protein. *ACS Omega*, *5*(41), 26382–26388.

Ramakrishnan, V., Ranbhor, R., Kumar, A., & Durani, S. (2006). The link between sequence and conformation in protein structures appears to be stereochemically established. *The Journal of Physical Chemistry B*, *110*(18), 9314–9323.

Ranbhor, R., Kumar, A., Tendulkar, A., Patel, K., Ramakrishnan, V., & Durani, S. (2018). IDeAS: Automated design tool for hetero-chiral protein folds. *Physical Biology*, *15*(6), 066005.

Ranbhor, R., Ramakrishnan, V., Kumar, A., & Durani, S. (2006). The interplay of sequence and stereochemistry in defining conformation in proteins and polypeptides. *Biopolymers: Original Research on Biomolecules*, *83*(5), 537–545.

Rodríguez-Llansola, F., Escuder, B., Hamley, I. W., Hayes, W., & Miravet, J. F. (2012). Structural and morphological studies of the dipeptide based L-Pro-L-Val organocatalytic gels and their rheological behaviour. *Soft Matter*, *8*(34), 8865–8872.

Rodriguez-Llansola, F., Escuder, B., & Miravet, J. F. (2009). Switchable perfomance of an L-proline-derived basic catalyst controlled by supramolecular gelation. *Journal of the American Chemical Society*, *131*(32), 11478–11484.

Rodríguez-Llansola, F., Escuder, B., & Miravet, J. F. (2009). Remarkable increase in basicity associated with supramolecular gelation. *Organic & Biomolecular Chemistry*, *7*(15), 3091–3094.

Rodríguez-Llansola, F., Miravet, J. F., & Escuder, B. (2009). A supramolecular hydrogel as a reusable heterogeneous catalyst for the direct aldol reaction. *Chemical Communications*, *47*, 7303–7305.

Rodríguez-Llansola, F., Miravet, J. F., & Escuder, B. (2010). Supramolecular catalysis with extended aggregates and gels: Inversion of stereoselectivity caused by self-assembly. *Chemistry–A European Journal*, *16*(28), 8480–8486.

Rufo, C. M., Moroz, Y. S., Moroz, O. V., Stöhr, J., Smith, T. A., Hu, X., . . . Korendovych, I. V. (2014). Short peptides self-assemble to produce catalytic amyloids. *Nature Chemistry*, *6*(4), 303–309.

Saikia, J., Dharmalingam, K., Anandalakshmi, R., Redkar, A. S., Bhat, V., & Ramakrishnan, V. (2021). Electric field modulated peptide based hydrogel nanocatalyst. *Soft Matter*, *17*(42), 9725–9735.

Sambasivan, R., & Ball, Z. T. (2010). Metallopeptides for asymmetric dirhodium catalysis. *Journal of the American Chemical Society*, *132*(27), 9289–9291.

Sambasivan, R., & Ball, Z. T. (2012). Screening rhodium metallopeptide libraries "on bead": asymmetric cyclopropanation and a solution to the enantiomer problem. *Angewandte Chemie International Edition, 51*(34), 8568–8572.

Sambasivan, R., & Ball, Z. T. (2013). Studies of asymmetric styrene cyclopropanation with a rhodium (II) metallopeptide catalyst developed with a high-throughput screen. *Chirality, 25* (9), 493–497.

Sasaki, T., & Kaiser, E. T. (1989). Helichrome: Synthesis and enzymic activity of a designed hemeprotein. *Journal of the American Chemical Society, 111*(1), 380–381.

Schowen, R. (1978). *Transition states of biochemical processes*. New York: Plenum Press, by RD Gandour and RL Schowen.

Schüth, F., Ward, M. D., & Buriak, J. M. (2018). Common pitfalls of catalysis manuscripts submitted to Chemistry of Materials. ACS Publications.

Schwizer, F., Okamoto, Y., Heinisch, T., Gu, Y., Pellizzoni, M. M., Lebrun, V., ... Ward, T. R. (2018). Artificial metalloenzymes: Reaction scope and optimization strategies. *Chemical Reviews, 118*(1), 142–231.

Shao, Z., & Arnold, F. H. (1996). Engineering new functions and altering existing functions. *Current Opinion in Structural Biology, 6*(4), 513–518.

Shaw, D. E., Maragakis, P., Lindorff-Larsen, K., Piana, S., Dror, R. O., Eastwood, M. P., ... Shan, Y. (2010). Atomic-level characterization of the structural dynamics of proteins. *Science, 330*(6002), 341–346.

Shaw, W. J. (2012). The outer-coordination sphere: incorporating amino acids and peptides as ligands for homogeneous catalysts to mimic enzyme function. *Catalysis Reviews, 54*(4), 489–550.

Singh, N., Conte, M. P., Ulijn, R., Miravet, J. F., & Escuder, B. (2015). Insight into the esterase like activity demonstrated by an imidazole appended self-assembling hydrogelator. *Chemical Communications, 51*(67), 13213–13216.

Singh, N., & Escuder, B. (2017). Competition vs cooperation in catalytic hydrogelators for anti-selective Mannich reaction. *Chemistry–A European Journal, 23*(41), 9946–9951.

Singh, N., Tena-Solsona, M., Miravet, J. F., & Escuder, B. (2015). Towards supramolecular catalysis with small self-assembled peptides. *Israel Journal of Chemistry, 55*(6-7), 711–723.

Singh, S. B., & Tandon, P. K. (2014). Catalysis: A brief review on nano-catalyst. *Journal of Energy and Chemical Engineering, 2*(3), 106–115.

Sluka, J. P., Horvath, S. J., Bruist, M. F., Simon, M. I., & Dervan, P. B. (1987). Synthesis of a sequence-specific DNA-cleaving peptide. *Science, 238*(4830), 1129–1132.

Soares, B., Aguilar, A., Silva, E., Coutinho-Neto, M., Hamley, I., Reza, M., ... Alves, W. (2017). Chiral organocatalysts based on lipopeptide micelles for aldol reactions in water. *Physical Chemistry Chemical Physics, 19*(2), 1181–1189.

Somorjai, G. A., Frei, H., & Park, J. Y. (2009). Advancing the frontiers in nanocatalysis, biointerfaces, and renewable energy conversion by innovations of surface techniques. *Journal of the American Chemical Society, 131*(46), 16589–16605.

Somwanshi, S. B., Somvanshi, S. B., & Kharat, P. B. (2020). Nanocatalyst: A brief review on synthesis to applications. In *Journal of Physics: Conference Series*. IOP Publishing Ltd.

Span, E. A., & Marletta, M. A. (2015). The framework of polysaccharide monooxygenase structure and chemistry. *Current Opinion in Structural Biology, 35*, 93–99.

Tiltscher, H., Wolf, H., & Schelchshorn, J. (1981). A mild and effective method for the reactivation or maintenance of the activity of heterogeneous catalysts. *Angewandte Chemie International Edition in English, 20*(10), 892–894.

Umpierre, A. P., de Jesús, E., & Dupont, J. (2011). Turnover numbers and soluble metal nanoparticles. *ChemCatChem, 3*(9), 1413–1418.

Vaaje-Kolstad, G., Forsberg, Z., Loose, J. S., Bissaro, B., & Eijsink, V. G. (2017). Structural diversity of lytic polysaccharide monooxygenases. *Current Opinion in Structural Biology*, *44*, 67–76.

Vallee, B. L., & Williams, R. (1968). Metalloenzymes: The entatic nature of their active sites. *Proceedings of the National Academy of Sciences of the United States of America*, *59*(2), 498.

Walton, P. H., & Davies, G. J. (2016). On the catalytic mechanisms of lytic polysaccharide monooxygenases. *Current Opinion in Chemical Biology*, *31*, 195–207.

Wang, M., Lv, Y., Liu, X., Qi, W., Su, R., & He, Z. (2016). Enhancing the activity of peptide-based artificial hydrolase with catalytic Ser/His/Asp triad and molecular imprinting. *ACS Applied Materials & Interfaces*, *8*(22), 14133–14141.

Wang, M., & Qi, W. (2020). Assembled peptides for biomimetic catalysis. *Artificial Protein and Peptide Nanofibers: Design, Fabrication, Characterization, and Applications*, 383.

Wong, Y.-M., Masunaga, H., Chuah, J.-A., Sudesh, K., & Numata, K. (2016). Enzyme-mimic peptide assembly to achieve amidolytic activity. *Biomacromolecules*, *17*(10), 3375–3385.

Xiang, S.-H., & Tan, B. (2020). Advances in asymmetric organocatalysis over the last 10 years. *Nature Communications*, *11*(1), 1–5.

Yamada, Y. M., Yoshikawa, N., Sasai, H., & Shibasaki, M. (1997). Direct catalytic asymmetric aldol reactions of aldehydes with unmodified ketones. *Angewandte Chemie International Edition in English*, *36*(17), 1871–1873.

Zaramella, D., Scrimin, P., & Prins, L. J. (2012). Self-assembly of a catalytic multivalent peptide–nanoparticle complex. *Journal of the American Chemical Society*, *134*(20), 8396–8399.

Zastrow, M. L., Peacock, A. F., Stuckey, J. A., & Pecoraro, V. L. (2012). Hydrolytic catalysis and structural stabilization in a designed metalloprotein. *Nature Chemistry*, *4*(2), 118–123.

Zhang, C., Xue, X., Luo, Q., Li, Y., Yang, K., Zhuang, X., … Zou, G. (2014). Self-assembled peptide nanofibers designed as biological enzymes for catalyzing ester hydrolysis. *Acs Nano*, *8*(11), 11715–11723.

Zhang, H., Jin, M., Xiong, Y., Lim, B., & Xia, Y. (2013). Shape-controlled synthesis of Pd nanocrystals and their catalytic applications. *Accounts of Chemical Research*, *46*(8), 1783–1794.

Zhu, M., Wang, M., Qi, W., Su, R., & He, Z. (2019). Constructing peptide-based artificial hydrolases with customized selectivity. *Journal of Materials Chemistry B*, *7*(24), 3804–3810.

Zozulia, O., Dolan, M., & Korendovych, I. (2018). Catalytic peptide assemblies. *Chemical Society Reviews*, *47*(10), 3621–3639.

Chapter 8

Bioinspired functional molecular constructs

Vivek Prakash and Vibin Ramakrishnan
Department of Biosciences and Bioengineering, Indian Institute of Technology Guwahati, Guwahati, Assam, India

8.1 Introduction

Supramolecular chemistry is an interdisciplinary field referring to the physical, chemical, and biological aspects of molecular assemblies having more complexity than their respective individual molecules (Jayawickramarajah & Wilson, 2017). According to Nobel laureate Lehn, supramolecular chemistry refers to the domain of chemistry beyond the realm of the normal molecules (Lehn, 1993). Supramolecular chemistry of biomolecules and bioinspired molecules has attracted tremendous interest due to their acknowledged importance in the construction of novel functional materials and in revealing the mechanisms of birth and evolution of natural living organisms. Bioinspired materials are synthetic materials whose structure, properties, or function mimic those of natural materials or living matter (McConney, Anderson, Brott, Naik, & Tsukruk, 2009). Molecular self-assembly is a spontaneous organization of molecules into a stable and well-defined structure under equilibrium conditions through noncovalent interactions (Bromley, Channon, Moutevelis, & Woolfson, 2008). It is a powerful tool in the synthesis of functional nanostructures as a bottom-up fabrication approach (Whitesides & Grzybowski, 2002). Top-down approaches are concerned with the protein engineering and modifications of already existing proteins, whereas the bottom-up approach allows for complete design and control over size, shape, folding, and assembly of the targets (Boyle & Woolfson, 2012).

Self-assembly is a spontaneous process of organizing molecules into ordered structures resulting from intramolecular and intermolecular interactions. The self-assembly process is controlled by the balance of attractive or repulsive forces within and between molecules. In most cases, a thermodynamically stable structure is formed through enthalpic and entropic interactions that involve the basic assembling units and the reacting solvent

De Novo Peptide Design. DOI: https://doi.org/10.1016/B978-0-323-99917-5.00003-2

molecules (Stephanopoulos, Ortony, & Stupp, 2013). Electrostatic interactions, hydrophobic interactions, hydrogen bonding, and $\pi - \pi$ stacking, together contribute to the overall stability of the assembly (Stupp, 2010).

The self-assembly process results in the generation of many biological molecules, such as DNA double helix, protein's secondary and tertiary structures, and cell membrane formation after phospholipid assembly (Yadav, Sharma, & Kumar, 2020). Peptides have drawn significant attention for self-assembly due to their simple structure, relative chemical and physical stability, diversity in sequences and shapes, and the feasibility to synthesize them in large amounts. In addition, peptides are known as useful building blocks for creating self-assembled nanomaterials due to their intrinsic biocompatibility and biodegradability.

This chapter focuses on the de novo design of peptide-based systems that are directed toward the self-assembly of biomaterials (specifically fibrous biomaterials and hydrogels) and discrete nanoscale objects.

8.2 Peptide-based functional materials

A good peptide design strategy starts with identifying potential bioactive candidates, followed by optimizing the sequence to improve the molecules' stability and reactivity. Several points should be considered during peptide design such as peptide length, solubility, cross-linking, and stability. The native peptide sequences can be modified or conjugated with other molecules to increase stability and activity. Random or site-specific modifications can also be implemented to increase the specificity of peptides toward the target (Trasatti, Woo, Ladiwala, Cramer, & Karande, 2018). These strategies, often based on the deep learning algorithms (Tran, Zhang, Xin, Shan, & Li, 2017), intensify the hope to generate new *de novo* designed peptides with targeted applications.

Diversification of chain stereochemistry offers tremendous increase in the peptide or protein design space (Kumar & Ramakrishnan, 2010; Ramakrishnan, Ranbhor, & Durani, 2005). Tacticity refers to the arrangement of successive chiral centers in a polymer (Ramakrishnan, Ranbhor, Kumar, & Durani, 2006). Peptides and proteins are polymers of amino acids with an immutable main chain and variable side chain. The C-alpha carbon in the amino acid is the stereo-center in the peptide main chain. The allowed region in the Ramachandran map for L-amino acid is restricted owing to the stearic hindrance of different side chains. This restriction limits the protein folding space accessible on the Ramachandran map to design the new peptide sequences. The incorporation of D-amino acids in the peptide sequence helps in accessing the "forbidden" region of the Ramachandran map for new peptide sequences composed of L and D-amino acids (Fig. 8.1). An isotactic polypeptide sequence is characterized by amino acids having L or D-stereochemistry, whereas a syndiotactic sequence has a stereo-regular arrangement with alternate L and D-amino acids (or vice versa) in succession, and a heterotactic polymer is a random distribution of L and D-amino acids in a sequence. Syndiotacticity in polypeptide was reported in Gramicidin, an antibacterial peptide (Urry, Goodall, Glickson, & Mayers, 1971), which leads to the formation of $\Pi_{(L,D)}$ helical structure. The formed helix is thermodynamically more stable than the isotactic alpha helix with (i, to i + 5) hydrogen bonding pattern. Therefore the syndiotactic backbone in Gramicidin leads to the formation of stable $\Pi_{(L,D)}$ helices. This design template, in theory, provides the template needed to design peptides with tailored properties such as charge distribution and structural amphipathicity. Heterochiral molecular designs, such as self-assembling organic nanotubes, make use of short cyclic peptides with alternating L and D-amino acid (Ghadiri, Granja, Milligan, McRee, & Khazanovich, 1993). Using stereochemically constrained unnatural amino acids, it is possible to stabilize secondary structures or design new conformations (Kumar, Ranbhor, Patel,

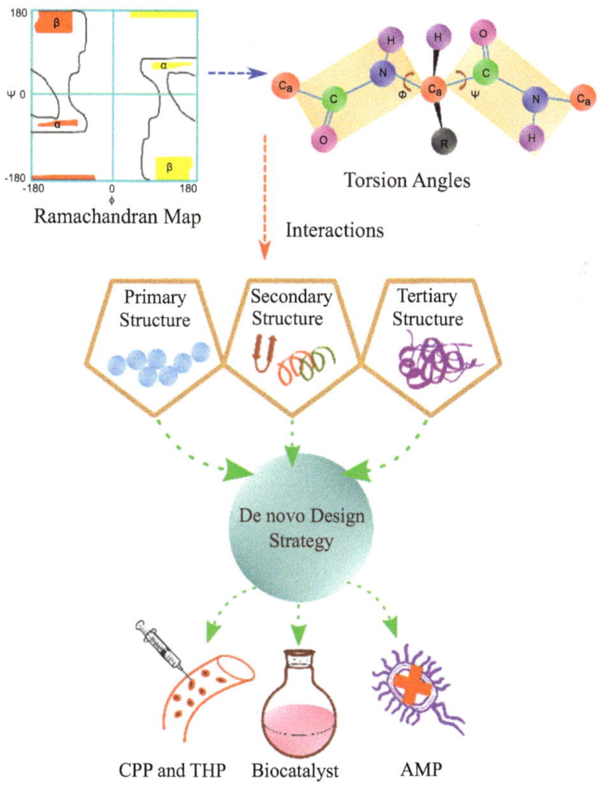

FIGURE 8.1 Schematic representation of Ramachandran map and torsion angles leading to the formation of structural organization of peptides.

Ramakrishnan, & Durani, 2017; Ranbhor, Kumar, Patel, Ramakrishnan, & Durani, 2018; Ranbhor, Kumar, Tendulkar, et al., 2018).

8.2.1 *De novo* designed fluorescent peptide

Synthetic peptides are popular choices as targeting molecules for bio-imaging or diagnostic applications. Unfortunately, broad utilization of linear, naturally occurring peptides is severely limited by their (1) inherent flexibility, which often lowers targeting capability, (2) charged N- and C-terminal groups, which inhibit cell permeation, and (3) sensitivity to proteolytic degradation (Zhai, Schreiber, Padilla-Coley, Oliver, & Smith, 2020). Efforts to obviate these drawbacks have led to a wide range of molecular design strategies in peptidomimetics with improved properties (Walport, Obexer, & Suga, 2017). One approach is to design structures that incorporate amino acids with unnatural side chains or opposite chirality, or alternatively use an unnatural polymer backbone (Nair, Baravkar, Ingole, & Sanjayan, 2014). By doing so, different strategies have been applied to design the peptide- based fluorophore for various applications such as cancer therapeutics (Fan et al., 2018), chemical detections (Wang et al., 2020), and heavy metal detections (Siepi et al., 2018). A key decision in the development of a new fluorescent peptide probe is determining the optimal sensing mechanism. It has to be considered that of the 20 natural amino acids, only Tyrosine (Tyr) and Tryptophan (Trp) display significant emission, and among them, only the Trp fluorescence is sensitive to its immediate environment (Masters, 2013). For this reason, the introduction of extrinsic fluorophores to the peptide probe is usually necessary, in order to obtain a good fluorescence. Moreover, Trp suffers from strong photobleaching, and its excitation and emission are masked by the strong background typical of biological media. Therefore applications that require measuring fluorescence in complex mixtures will demand the introduction of extrinsic fluorophores with long-wavelength excitation and emission.

Except for peptide-functionalized fluorescent platforms, selected peptides could also be self-assembled into fluorescent nanoarchitectures without introducing organic dyes. Inspired by the fluorescence mechanism in natural fluorescent proteins, Zhang and co-workers have reported Trp-Phe dipeptide nanoparticles (DNPs) that shifted the intrinsic peptide fluorescence from ultraviolet to the visible range (Fan, Sun, Huang, Wang, & Zhang, 2016). Similar to natural florescent proteins, DNPs were prepared through self-assembly with Zn^{2+}, which shift peptide's intrinsic fluorescent signal from the ultraviolet to the visible range.

Narrow emission bandwidth in visible range and remarkable photostability of DNPs were observed, which could be an effective agent for fluorescent biomedical imaging. Subsequently, Li and co-workers have reported the quantum confined self-assembly of aromatic cyclic dipeptides cyclo-Phe-Trp

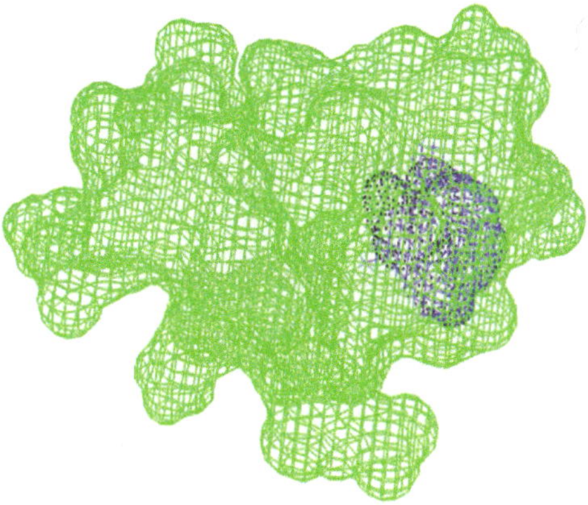

FIGURE 8.2 Surface mesh model of BP02 mini-protein. *Blue color* represents β-(1-azulenyl)-L-alanine.

and cyclo-Trp-Trp (Tao et al., 2018). The cyclic dipeptides were designed as quantum dots (QDs) to self-assemble into quantum-confined nanostructures, which dimerized to form nanostrucutres. The emission of QDs can be tuned from the visible region to the near-infrared (NIR) region (420−820 nm) by modulating self-assembly. Ramakrishnan and co-workers have designed a heterochiral de novo minimal fluorescent protein that can be selectively excited at 342 nm (Prakash, Ranbhor, & Ramakrishnan, 2020). β-(1-azulenyl)-L-alanine, an unnatural amino acid, was impregnated into the hydrophobic core of a heterotactic protein scaffold, which gives emission in visible region (Fig. 8.2).

8.2.2 De novo designed antimicrobial peptides

The emergence of drug-resistant bacteria has restricted the use of many conventional antibiotics, leading to the new research to develop antibiotics, specifically antimicrobial peptides (AMPs), which generally targets the bacterial cell membranes, possibly decreasing the risk of resistance (Bishop & Finlay, 2006). AMPs are typically 20−50 residues and generally tend to adopt highly amphiphilic structure. The electrostatic interactions between the positively charged AMPs and the negatively charged bacterial phospholipids provide an initial mode of interaction, whereas hydrophobic interactions allow the peptides to penetrate the cell membrane, in some cases leading to depolarization of the bacterial membrane and cell death (Fig. 8.3) (Boman, 1995). Although natural peptides are excellent candidates as antimicrobial agents, they have limitations as well, such as reduced bioavailability, protease-mediated degradation, and instability in pH environment (Hazam, Singh, Chaudhary, & Ramakrishnan, 2019). To counter these limitations, tailor-made approach to design the AMPs has increased in recent decades

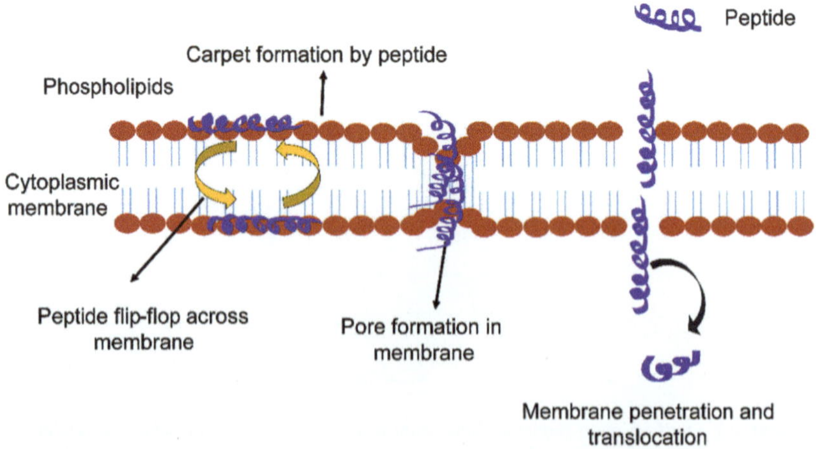

FIGURE 8.3 Schematic representation of mechanism of antimicrobial peptides.

(Hazam, Goyal, & Ramakrishnan, 2019). Modification of N and C-terminus by acetylation and amidation has reported to increase the peptide stability (Arnesen, 2011). Altering peptide chirality by insertion of D-amino acids has also resisted degradation of peptides with increased half-life. Many studies have reported the de novo design of AMPs, mimicking natural AMPs (Hazam, Akhil, Jerath, Saikia, & Ramakrishnan, 2019). Hodges and co-workers have utilized the structural framework of an amphipathic α-helical AMP (V_{681}) with the mutation at 11th and 13th position by L and D-amino acid (Chen et al., 2005). They have observed the improved therapeutic activity against Gram-negative and Gram-positive bacteria. Ghadiri and co-workers have designed a series of cyclic D,L-alpha peptides as antibacterial agents, which are resistant to proteolytic degradation (Fernandez-Lopez et al., 2001). Ramakrishnan and co-workers have designed a syndiotactic peptide based on "peptidomimetic" approach that mimics stereochemistry of Gramicidin, a naturally occurring AMP (Hazam, Jerath, Chaudhary, & Ramakrishnan, 2018; Hazam, Jerath, Kumar, Chaudhary, & Ramakrishnan, 2017). They have also examined the acetylated peptide as a bactericidal agent (Hazam et al., 2018). The redesigned peptide sequence retained its stereochemistry and also showed less toxicity as compared to Gramicidin, while maintaining good antimicrobial activity. Metzler-Nolte and co-workers have reported a 4–eightfold decrease in MIC values by adopting a specific combination of L and D-amino acids (Albada, Prochnow, Bobersky, Bandow, & Metzler-Nolte, 2014). Ramakrishnan and co-workers have synthesized the syndiotactic peptide, which showed antimicrobial activity against antibiotic resistant strains such as Gentamicin-resistant MRSA (methicillin-resistant *Staphylococcus aureus*), Escherichia *coli*, and Mycobacterium (Hazam, Phukan, Akhil, Singh, & Ramakrishnan, 2021). The synthesized peptides also retained their activity in serum, which was the principal drawback of poly L-peptide sequences due to enzyme-mediated degradation.

Cationic antimicrobial peptides (CAMPs), a class of AMPs, possess a highly amphiphilic conformation in which cationic hydrophilic and hydrophobic residues are distributed throughout the folded conformation (Wimley & Hristova, 2011). This amphiphilic topology helps to disrupt the bacterial membrane integrity, leading to cell content leakage or the infiltration of extracellular materials and eventually cell death (Hancock & Sahl, 2006). However, limitations of CAMPs are poor oral bioavailability, low stability, side effects, and a high production cost (Sgolastra, Deronde, Sarapas, Som, & Tew, 2013). To address these shortcomings, small-molecule-based peptidomimetics have been designed as antimicrobials to mimic CAMPs (Lin et al., 2020). Maffia and co-workers have synthesized CAMPs, which have antibacterial activity against multiresistant clinical isolates, and have different alpha helical content and different amphipathicity (Faccone et al., 2014).

8.2.3 *De novo* designed cell-penetrating and tumor homing peptide

The most crucial factor in successful drug development is insuring that the drug molecule interacts with the target efficiently. For intracellular drug targets, attaining considerable intracellular delivery is crucial. Several drug delivery strategies have been proposed to increase the bioavailability of the drugs, such as electroporation, liposomal formulation, and micro injection. However, each of these has its limitations, mainly in terms of toxicity and therapeutic feasibility.

Peptide molecules can be an alternative delivery strategy that can translocate the cytoplasmic membrane (Langel, 2002). This strategy emerged in 1994, after discovery of third helix of Antennapedia homeodomain, pAntp (43–58), which can translocate the cell membrane (Derossi, Joliot, Chassaing, & Prochiantz, 1994). Since then many cell-penetrating peptides (CPPs) have been observed. The most common feature among CPPs is the presence of positive charges, and hydrophobicity (Mitchell, Steinman, Kim, Fathman, & Rothbard, 2000).

Majority of CPPs reported till date have been identified or developed from natural protein sources of different origin and mainly have L-amino acids. The uptake of poly L-CPPs is either through the direct plasma membrane translocation or receptor-based mechanism of endocytosis (He, Sayers, Watson, & Jones, 2018). The recognition of such CPPs by receptors not only helps in their uptake but also results in their entrapment in endosome and subsequent degradation in lysosomes (Ruseska & Zimmer, 2020). Another problem in peptide-based drug delivery vectors is their susceptibility to proteolytic cleavage (Patel et al., 2019). The incorporation of D-amino acids in CPP sequences increased peptide stability in biological fluids, but often have lower cellular uptake than poly L-amino acid CPP sequences (Pujals, Fernández-Carneado, Ludevid, & Giralt, 2008) However, the use of D-proline in CPP sequences provides the much needed proteolytic stability. Ramakrishnan and co-workers have designed a series of 12-mer peptides to address the instability issues of the CPPs, by reverse engineering the syndiotactic backbone (Jerath, Goyal, Trivedi, Santhoshkumar, & Ramakrishnan, 2019). The synthesized peptides were stable in both human plasma and bovine serum. Moreover, the designed peptides have higher rates of cellular uptake than the model Tat (48–60) peptide in breast and cervical cancer cells.

Poor penetration of antitumor drugs in the extravascular tumor tissue is often a major factor limiting the efficacy of cancer treatments. Tumor homing peptides (THPs) are the molecules engineered for the targeted drug delivery in cancerous cells, and can be used to deliver drugs into tumors (Goyal, Jerath, Akhil, et al., 2021). Receptor-mediated interactions of THPs favor high specificity, and low cross-reactivity in recognition of cancer cells due to the presence of frequently occurring RGD (Arginine–Glycine–Aspartic acid) and NGR (Asparagine–Glycine–Arginine) peptide motifs. RGD is the most studied THP motif that targets integrins, overexpressed on tumor blood vessels (Goyal, Jerath, Akhil, et al., 2021).

Kessler and co-workers introduced the D-amino acids in the RGD motif and synthesized the cyclic RGD peptide (Mas-Moruno, Rechenmacher, & Kessler, 2010). The peptide cyclo(RGDfV), the most studied version of cyclic RGD, was commercialized for research use under the name Cilengitide, which is proved to be effective in reducing glioblastoma in mice (Stupp et al., 2010). Saviano and co-workers have rationally designed an RGD-containing peptide that binds to $\alpha_v\beta_3$ integrin, which is highly expressed on activated endothelial and tumor cells but is not present in resting endothelial cells. They named this new peptide RGDechi (Del Gatto et al., 2006). Salvaotore and co-workers conjugated the RGDechi with positron emission tomography (PET) imaging probe to test whether RGDechi can discriminate $\alpha_v\beta_3$ from $\alpha_v\beta_5$ integrin, thus allowing the *in vivo* selective visualization of $\alpha_v\beta_3$ expression by PET imaging (Zannetti et al., 2009). They concluded that it selectively binds to $\alpha_v\beta_3$ integrin and does not cross-react with $\alpha_v\beta_5$ integrin.

A tumor penetrating variant of RGD motif was identified in 2009 (denoted as iRGD) on prostate cancer cells (Teesalu, Sugahara, Kotamraju, & Ruoslahti, 2009). Different research groups over the last 10 years have shown that iRGD enhances drug accumulation and efficacy in a number of tumor models (Akashi et al., 2014). Ruoslahti and co-workers have described a strategy to enhance tumor penetration of chemotherapeutic drugs, by the use of iRGD peptide (CRGDK/RGPDC) (Sugahara et al., 2009). The peptide comprises two sequence motifs: RGD, which binds to $\alpha_v\beta_{3/5}$ integrins on tumor endothelial and tumor cells, and a cryptic CendR motif (R/KXXR/K-OH).

Once integrin binding has brought iRGD to the tumor, the peptide is proteolytically cleaved to expose the cryptic CendR motif. The truncated peptide loses affinity for its primary receptor and binds to neuropilin-1, activating a tissue penetration pathway that delivers the peptide along with attached or coadministered payload into the tumor mass. Similar to RGD, another tripeptide NGR is associated with tumor vasculature homing (Svensen, Walton, & Bradley, 2012). An NGR containing phage clone was first isolated by Koivunen and Ruoslahti during in vitro biopanning on $\alpha_5\beta_1$ integrin (Koivunen, Gay, & Ruoslahti, 1993).

The NGR motif binds to integrins with low affinity, but its receptor is aminopeptidase N (APN/CD13), a protein that is detectable in tumor vasculature and in angiogenic blood vessels (Pasqualini et al., 2000). In THPs, RGD and NGR motifs are known to impart selective targeting of tumor vasculatures by recognition of their cognate receptors. Methylated NGR motif can be used as a cargo for radiolabelled compound delivery (Corti et al., 2017).

RGD derivatives have been used for targeting tumor-associated vasculature (Goyal, Jerath, Chandrasekharan, et al., 2021). The basic principles of conformational biasing through cyclization, turn-inducers, and rigid cores have been utilized to explore RGD mimetics (Goyal, Jerath, Chandrasekharan, Kumar, & Ramakrishnan, 2020). Baiula and co-workers have used the β-turn mimetic approach by introducing D-proline in RGD tripeptide (Gentilucci et al., 2007). Ramakrishnan and co-workers have designed a library of RGD and NGR/QGR (Glutamine−Glycine−Arginine) mimics, exploring the concept of informed walk across the Ramachandran map (Goyal, Jerath, Akhil, et al., 2021). They have designed several 11-mer peptides by "locking" glycine residue conformation to specific basins in the D-region of Ramachandran map, with the desired unnatural amino acid substitutions: D-Pro, D-Ala, Aib, and D-Val, to look into their corresponding effect in homing tumors with better specificity and across cancer cell types with better selectivity. The synthesized peptides retained their functional activity in serum, indicating their stability against serum proteases. Further, they have conjugated the peptides with methotrexate, and reported that the designed peptides mediate drug delivery in cancer cells.

8.2.4 *De novo* designed biocatalysts

The efficiency and selectivity of enzymes have encouraged the researchers to find out the approaches to design the catalysts for the reactions not found in nature, or to come up with the new approach for catalyzing the reactions already known, with better performance. The computational approach is prominently successful in this effort owing to the high complexity of protein structures and our limited understanding of the protein structure − function relationships (Damborsky & Brezovsky, 2009). Computational *de novo* design relies on placing the designed catalytic active sites into the existing scaffolds. The idea behind this is that enzymes enhance chemical reactions by lowering of activation energy, by designed active site (Garcia-Viloca, Gao, Karplus, & Truhlar, 2004). The *de novo* design of enzymes has been supported by the computational tools such as DEZYMER, ORBIT, and ROSETTA.

The main drawback of many enzymes is their size as they are either large polypeptide or fragments of proteins. As even short peptides can catalyze many chemical reactions (Kolundzic, Noshi, Tjandra, Movassaghi, & Miller, 2011), it seems logical that their subsequent improvement can yield incredibly efficient enzymes. The supramolecular self-assembly of peptides yields an array of nanostructures such as amyloid fibers (Cherny & Gazit, 2008), hydrogels (Saikia et al., 2021), and coil-coiled bundles (Woolfson, 2005). Synthetic modifications of the peptide such as N and C-terminus modification, introduction of unnatural amino acids, and aromatic $\pi - \pi$ stacking, provide additional opportunity to tune the large variety of supramolecular structures including twisted ribbons (Fishwick et al., 2003), collagen mimicking peptides (Pires, Przybyla, & Chmielewski, 2009), and nanobelts (Cui, Muraoka, Cheetham, & Stupp, 2009). In one of the earliest attempts Benner and co-workers have rationally designed a 14-mer alpha-helical peptide forming helical bundle, which converts oxaloacetate to pyruvate (Johnsson, Allemann, Widmer, & Benner, 1993). Baltzer and co-workers have designed KO-42, a 42-residue polypeptide, that folded into a helix-loop-helix hairpin motif (Broo, Brive, Ahlberg, & Baltzer, 1997). KO-42 peptide catalyzes hydrolysis of p-nitrophenyl esters. Further they introduced aldimine binding sites into KO-42 peptide to catalyze conversion of aldimine into ketamine, emulating the biosynthesis of amino acids. The resulting catalysts bound the aldimines and catalyzed conversion of aldimine into ketamine three times faster than imidazole (Allert & Baltzer, 2003). Inoue described that the aggregation of a peptide-based catalyst notably improved the stereoselectivity of the asymmetric addition of hydrogen cyanide to m-phenoxybenzaldehyde (Tanaka, Mori, & Inoue, 1990). Recently, Das and co-workers have developed KLVFF-based nanotube peptide catalysts and demonstrated catalysis of aldol and retro-aldol condensation reactions (Reja, Afrose, & Das, 2020).

8.2.4.1 *Proline-based supramolecular catalyst*

The extensive work suggests great promise for the proline-based self-assembled peptide as a potential catalyst (Saikia et al., 2021). It was first reported in early 1970s that amino acids, such as proline, could catalyze asymmetric aldol cyclizations (Eder, Sauer, & Wiechert, 1971). Clark and co-workers designed minimal peptide pro-nap, by tagging proline to amido naphthyridine (Nap), and provided a library of asymmetric assemblies capable of enamine catalysis (Fuentes, Lebl, Slawin, & Clarke, 2011). Miravet and co-workers have designed Pro-Val dipeptide hydrogel for enamine-based catalysis (Rodríguez-Llansola, Escuder, Hamley, Hayes, & Miravet, 2012).

The formed hydrogel showed dual catalytic behavior as free molecules and as aggregates. In solution phase dipeptides behave as moderately active and stereo-selective L-Pro-based organocatalysts participating in the aldol reaction, whereas in the gel phase they turned into basic catalytic residues inactive in the aldol reaction, but active in the nonstereospecific deprotonation of the aldol product (Rodríguez-Llansola, Escuder, & Miravet, 2009). Recently, Hamley and co-workers developed a series of proline-based amyloid peptides and lipopeptides, which form several nanostructures and show excellent catalytic activity and high selectivity for aldol reactions (Pelin et al., 2020).

8.2.4.2 *Histidine-based supramolecular catalyst*

Much attention has been given to develop self-assembling catalysts for hydrolytic reactions. A single amino acid histidine (His) on its own is capable of promoting hydrolysis, although the catalytic efficiency is very low. However, incorporation of His into self-assembling structures has emerged as a powerful strategy for creating artificial hydrolases (Fig. 8.4). Guller and Stupp have demonstrated the hydrolysis of para-nitrophenyl acetate (PNPA) on the surface of nanofibers formed by the assembly of peptide amphiphiles (Guler & Stupp, 2007). They have incorporated the two histidine residues in the peptide amphiphiles. The resulting nabofiber hydrolyzed the PNPA, while nonfibrils structure showed poor catalytic activity. Leu and co-workers have synthesized the short amphiphilic peptide Fmoc-FFH-$CONH_2$, which self-assembles to form nanotube and hydrolyzed PNPA (Huang et al., 2013). The resulting hydrolase model shows typical saturation kinetics behavior to that of natural enzymes and the catalytic efficiency of a single catalytic center is 519-fold higher than the one without catalysts. To develop further the hydrolase model, Zhimin He and co-workers formed the catalytic Ser/His/Asp triad by mixing varying ratio of Fmoc-FFH-$CONH_2$ with Fmoc-FFS-$CONH_2$ and Fmoc-FFD-$CONH_2$ peptide. At the optimal ratio of 40:1:1 (Fmoc-FFH-$CONH_2$: Fmoc-FFS-$CONH_2$: Fmoc-FFD-$CONH_2$) catalytic efficiency was observed to be twofold higher than the activity of the fibrils formed by Fmoc-FFH-$CONH_2$ (Wang et al., 2016). Dejugnat and co-workers have designed lipopeptides by joining tripeptides with lipophilic molecules, which catalyzed the hydrolysis of PNPA upon self-assembly (Bélières, Chouini-Lalanne, & Déjugnat, 2015). They concluded that the lipopeptides are more efficient catalysts as monomers than in aggregates, showing that a subtle balance of the hydrophobic part could be used to orient it toward activated hydrolysis or rather protection of the substrate.

8.2.4.3 *Peptide catalyst with metal ions*

Metal ions are ubiquitous in nature, playing structural or catalytic roles in nearly half of all proteins. They play a key role in regulating the activity of proteins both at allosteric and at catalytic sites. Natural amino acids comprise

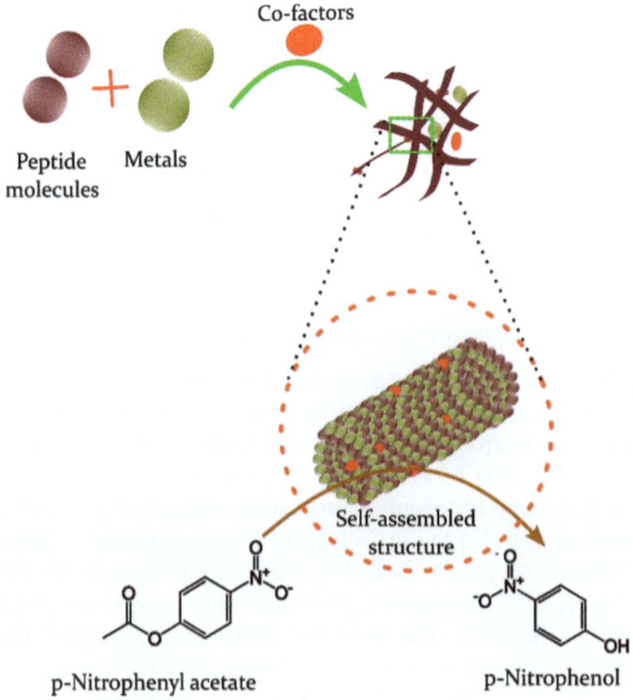

FIGURE 8.4 Schematic representation of ester hydrolysis by self-assembled nanostructures.

several functional groups in the side arms, which act as binding sites for metal ions (Yamashita, Wesson, Eisenman, & Eisenberg, 1990). This can occur for peptides of amino acids without coordinating side chains, such as alanine or phenylalanine, or with coordinating side chains, such as histidine or cysteine (Sóvágó, Kállay, & Várnagy, 2012). Kaplan and DeGrado developed a heterotetrameric coiled-coil "DFtet" peptide, which self-assembled to bind to two Fe (II) in active sites. The structure of DFet comprised of two different "A" and "B" subunits, which helps in binding with iron (Marsh & DeGrado, 2002). The resulting co-assembly with iron catalyzes oxidation of 4-aminophenol to the corresponding quinone. Serpell and co-workers have designed the peptides that form an amyloid-like architecture and reveal their capability to mimic carbonic anhydrase (Al-Garawi et al., 2017). These amyloid fibril structures can bind the metal ion Zn^{2+} by a three-dimensional arrangement of His residues created by the amyloid architecture. The peptide catalyzed the PNPA molecules in the presence of Zn^{2+} ion. Laungani and co-workers have designed C- and N-terminal phosphane-functionalized peptides to bind noble metals (Pt/Rh) (Laungani, Slattery, Krossing, & Breit, 2008). Upon mixing, the peptides and appropriate metal salts form β-sheet assemblies that can do asymmetric hydroformylation of styrene with quantitative yield and excellent regioselectivity but modest enantioselectivity.

8.2.5 Peptide-based smart materials

With the technological advancement in imaging, peptides have recently been considered for their potential use as biomaterials. Researchers are now exploring the ability of peptide-based biomaterials to respond to external cues; this responsiveness has been collectively referred to as "smart behavior" (Rajagopal & Schneider, 2004). Responsiveness can be defined either at the structural or at the functional level (Rajagopal & Schneider, 2004). For structural smartness, assemblies built from α-helical units tend to be more readily reversible than structures dominated by β-strand units. This difference arises principally from the relative importance of hydrogen bonding and hydrophobic contributions to structural stability.

The ability of peptides and proteins to change the conformations in response to the external stimuli such as pH, temperature, and the presence of specific small molecules is ubiquitous in nature. This property of peptide has been exploited to engineer several stimulus-responsive systems with potential applications in biomaterials, nanodevices, biosensors, bio-separations, tissue engineering, and drug delivery. This section will highlight some recent advances in the field of self-assembling peptides used in biomaterials design. Emphasis will be placed on work that either implicitly or explicitly incorporates concepts of smartness.

8.2.5.1 Stimulus-responsive materials

Several peptide-based systems have been designed to respond to various stimuli. In a work, temperature-dependent 12 and 16 residue peptide were designed. At the temperature more than 70°C, the designed peptides showed conformational change from β-sheet formation to an α-helical structure (Zhang & Rich, 1997). With cooling, these peptides retained the a-helix structure, and took several weeks at room temperature to partially return to the β-sheet form. Peptides whose bulk behavior can be controlled by metal ions have also been successfully engineered (Ghadiri, Soares, & Choi, 1992).

An emerging area of interest in the field of stimulus-responsive peptides is the concept of enzyme-responsive peptides. Gao and co-workers have designed an enzyme-responsive zwitter ionic peptide coating capable of responding to matrix metalloproteinase-9 (MMP-9), which is overexpressed in tumor microenvironment (Wu et al., 2019). The repetitive pentameric sequence of peptide VPGVG is one of the most studied stimulus-responsive peptide, which is naturally found in the polymeric elastin-like polypeptide (ELP) of the mammalian elastin protein (Meyer & Chilkoti, 2004). The physiochemical properties of ELP can be altered by mutating fourth residue of this sequence. The highly multimeric form of peptides (20–300 pentameric repeats) induces a sharp reversible hydrophilic − hydrophobic phase transition in ELP, which is triggered by changes in pH, temperature, or ionic strength. Below transition temperature, the ELP exists in an extended state and is soluble in water, while at above transition temperature, the ELP collapses into an ordered b-spiral that aggregates and precipitates out of solution (Meyer & Chilkoti, 2004). The stimulus-responsive properties of ELPs fused to appropriate peptides or proteins have been used for remediation of toxic metals (Prabhukumar, Matsumoto, Mulchandani, & Chen, 2004) and targeted drug delivery (Sreekumar et al., 2018).

8.2.5.2 Peptide-based molecular shuttles

The inherent restrictions in rotational and translational freedom imposed on the components of mechanically interlocked molecules make them particularly attractive architectures for precisely controlling the positioning of functional units or substituents with the possibility of switching their relative separation and orientation (Anelli et al., 1997). Such control has been elegantly demonstrated through π − donor − acceptor interactions in the "molecular shuttles"(Anelli et al., 1997), metal ion coordination in the catenates (Cárdenas, Livoreil, & Sauvage, 1996), and in pseudorotaxanes, prepared by Sauvage and co-workers(Collin, Gavinã, & Sauvage, 1996). Molecular shuttles are rotaxanes in which the macrocycles are translocated from one binding site ("station") on the thread to the second station in response to the external stimuli (e.g., chemical fuels, light, pH, or temperature) that invert the initial relative affinity of each of the sites for the macrocycle (Villalva, Nieto-Ortega, Melle-Franco, & Perez, 2020). Some of the most advanced synthetic molecular machines described to date are based on the molecular shuttle concept (Qiu et al., 2020). In particular, the unique flipping motion between multiple recognition sites in a rotaxane molecular system makes it a highly suitable choice for developing molecular devices (Davidson, Sharma, & Loeb, 2010). Gariepy and co-workers have designed loligomers, which are peptide-based intracellular vehicles, that can penetrate cells and self-localize into distinct cellular compartments (Singh, Bisland, Kawamura, & Gariépy, 1999). The loligomers were able to transfer the gene into mammalian cells, which was a nonviral gene delivery system (Singh et al., 1999). As is well-known, ion passive transport is essential for maintaining normal cell function and biological processes. Tian and co-workers have designed an artificial molecular shuttle that operates in lipid bilayers for ion transport (Chen et al., 2018). Recently, peptide-based molecular shuttle system is explored for applications such as gene therapy(Garcia, Arranz-Gibert, Sánchez-Navarro, Giralt, & Teixidó, 2020) and drug delivery (Bukchin et al., 2021).

8.2.5.3 Peptide-based metal organic frameworks

Meta organic frameworks (MOFs) have been used extensively for different applications such as gas storage (Farha et al., 2010), catalysis (Lee et al., 2009), and chemical separation (Qiu, Xue, & Zhu, 2014). The incorporation of biomolecules as organic linkers in MOF synthesis makes them one of the most exciting and rapidly growing areas of modern chemistry research. Ferry G and co-workers first reported the loading and release kinetics of the analgesic drug, ibuprofen from the first MOF family, Materials Institute Lavoisier-100 (MIL-100) and MIL-101 (Horcajada et al., 2006). Both MIL structures are composed of trivalent metal centers with carboxylic bridging ligands and possess large pores with large surface areas. Several theoretical and experimental researches demonstrated the potential application of MOFs as an ideal drug delivery system for various therapeutic drugs (Cui et al., 2021). Zhang and co-workers have reported the RGD peptide functionalized MIL-101 for targeted release of the drug doxorubicin with effective cancer cell inhibition and reduced side effect (Wang et al., 2015). Jimenez and co-workers have reported the use of an MOF to load, protect, and deliver small interfering ribonucleic acids (siRNAs) by avoiding enzymatic degradation (Teplensky et al., 2019). They have encapsulated the siRNAs in the MOF with KALA peptide and ammonium chloride, to evade the endosomal retention and ensure that gene knockdown is efficacious.

Surface functionalization of MOFs are important for biomedical application because this interface controls the stability of the framework, the rate of drug release, cell uptake, and the overall biological response (Horcajada et al., 2012). Short peptides functionalized MOFs have been used as electrochemical devices, and biosensors (Sun et al., 2021).

8.2.5.4 Peptide-based vaccines

Classical whole organism vaccines may have several pitfalls such as the potential risk of inducing autoimmune and allergic responses (Skwarczynski & Toth, 2016). The "peptide vaccines" containing only epitopes capable of inducing positive, desirable T-cell and B-cell-mediated immune response may be a substitute for this (Sesardic, 1993). Since epitopes are the antigenic determinants within larger proteins, these peptides are considered sufficient for activation of the appropriate cellular and humoral responses, while eliminating allergenic and reactogenic responses (Bijker, Melief, Offringa, & Van Der Burg, 2007). Peptide-based vaccines are built of defined, small-peptide antigens engineered to induce the desired immune response.

Several factors should be considered in formulating an effective peptide vaccine including inclusion of a universal T-helper epitope, and the necessity to mimic the structure of the parent antigen to generate high-affinity antibodies (Steele, Allen, & Kaumaya, 2006). On the other hand, owing to the relatively small size of peptides, they are often weakly immunogenic by themselves and therefore require carrier molecules, to add chemical stability and adjuvant, for the induction of a robust immune response (Purcell, McCluskey, & Rossjohn, 2007). There are many peptide vaccines under

TABLE 8.1 List of some peptide-based vaccines.

Disease	Vaccine	Clinical phase	Reference
Alzheimer's disease	CAD106	Phases II/III	Wiessner et al. (2011)
	ACI-35	Phase I	Theunis et al. (2013)
	AADvac-1	Phase II	Novak et al. (2018)
Cancer	Gp100:209–217 (210 M)	Phase III	Rosenberg et al. (1998)
	AE37	Phase II	Voutsas et al. (2007)
Malaria	MSP1 with GLA-SE adjuvant	Phase I	Blank et al. (2020)
Hepatitis C	IC41	Phase II	Firbas et al. (2006)
HIV	Rv144	Phase III	Malonis, Lai, and Vergnolle (2019)

development, such as vaccine for human immunodeficiency virus (Liu et al., 2007), malaria (Epstein, Giersing, Mullen, Moorthy, & Richie, 2007), and anticancer vaccine (Greten et al., 2010) (Table 8.1).

8.3 *De novo* designed peptide nano-assemblies

Peptides self-assemble into different nanostructures, playing crucial role in fabricating hierarchical architectures that have good biocompatibility and functionalities (Zhao et al., 2010). On the basis of their chemical structure, peptides can be classified as linear (Blanco, Rivas, & Serrano, 1994), cyclic (Zorzi, Deyle, & Heinis, 2017), and branched (Guler, Soukasene, Hulvat, & Stupp, 2005). Some self-assembled nanostructures are summarized in Fig. 8.5.

8.3.1 Nanotubes

Among the notable supramolecular self-assembly, peptide nanotube is a well-studied architecture with diversified applications. Alternating D-octapeptide and L-octapeptide, upon cyclization, has been shown to stack through hydrogen bonding to form a nanotubular assembly (Ghadiri et al., 1993). One of the most well-studied building blocks for supramolecular structures is diphenylalanine (FF) peptide. The nanotube formed by FF molecules has shown remarkable rigidity along with various physical and chemical functions such as semiconductivity (Hauser & Zhang, 2010), luminescence (Amdursky et al., 2009), and piezoelectricity (Bosne et al., 2013).

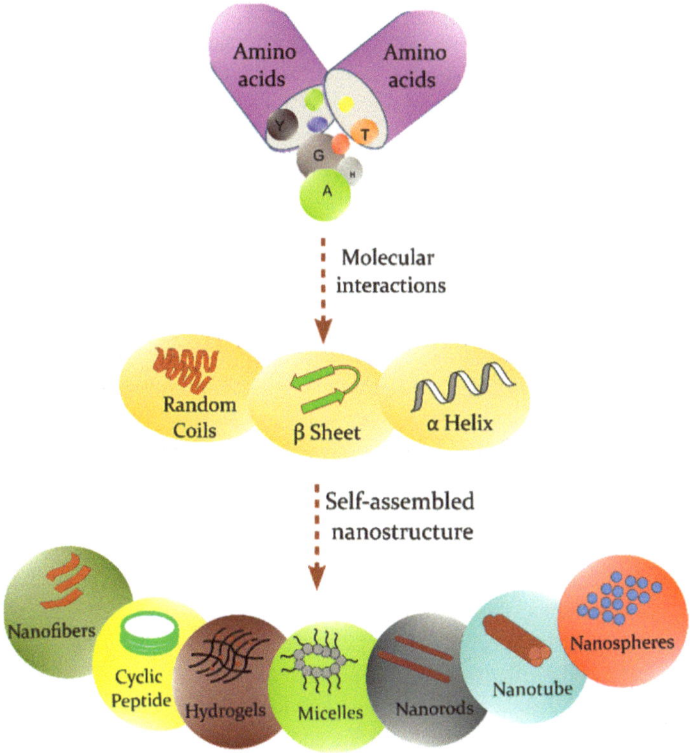

FIGURE 8.5 Peptide-based self-assembled nanostructures.

8.3.2 Nanosheets

Surfactant-like peptides were known to form nanotubes (Bucak, Cenker, Nasir, Olsson, & Zackrisson, 2009), but Castelletto and co-workers first reported the nanosheet formation with surfactant-like peptide, A_6R (Hamley, Dehsorkhi, & Castelletto, 2013). Further, they have reported that the RFL_4FR peptide aggregates into ultrathin flat sheets because of β-sheet assemblies (da Silva et al., 2015). Liu and co-workers have first reported the amyloid-like nanosheets formed by the KLVFFAK segment (Dai et al., 2015). They have also shown its application as retroviral transduction.

8.3.3 Cyclic peptides

Cyclic peptides have promising potential applications in biologicals and materials (Fernandez-Lopez et al., 2001). Rationally designed cyclic peptides with flat conformations can self-assemble into tubular nanostructures mainly stabilized by intercycle hydrogen bonding (Ghadiri et al., 1993). Generally, when the assembly of cyclic peptides occurs in solution, which can be triggered by cooling the solution, they can stack into nanotubes that, in turn, organize into a compact crystal structure (Bélanger, Tong, Soumaré, Dory, & Zhao, 2009). Cyclic peptide-based molecular constructs have been utilized as enzyme inhibitors (Colgrave, Korsinczky, Clark, Foley, & Craik, 2010), gene delivery (Oba et al., 2007), organic electronics (Ardoña & Tovar, 2015), and antibacterial agents (Brown et al., 2020).

8.3.4 Nanofibers

Among the various constructs of self-assembly, nanofibers are considered to be the most promising structure with various applications. Fibrous structures have large length to diameter ratio. Hierarchical architectures with self-assembled peptide nanofibers may be constructed through entanglement and noncovalent binding between peptide nanofibers. Self-assembled fibrous materials have broad range of biomedical application such as scaffold for three-dimensional cell culture, biofilm (Cherny & Gazit, 2008), and tissue engineering (Yanzhong Zhang, Ouyang, Lim, Ramakrishna, & Huang, 2005). Factors promoting the nanofiber formation are aromatic stacking, metal $-\pi$ interaction, and hydrogen bonding (Sasidharan, Hazam, & Ramakrishnan, 2017). It has been observed that the high content of β-strand could enhance the rigidity of nanofibers (Xiao, Xiao, & Gräter, 2013). pH also modulates the mechanical property of nanofibers. Dong and co-workers have investigated the effect of pH on peptide sequence SSSSFAFAC (Bortolini et al., 2015). At pH 2, this peptide formed amyloid-like fibers with relatively higher amount of ordered random coil (38%) and β-sheets (37%), but no α-helices, while more flexible fibrils that are rich in α-helices were formed at pH 7. Furthermore, they confirmed that a high amount of ordered β-sheets contributed to improve the rigidity of the formed fibrils (Bortolini et al., 2015).

8.3.5 Hydrogels

Peptide-based hydrogels are a class of soft materials that typically have three-dimensional fibrous networks cross-linked through physical or chemical bonding. They are characterized by their high-water content, microporous structure, tuneable mechanical stability, good biocompatibility, and elasticity (Bahram, Mohseni, & Moghtader, 2016). In addition, hydrogels are readily decorated or functionalized through side-chain modifications to realize specific applications (Du, Zhou, Shi, & Xu, 2015). The main factor responsible for the gelation is aromatic $\pi-\pi$ interaction, which also increases the hydrogen bonding and other nonbonding interactions in water, usually leading to supramolecular hydrogels.

8.3.5.1 Single amino acid-based hydrogels

There are several reports on hydrogels formed by dipeptides and tripeptides with different N-terminally conjugated aromatic groups. However, single amino acid containing hydrogels are relatively much less reported

(Chakraborty & Gazit, 2018). Nilsson and co-workers have reported that halogenation of the phenyl side chain of Fmoc-Phe resulted in rapid self-assembly, and hydrogelation relative to Fmoc-Phe or Fmoc-Tyr (Ryan, Anderson, & Nilsson, 2010). Hydrogels formed from the self-assembly of Fmoc-F_5-Phe-OH or monohalogenated derivatives were mechanically rigid, but unstable in solvent (Ryan, Doran, & Nilsson, 2011). To increase the stability of the designed peptide, they have conjugated it with polyethylene glycol. Banerjee and co-workers have reported pyrene conjugated phenylalanine derivate superhydrogel with minimum gelation concentration less than 0.1% w/v, showing thixotropic property (Nanda, Biswas, & Banerjee, 2013). Thakur and co-workers have reported the Fmoc-Phe hydrogel with antimicrobial effect (Gahane et al., 2018). Marchesan and co-workers have also synthesized phenylalanine derivative *N*-(4-Nitrobenzoyl)-Phe, which self-assembled to form hydrogel showing significant antimicrobial property (Garcia et al., 2020).

8.3.5.2 *Short and ultrashort peptide-based hydrogel*

Peptide sequences having amino acids chain length less than 20 are termed as short peptides, while sequences having less than 7 amino acids are ultrashort peptide (Seow & Hauser, 2014). Zhang and co-workers have designed several self-assembling short peptides (Zhang, Holmes, Lockshin, & Rich, 1993). Features of these peptides include regularly alternating hydrophilic and hydrophobic amino acids, where positive charges are juxtaposed to negative ones (Yokoi, Kinoshita, & Zhang, 2005). Due to ionic self-complementarity encouraged by the presence of salts, these peptides self-assemble to form nanofibers that interweave to finally form β-sheet hydrogels (Yokoi et al., 2005). FF is frequently incorporated as a backbone to generate short peptide-based hydrogelators because it promotes $\pi-\pi$ stacking interactions (Marchesan, Vargiu, & Styan, 2015). The N-terminal capping groups also enhance intramolecular interactions, which can then induce hydrogel formation. The protecting groups that are generally used as capping groups such as fluorenylmethyloxycarbonyl (Fmoc) (Orbach et al., 2012), naphthalene (Laverty et al., 2014), carbazole (Kubiak et al., 2015), and indole (Martin, Robinson, Mason, Wojciechowski, & Thordarson, 2014) are reported to induce gelation. Naresh Kumar and co-workers have designed an anthranilamide-capped short peptide hydrogel having antimicrobial property (Aldilla et al., 2020).

8.3.5.3 *Stimulus-responsive peptide hydrogel*

In stimulus-responsive hydrogel the polymer networks comprising of stimuli-responsive basic building blocks, for example, photoisomerizable units (Huang et al., 2011) or redox groups (Yoshii, Onogi, Shigemitsu, & Hamachi, 2015), allow the triggered and reversible transitions between hydrogel and solution phases or hydrogel and solid states (Echeverria, Fernandes, Godinho, Borges, & Soares, 2018). The conjugation of p-boronophenylmethoxycarbonyl (BPmoc) or p-nitrophenylmethoxycarbonyl (NPmoc) to the FF peptide yielded redox-active hydrogels (Ikeda, Tanida, Yoshii, & Hamachi, 2011). Incorporation of fumaric amide as a conformational switch has proved to be an effective strategy to construct photo-responsive supramolecular hydrogels with potential applications (Huang et al., 2011). pH-responsive hydrogels have been designed as drug delivery systems. Zhong and co-workers have synthesized an RGD derived peptide conjugate, which entraps doxorubicin and releases it under mild acidic conditions paving ways for the controlled release of drug cargos in a typical tumor microenvironment (Mei et al., 2019). Ge and co-workers have designed a paclitaxel-loaded peptide hydrogel for the localized delivery of paclitaxel *via* intratumoral injection to enhance the therapeutic efficacy against the tumor (Raza et al., 2019). Apart from drug delivery systems, stimulus-responsive hydrogels have been explored for tissue engineering (Tang, Mura, & Lampe, 2019), self-healing materials (Basak, Nanda, & Banerjee, 2014), and sprays (Shundo et al., 2014).

8.3.5.4 Amyloid-based hydrogels

Amyloid-based hydrogels are water-laden, three-dimensional materials formed by cross-linking of protein fibrils. They hold large amounts of water because they have a large number of hydrophilic functional groups present in their polymeric chain (Kumari & Ahmad, 2019). Several proteins and peptides are known to form hydrogels through the cross-linking of amyloid fibrils (Hu et al., 2018). Amyloid-based hydrogels are useful in the development of materials such as drug delivery systems and tissue engineering (Fichman & Gazit, 2014).

8.3.5.5 Fmoc-based hydrogels

Janmey and co-workers reported for the first time the formation of fibrous networks with fluorenylmethoxycarbonyl (Fmoc)-protected amino acids and dipeptides (Vegners, Shestakova, Kalvinsh, Ezzell, & Janmey, 1995), while Xu and co-workers were the first to report that certain Fmoc-protected amino acids and dipeptides spontaneously formed fibrous scaffolds (Zhang, Gu, Yang, & Xu, 2003). Fmoc-modified diphenylalanine (Fmoc-FF) has been widely studied as the simplest and most effective hydrogelator (Rajbhandary & Nilsson, 2017; Zhang et al., 2003). Recently, Fmoc-FF-based hybrid gels were also used as drug delivery systems (Abbas, Xing, Zhang, Zou, & Yan, 2017). Ikeda and co-workers have synthesized a new Fmoc-dipeptide, comprising α-methyl-L-phenylalanine (Arakawa et al., 2020). Few examples of Fmoc-based peptide hydrogels are listed in Table 8.2.

TABLE 8.2 Fmoc-based peptide hydrogels.

S. no	Sequence	References
1	Fmoc-FF	Mahler, Reches, Rechter, Cohen, and Gazit (2006)
2	Fmoc-F, Fmoc-W, Fmoc-Y	Tao, Levin, Adler-Abramovich, and Gazit (2016)
3	Fmoc-FRGD-OH, Fmoc-RGDF-OH, Fmoc-FG-OH	Orbach et al. (2012)
4	Fmoc-VLK(Boc), Fmoc-K(Boc)LV	Cheng, Castelletto, Moulton, Newby, and Hamley (2010)
5	Phosphorylated Fmoc-FFpY	Criado-Gonzalez, Wagner, et al. (2020)
6	Fmoc-FFY	Criado-Gonzalez, Iqbal, et al. (2020)
7	Fmoc-RGDS, Fmoc-RGDS, Fmoc-GRDS, and Fmoc-GRDS	Castelletto et al. (2011)

8.4 Properties of self-assembled nanostructures

8.4.1 Mechanical properties

Several previous studies reported the mechanical strength of the diphenylalanine (FF) peptide nanostructure (Sedman et al., 2013). Atomic force microscopy measurements have indicated an average point stiffness of 160 N/m and a calculated Young's modulus of about 20 GPa for nanotubular peptide assemblies (Adler-Abramovich et al., 2010). Density functional theory calculations show that the high mechanical strength of FF nanostructure was the result of an aromatic zipper present in the nanotubes (Azuri, Adler-Abramovich, Gazit, Hod, & Kronik, 2014). Gazit and co-workers have designed dipeptide based on β,β-diphenyl-Ala-OH (Dip), Dip-Dip, cyclo-Dip-Dip, and *tert*-butyloxycarbonyl (Boc)-Dip-Dip (Basavalingappa et al., 2020). The phenyl-enriched structures showed material performance superior to that of FF nanotubes, with improved mechanical rigidity, thermal stability, and piezoelectricity. The mechanical properties of the nanostructure can also be tuned by using two peptides of different stereochemistry. Elevating the level of FF peptide in the peptide mixture decreases the nanostructure's stiffness (Sedman et al., 2013).

8.4.2 Electrical properties

Among the amino acids, some side chains include hydrogen donor and acceptor groups. These side chains are commonly involved in proton transfer and proton-coupled electron transfer processes (Migliore, Polizzi, Therien, & Beratan, 2014). In the context of protein and peptide-based films, these side chains give rise to long-range proton transport (Amit et al., 2014). Nakayama and co-workers have shown conductivity in the designed peptide, GFPRFAGFP, which self-assembled to form fibers (Creasey, Shingaya, & Nakayama, 2015). Ashkenasy and co-workers have demonstrated that the aromatic stacking in D,L, α-cyclic peptide self-assembled nanotubes was found to promote long range conductance in peptide self-assembly, and hence could increase the conductivity (Yardeni, Amit, Ashkenasy, & Ashkenasy, 2016). However, due to the inherently low dielectric constant, energy translocation through peptide architectures is inefficient, which severely hinders their practical application (Knowles & Buehler, 2011). Incorporation of water molecules can improve the conductivity of bioinspired supramolecular structure (Amit et al., 2014). Another way to improve charge transfer inside peptide assemblies is by the utilization of doping, such as through chemical conjugation or co-assembly with functional moieties (Erwin et al., 2005). By concerning these points, Gazit and co-workers have designed L,D-amino acid containing Trp dipeptide (Ww), which crystallized into a layered organic-water structure (Tao et al., 2020).

8.4.3 Optical properties

The molecular arrangement during self-assembly determines the optical properties of peptide nanostructures. The elementary building block of FF-nanotubes or Boc-FF-nanospheres was shown to have the optical properties of zero dimensional QD (Amdursky, Molotskii, Gazit, & Rosenman, 2009). However, FF was also reported to form two-dimensional quantum wells during thermal deposition at high temperature (Adler-Abramovich & Gazit, 2014). The change in optical property was attributed to the alteration of their nanoscale packing, during vapor deposition process. Besselievre and co-workers have used the lanreotide octapeptide cyclized peptide scaffold to form nanotubes (Tarabout et al., 2011). The high resolution structure of the nanotube walls shows that the cyclic peptides are stacked in a two-dimensional crystal, which resembles a quantum well (Alivisatos, 1996).

Some of the most promising quantum-sized physical effects, found in these nanomaterials, are the electron-optical spectral properties such as light absorption and luminescence (Harrison, 2016). The nanoscale architecture comprising the supramolecular structure revealed pronounced electron-optical effects, which are highly specific for different quantum confinement ordering and could be recognized optically for understanding of quantum confined shapes and dimensions (Amdursky, Molotskii, Gazit, & Rosenman, 2010).

8.5 Factors affecting nanoassemblies

Peptide self-assembly occurs due to the association of two or more molecules mediated by physical interactions. In aqueous solution and in equilibrium, a peptide molecule adopts a conformation that minimizes its free energy. The main driving forces in the intramolecular interactions include hydrophobic interactions, hydrogen bonding (intermolecular and intramolecular), Van der Waals, aromatic, and electrostatic interactions (Wang, Liu, Xing, & Yan, 2016). These noncovalent interactions are relatively insignificant for isolated peptides because they are generally weaker than covalent interactions. pH (Martins et al., 2011), electric field,(Pandey & Ramakrishnan, 2020; Pandey, Morla, Nemade, Kumar, & Ramakrishnan, 2019; Saikia et al., 2019), magnetic field (Pandey et al., 2017), and temperature (Ciani, Hutchinson, Sessions, & Woolfson, 2002) also modulate nanoassemblies.

8.5.1 Aromatic $\pi - \pi$ interactions

Burley and Petsko first reported the aromatic interactions in proteins (Burley & Petsko, 1985). They have reported that phenyl ring centroids are separated by a preferential distance, between 4.5 and 7.0 Å, and dihedral angles of 90 degrees are commonly found to have aromatic $\pi - \pi$ interaction (Burley & Petsko, 1985). In supramolecular chemistry, aromatic $- \pi$ interaction is considered as a noncovalent interaction between the organic molecules having aromatic moieties (Hill & DeGrado, 1998; Sasidharan & Ramakrishnan, 2022).

On the basis of interactions between π electron-rich aromatic species, aromatic interactions can be classified as sandwich, parallel displaced, and face-to-face stacking (Wheeler, 2011) (Fig. 8.6). The small, unsubstituted aromatic compounds prefer edge-to-face geometry, whereas substituted and large multiring aromatic compounds prefer parallel displaced geometry (Thakuria, Nath, & Saha, 2019). Chakrabarti and co-workers examined the specific interactions and packing preferences of tryptophan (Trp) with other aromatic amino acids. They reported that interaction of Trp with Phe was more favorable in face-to-edge (FtE) geometry as compared to face-to-edge (FtE) geometry (Samanta, Pal, & Chakrabarti, 1999). The bulky indole side chain of Trp amino acid is capable of establishing multiple interactions, including $\pi \cdots \pi$, $C-H \cdots \pi$, band $N-H \cdots \pi$, in which the indole nitrogen can also act as a hydrogen bond donor (Mahalakshmi, Sengupta, Raghothama, Shamala, & Balaram, 2005). Aromatic interactions are also crucial for the stability of the helix bundle proteins. The different nanostructures formed by the pi $-$ pi interactions are nanotube (Görbitz, 2001), nanofibers (H. Zhang, Lou, & Yu, 2019), nanoflowers

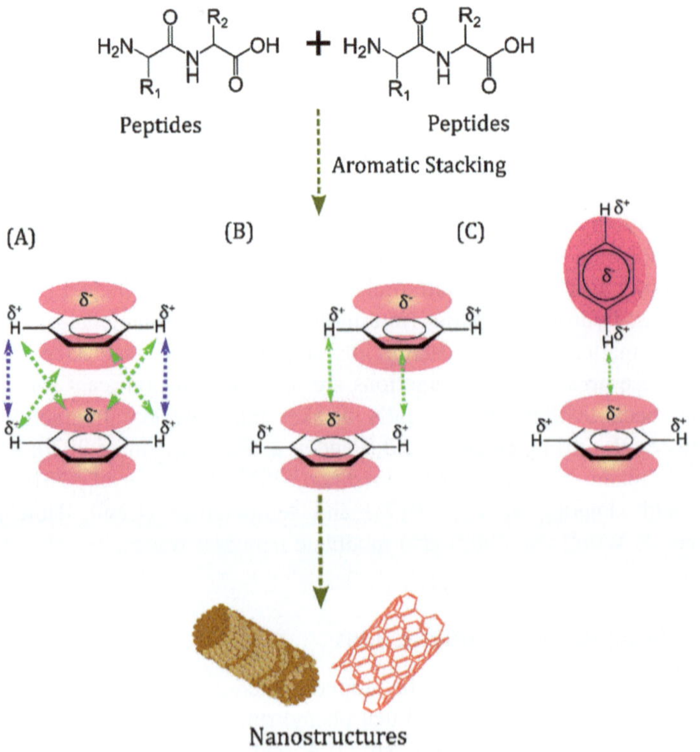

FIGURE 8.6 Schematic representation of aromatic interactions; (A) sandwich, (B) parallel displaced, and (C) edge-to-face stacking leading to the formation of nanostructure. *Blue line* represents repulsive interactions, and *green line* represents attractive interactions.

(Sasidharan, Shyni, Chaudhary, & Ramakrishnan, 2017), and nanorods (Guo, Luo, Zhou, & Wei, 2014).

8.5.2 Hydrogen bonding

Hydrogen bonds, are the interactions between hydrogen atoms and the electronegative atoms (Bartocci et al., 2018). The selectivity and high directionality of hydrogen bonds can convert short peptides into diverse one-dimensional, two-dimensional, and three-dimensional nanostructures (Kim et al., 2010; Krone et al., 2008). Li and co-workers have demonstrated the formation of FF microrods by hydrogen bond-based self-assembly (Li, Jia, Dai, Yang, & Li, 2015). The characterizations of the self-assembled microrods indicated that 1,1,3,3,6,6-Hexafluoro-2-propanol formed stable intermolecular hydrogen bonds with an FF peptide, leading to the solvation of peptide molecules. The intermolecular hydrogen bond strength modulated by elevated temperature and sonication enables structural transitions between nanowires and nanotubes (Ziserman, Lee,

Raghavan, Mor, & Danino, 2011). Hydrogen bonding between peptide backbones is unique and plays a critical role in the one-dimensional growth of many peptide aggregates (Paramonov, Jun, & Hartgerink, 2006). Lu and co-workers have compared the self-assembly of I_3K and L_3K peptide, and found that replacing isoleucine residues with leucine leads to a morphological change from long nanotubes to spherical aggregates, with the corresponding secondary structure transition from β-sheet to random conformation (Xu et al., 2010).

8.5.3 Hydrophobic interactions

Peptides resemble conventional surfactants in terms of possessing hydrophobic and hydrophilic moieties (Ghadiri et al., 1993). Stupp and co-workers have shown that the self-assembly of amphiphilic peptide is prominently governed by the relative strength of hydrogen bonding, compared with the hydrophobic interactions (Tsonchev, Niece, Schatz, Ratner, & Stupp, 2008). The aromatic residue of peptide plays a crucial role in determining the interactions between molecules. The organizational mode of the aromatic residues in the hydrophobic interactions is commonly disordered, while in $\pi - \pi$ interactions, it is well-ordered (Hu et al., 2020).

8.5.4 Electrostatic interactions

Electrostatic interactions play a significant role in the self-assembly of biomolecules (Kundu, Eltohamy, Yadavalli, Kundu, & Kim, 2016). The strength of an ionic bond is dependent on the solvent (particularly its dielectric constant) and the presence of mobile ions. Ionic bonds are generally stronger and of longer range than hydrogen bonding (Faul & Antonietti, 2003). Electrostatic bonds based on the Coulombic attractions between opposite charges lead to the formation of ion-pairs. Bernier and co-workers have described the self-assembly of a crown ether functionalized peptide through the mediation of electrostatic interactions (Voyer, Roby, Deschĕnes, & Bernier, 1995). Su and co-workers have fabricated an artificial peptide nanofiber with peptide motif, VIAGASLWWSEKLVIA, further metallized to synthesize nanofiber-based silver nanowires (Wang et al., 2014). A novel hybrid nanomaterial was obtained successfully by assembling the prepared silver nanowires on graphene nanosheets. In another study, Liu and coworkers fabricated micelle-induced protein nanowires through an electrostatic interaction when the electronegative cricoid stable protein one (SP1) assembled with positively charged cross-linked micelles (Sun et al., 2016).

8.5.5 Van der Waals interactions

Van der Waals forces, such as the interaction between aliphatic tails in peptide amphiphiles, provide a crucial contribution to various noncovalent interactions and are ubiquitous in assembly systems (Tahara, Lei, Adisoejoso, De Feyter, & Tobe, 2010). There are, however, only a few examples that employ Van der Waals interaction as a dominant force for the control of a peptide nanostructure (Tahara et al., 2010).

8.5.6 Solvent effects

Solvent affects the self-assembled nanostructure owing to different polarity, dielectric constant, and hydrogen bonded network. Murphy and co-workers investigated the effect of solvent system on the self-assembly of beta amyloid peptide (Shen & Murphy, 1995). In pure dimethylsulfoxide (DMSO), Aβ peptide had no detectable β-sheet content. In 0.1% trifluoroacetate, however, content went upto 33% β-sheet, and in 35% acetonitrile/0.1% trifluoroacetate, Aβ peptide had two-third amino acids in β-sheet conformation, equivalent to the fibrillar peptide in physiological buffer. Hamley and co-workers have demonstrated that the short peptide fragments, derived from β-amyloid peptide, AAKLVFF exhibited distinct structural differences depending on solvent polarity (Hamley et al., 2010). Zhimin and co-workers examined the effect of solvent system on the FF peptide to form either nanofibers or nanotubes after heating to 95°C and subsequent cooling in solution, or alternatively, deposition on the surfaces (Huang, Qi, Su, Zhao, & He, 2011). Interestingly, they have reported that upon cooling from mixed water−methanol systems, nanofibers can form in solution, but no formation is observed when the percentage of methanol exceeds 70%. In these conditions, nanofibers are formed only upon drying on glass surfaces. Authors attributed this behavior to the ability of methanol to form hydrogen bonds with the diphenylalanine molecules, leading to good solvation of peptide molecules (Huang et al., 2011). Roy and co-workers have investigated the solvent effect on the gelation of laminin-derived peptide (Jain & Roy, 2020). They have chosen the solvents of different polarity varying from highly polar phosphate buffer saline (pH 6) to 10% DMSO/water (0.1% TFA) and comparatively weak polar 50% ACN/ water (0.1% TFA). The gelation results clearly revealed that aqueous-organic mixtures are more favorable for inducing hierarchical self-assembly in more hydrophobic peptide derivatives by promoting hydrophobic and aromatic interactions between gelator moieties, which otherwise could not have been possible in aqueous solvents due to their poor solubility (Liu et al., 2017).

8.6 Applications of de novo designed peptide constructs

Peptide-based nanomaterials have been used for a myriad of biological applications, leveraging in large part their property to self-assemble (Fig. 8.7).

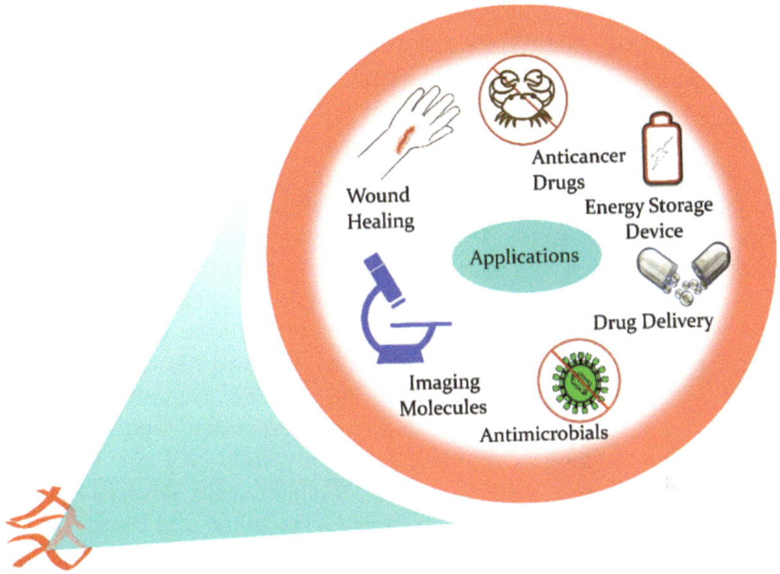

FIGURE 8.7 Applications of self-assembled nanostructures.

8.6.1 Tissue engineering

Bioinspired materials for tissue repair have been among the most exhaustively explored fields in biomaterials research (Panda & Chauhan, 2014). Pedone and co-workers have designed the peptide, QK, mimicking the vascular endothelial growth factor (VGEF) (D'Andrea et al., 2005). VGEF is the principle regulator of angiogenesis. Hartgerink and co-workers have incorporated the QK peptide to the nanofiber forming peptide sequence, which self-assembled to form hydrogel (Kumar et al., 2015). The conjugated peptide sequences potentially promoting VEGF receptor activation, dimerization, clustering, and intracellular angiogenic signaling at the site of nanofibrous hydrogel delivered (Kumar et al., 2015). Stupp and co-workers have designed a peptide nanoribbon mimicking the activity of fibroblast growth factor-2 (FGF-2) (Rubert Pérez, Álvarez, Chen, Aytun, & Stupp, 2017), which is a regulating factor in angiogenesis, cell differentiation, and wound healing (Rubert Pérez et al., 2017). Peptide sequences have also been designed as antiinflammatory motifs (Niyonsaba et al., 2013), surface adherence molecules (Shin, Jo, & Mikos, 2003), and corneal tissue repair (Islam et al., 2016).

8.6.2 Drug delivery systems

The clinical efficacy and effect of conventional chemotherapeutics against tumors are compromised by several factors including poor solubility, short half-life in vivo, the weak penetration capability, and the low specificity (Luo, Dai, & Gao, 2019). Ingeniously designed drug delivery systems can improve the bioavailability of drugs or minimize the adverse effects of drugs. Among the different drug delivery systems, peptide-based drug delivery systems are important because (1) the unique biochemical functionality encoded by peptide sequences enables an active targeting (Kebebe et al., 2018), (2) the structure of peptide assembly can be programmatically modulated by intrinsic and external stimuli to achieve a controllable release of the payload into the target region (Raza et al., 2019), (3) peptides are biocompatible compared to the synthetic organic compounds, and (4v) the reactive terminus and side chains of peptides can be used as a reactive site to conjugate chemotherapeutics (Meng et al., 2019). Peptides as a drug delivery vehicle can be utilized as targeting peptides, CPPs (Gaurav & Santhoshkumar, 2020; Goyal & Ramakrishnan, 2019), and stimulus-responsive peptides. Peptides have been *de novo* designed to target the blood vessels. NGR and RGD motif peptides can bind to the blood vessels of tumor areas (Ruoslahti, 2002). Deutscher and co-workers have synthesized a six amino acid containing peptide, KCCYSL, which was used for in vivo imaging of EGFR-2 expressing breast tumors by PET (Kumar et al., 2010). In another work, Sofou and co-workers have used the same KCCYSL sequence for the targeted delivery of doxorubicin drug (Bandekar et al., 2013).

8.6.3 Antimicrobial peptide hydrogel for wound healing

Despite advancements in medical technology, infectious diseases caused by microorganisms such as bacteria, viruses, and fungi remain a major threat to public health. Furthermore, overprescription and misuse of antibiotics also have escalated the drug resistance. Chronic wound infections and development of antibiotic resistance are affecting people worldwide. Antimicrobial peptides kill bacteria effectively, but their susceptibility to degradation limits their use for topical application only. In recent years, tremendous research has been conducted to utilize hydrogels either as an antimicrobial agent (Moorcroft, Roach, Jayne, Ong, & Evans, 2020) or as a drug carrier (Atefyekta et al., 2021). Chitosan, a natural linear polysaccharide, was investigated to form hydrogels and subsequently as a potential antimicrobial hydrogel. Hsieh and co-workers have reported a chitosan-γ-poly(glutamic acid) polyelectrolyte complex hydrogel possessing antimicrobial property against *Staphylococcus aureus* and *E. coli*. (Tsao et al., 2010). Interestingly, the formed hydrogels also promote the 3T3 fibroblast cell proliferation, and hence induced wound healing. In addition to polymer-based hydrogels, peptide-based antimicrobial hydrogels are also reported recently. Schneider and co-workers have designed a 20-residue β-hairpin peptide hydrogel scaffold for tissue regeneration, which also has an inherent antimicrobial property (Salick, Kretsinger, Pochan, & Schneider, 2007).

TABLE 8.3 Peptide-based hydrogels for diabetic foot ulcer.

S. no.	Hydrogels	Compositions	Reference
1	Chitosan-peptide-collagen hydrogel	Chitosan-collagen hydrogel with peptide: QHREDGS	Xiao et al. (2016)
2	Hydroxyethyl cellulose peptide hydrogel	Peptide: RQPKIWFPNRRKPWKKRPRPDDLEI entrapped in cellulose	Grek et al. (2015)
3	Multidomain peptide hydrogel	Peptide: $K_2(SL)_6K$	Carrejo et al. (2018)
4	Dispersin B-KSL-W hydrogel	Dispersin BPeptides: KKVVFWVKFK, and KKVVFKVKFKPluronic F-127	Gawande, Leung, and Madhyastha (2014)
5	ODEX/HA-AMP/PRP hydrogel	Oxidized dextranPlatelet-rich plasmaPeptide: SWLSKTAKKLFKKIPKKIPKKRFPRPRPWPRPNMI-NH2	Wei et al. (2021)
6	Chitosan-peptide hydrogel	ChitosanPeptide: LLGDFFRKSKE-KIGKEFKRIVQRIKDFLRNLVPRTES	Yang et al. (2020)
7	Glucose oxidase loaded antimicrobial peptide hydrogels	Glucose oxidePeptide: IKYLSVN	Zhao et al. (2020)

Diabetic foot ulceration and infections are one of the prominent causes of mortality and morbidity in developing countries. The microbiological profiling of diabetic foot infection shows the presence of pathogens such as *Pseudomonas aeruginosa*, *S. aureus*, and *Enterobacter* species (Gadepalli et al., 2006). Peptide hydrogels and its composite with chitosan have been investigated as an antimicrobial agent useful in the diabetic foot ulcer (Table 8.3).

8.6.4 Biosensors

Peptides have also been employed as recognition elements in bio-sensing (Liu, Wang, & Boyd, 2015). For instance, peptides with short chains of amino acids generally have better chemical and conformational stability than proteins. Peptide-based biosensors have been fabricated for the detection of several analytes, including cell proteins, metal ions, proteases, kinases, *Bacillus* species, nucleic acids, and antibodies (Liu et al., 2015). Electrochemical biosensors based on screen-printed electrodes and peptides are interesting alternatives for molecular diagnosis (Hasanzadeh & Shadjou, 2017).

Chai and co-workers have developed an electrochemical peptide cleavage-based biosensor for the detection of prostate-specific antigen using silver deposition on electrode (He, Xie, Yang, Yuan, & Chai, 2015). Other recently explored applications of peptide-based molecular constructs are in energy storage devices (Colherinhas, Malaspina, & Fileti, 2018), bioimaging (Liu, Sun, & Liang, 2020), antifouling agents (Sakala & Reches, 2018), and MOF (Schnitzer, Paenurk, Trapp, Gershoni-Poranne, & Wennemers, 2021).

8.7 Conclusion and future directions

Extensive research has been performed in the field of supramolecular self-assembly of peptides in last 20 years. The wide range of self-assembled architecture, formed by these molecules, has been explored in various fields. The work has been conducted to address different problems, such as to deliver the small drug molecules to specific targets (Yunching Chen, Wu, & Huang, 2010), to increase the efficiency of anticancer drug, sustained release of drug molecules, find the replacement of conventional vaccine for reducing side effects, and replacement of hazardous batteries with biodegradable materials. Here we have provided an overview of different nanomaterials fabricated by peptide molecules with possible applications. The stimulus-responsive materials have opened a different field of research in the field of drug delivery systems where drugs can be released only by changing the pH. Peptides and small molecule-based nanostructures can be good alternatives as drug delivery systems as they possess certain characteristics such as good biocompatibility, ease of synthesis, and functionalization. The efforts to replace the conventional vaccines with peptide-based vaccines offers several advantages such as less toxicity, easy to synthesis protocol, and stability. Although several methods have been applied to control the formation of self-assembled structures, total control of the designed peptide conformation and assembled supramolecular structure still remains elusive.

References

Abbas, M., Xing, R., Zhang, N., Zou, Q., & Yan, X. (2017). Antitumor photodynamic therapy based on dipeptide fibrous hydrogels with incorporation of photosensitive drugs. *ACS Biomaterials Science & Engineering*, *4*(6), 2046–2052.

Adler-Abramovich, L., & Gazit, E. (2014). The physical properties of supramolecular peptide assemblies: From building block association to technological applications. *Chemical Society Reviews*, *43*(20), 6881–6893.

Adler-Abramovich, L., Kol, N., Yanai, I., Barlam, D., Shneck, R. Z., Gazit, E., & Rousso, I. (2010). Inside cover: Self-assembled organic nanostructures with metallic-like stiffness (Angew. Chem. Int. (Ed.) 51/2010). *Angewandte Chemie International Edition*, *49*(51), 9788.

Akashi, Y., Oda, T., Ohara, Y., Miyamoto, R., Kurokawa, T., Hashimoto, S., & Ohkohchi, N. (2014). Anticancer effects of gemcitabine are enhanced by co-administered iRGD peptide in murine pancreatic cancer models that overexpressed neuropilin-1. *British Journal of Cancer*, *110*(6), 1481–1487.

Albada, H. B., Prochnow, P., Bobersky, S., Bandow, J. E., & Metzler-Nolte, N. (2014). Highly active antibacterial ferrocenoylated or ruthenocenoylated Arg-Trp peptides can be discovered by an L-to-D substitution scan. *Chemical Science*, *5*(11), 4453–4459.

Aldilla, V. R., Chen, R., Martin, A. D., Marjo, C. E., Rich, A. M., Black, D. St. C., & Kumar, N. (2020). Anthranilamide-based short peptides self-assembled hydrogels as antibacterial agents. *Scientific Reports*, *10*(1), 1–12.

Al-Garawi, Z. S., McIntosh, B. A., Neill-Hall, D., Hatimy, A. A., Sweet, S. M., Bagley, M. C., & Serpell, L. C. (2017). The amyloid architecture provides a scaffold for enzyme-like catalysts. *Nanoscale*, *9*(30), 10773–10783. Available from https://doi.org/10.1039/C7NR02675G.

Alivisatos, A. P. (1996). Semiconductor clusters, nanocrystals, and quantum dots. *Science (New York, N.Y.)*, *271*(5251), 933–937.

Allert, M., & Baltzer, L. (2003). Noncovalent binding of a reaction intermediate by a designed helix-loop-helix motif—Implications for catalyst design. *Chembiochem: A European Journal of Chemical Biology*, *4*(4), 306–318.

Amdursky, N., Molotskii, M., Aronov, D., Adler-Abramovich, L., Gazit, E., & Rosenman, G. (2009). Blue luminescence based on quantum confinement at peptide nanotubes. *Nano Letters*, *9*(9), 3111–3115. Available from https://doi.org/10.1021/nl9008265.

Amdursky, N., Molotskii, M., Gazit, E., & Rosenman, G. (2009). Self-assembled bioinspired quantum dots: Optical properties. *Applied Physics Letters*, *94*(26), 261907.

Amdursky, N., Molotskii, M., Gazit, E., & Rosenman, G. (2010). Elementary building blocks of self-assembled peptide nanotubes. *Journal of the American Chemical Society*, *132*(44), 15632–15636.

Amit, M., Appel, S., Cohen, R., Cheng, G., Hamley, I. W., & Ashkenasy, N. (2014). Hybrid proton and electron transport in peptide fibrils. *Advanced Functional Materials*, *24*(37), 5873–5880.

Anelli, P. -L., Asakawa, M., Ashton, P. R., Bissell, R. A., Clavier, G., Górski, R., & Menzer, S. (1997). Toward controllable molecular shuttles. *Chemistry–A European Journal, 3*(7), 1113–1135.

Arakawa, H., Takeda, K., Higashi, S. L., Shibata, A., Kitamura, Y., & Ikeda, M. (2020). Self-assembly and hydrogel formation ability of Fmoc-dipeptides comprising α-methyl-L-phenylalanine. *Polymer Journal, 52*(8), 923–930.

Ardoña, H. A. M., & Tovar, J. D. (2015). Peptide π-electron conjugates: Organic electronics for biology? *Bioconjugate Chemistry, 26*(12), 2290–2302.

Arnesen, T. (2011). Towards a functional understanding of protein N-terminal acetylation. *PLoS Biology, 9*(5), e1001074.

Atefyekta, S., Blomstrand, E., Rajasekharan, A. K., Svensson, S., Trobos, M., Hong, J., & Andersson, M. (2021). Antimicrobial peptide-functionalized mesoporous hydrogels. *ACS Biomaterials Science & Engineering, 7*(4), 1693–1702.

Azuri, I., Adler-Abramovich, L., Gazit, E., Hod, O., & Kronik, L. (2014). Why are diphenylalanine-based peptide nanostructures so rigid? Insights from first principles calculations. *Journal of the American Chemical Society, 136*(3), 963–969.

Bahram, M., Mohseni, N., & Moghtader, M. (2016). *An introduction to hydrogels and some recent applications. Emerging concepts in analysis and applications of hydrogels*. IntechOpen.

Bandekar, A., Zhu, C., Gomez, A., Menzenski, M. Z., Sempkowski, M., & Sofou, S. (2013). Masking and triggered unmasking of targeting ligands on liposomal chemotherapy selectively suppress tumor growth in vivo. *Molecular Pharmaceutics, 10*(1), 152–160.

Bartocci, S., Berrocal, J. A., Guarracino, P., Grillaud, M., Franco, L., & Mba, M. (2018). Peptide-driven charge-transfer organogels built from synergetic hydrogen bonding and pyrene–naphthalenediimide donor–acceptor interactions. *Chemistry–A European Journal, 24*(12), 2920–2928.

Basak, S., Nanda, J., & Banerjee, A. (2014). Multi-stimuli responsive self-healing metallo-hydrogels: Tuning of the gel recovery property. *Chemical Communications, 50*(18), 2356–2359.

Basavalingappa, V., Bera, S., Xue, B., O'Donnell, J., Guerin, S., Cazade, P.-A., & Tao, K. (2020). Diphenylalanine-derivative peptide assemblies with increased aromaticity exhibit metal-like rigidity and high piezoelectricity. *ACS Nano, 14*(6), 7025–7037.

Bélanger, D., Tong, X., Soumaré, S., Dory, Y. L., & Zhao, Y. (2009). Cyclic peptide–Polymer complexes and their self-assembly. *Chemistry–A European Journal, 15*(17), 4428–4436.

Bélières, M., Chouini-Lalanne, N., & Déjugnat, C. (2015). Synthesis, self-assembly, and catalytic activity of histidine-based structured lipopeptides for hydrolysis reactions in water. *RSC Advances, 5*(45), 35830–35842.

Bijker, M. S., Melief, C. J. M., Offringa, R., & Van Der Burg, S. H. (2007). Design and development of synthetic peptide vaccines: Past, present and future. *Expert Review of Vaccines, 6*(4), 591–603.

Bishop, J. L., & Finlay, B. B. (2006). Friend or foe? Antimicrobial peptides trigger pathogen virulence. *Trends in Molecular Medicine, 12*(1), 3–6. Available from https://doi.org/10.1016/j.molmed.2005.11.001.

Blanco, F. J., Rivas, G., & Serrano, L. (1994). A short linear peptide that folds into a native stable β-hairpin in aqueous solution. *Nature Structural Biology, 1*(9), 584–590.

Blank, A., Fürle, K., Jäschke, A., Mikus, G., Lehmann, M., Hüsing, J., & Böhnlein, E. (2020). Immunization with full-length *Plasmodium falciparum* merozoite surface protein 1 is safe and elicits functional cytophilic antibodies in a randomized first-in-human trial. *NPJ Vaccines, 5*(1), 1–15.

Boman, H. G. (1995). Peptide antibiotics and their role in innate immunity. *Annual Review of Immunology*, *13*, 61–92. Available from https://doi.org/10.1146/annurev.iy.13.040195.000425.

Bortolini, C., Jones, N. C., Hoffmann, S. V., Wang, C., Besenbacher, F., & Dong, M. (2015). Mechanical properties of amyloid-like fibrils defined by secondary structures. *Nanoscale*, *7* (17), 7745–7752.

Bosne, E. D., Heredia, A., Kopyl, S., Karpinsky, D. V., Pinto, A. G., & Kholkin, A. L. (2013). Piezoelectric resonators based on self-assembled diphenylalanine microtubes. *Applied Physics Letters*, *102*(7), 073504.

Boyle, A L, & Woolfson, D.N. (2012). Rational design of peptide-based biosupramolecular systems. Supramolecular Chemistry: From Molecules to Nanomaterials.

Bromley, E. H. C., Channon, K., Moutevelis, E., & Woolfson, D. N. (2008). Peptide and protein building blocks for synthetic biology: From programming biomolecules to self-organized biomolecular systems. *ACS Chemical Biology*, *3*(1), 38–50.

Broo, K. S., Brive, L., Ahlberg, P., & Baltzer, L. (1997). Catalysis of hydrolysis and transesterification reactions of p-nitrophenyl esters by a designed helix – loop – helix dimer. *Journal of the American Chemical Society*, *119*(47), 11362–11372.

Brown, P., Abdulle, O., Boakes, S., Duperchy, E., Moss, S., Simonovic, M., & Dawson, M. J. (2020). Direct modifications of the cyclic peptide polymyxin B leading to analogues with enhanced in vitro antibacterial activity. *Bioorganic & Medicinal Chemistry Letters*, *30*(11), 127163.

Bucak, S., Cenker, C., Nasir, I., Olsson, U., & Zackrisson, M. (2009). Peptide nanotube nematic phase. *Langmuir: The ACS Journal of Surfaces and Colloids*, *25*(8), 4262–4265. Available from https://doi.org/10.1021/la804175h.

Bukchin, A., Sanchez-Navarro, M., Carrera, A., Resa-Pares, C., Castillo-Ecija, H., Balaguer-Lluna, L., & Carcaboso, A. M. (2021). Amphiphilic polymeric nanoparticles modified with a protease-resistant peptide shuttle for the delivery of SN-38 in diffuse intrinsic pontine glioma. *ACS Applied Nano Materials*, *4*(2), 1314–1329.

Burley, S. K., & Petsko, G. A. (1985). Aromatic-aromatic interaction: A mechanism of protein structure stabilization. *Science (New York, N.Y.)*, *229*(4708), 23–28.

Cárdenas, D. J., Livoreil, A., & Sauvage, J.-P. (1996). Redox control of the ring-gliding motion in a Cu-complexed catenane: A process involving three distinct geometries. *Journal of the American Chemical Society*, *118*(47), 11980–11981.

Carrejo, N. C., Moore, A. N., Lopez Silva, T. L., Leach, D. G., Li, I.-C., Walker, D. R., & Hartgerink, J. D. (2018). Multidomain peptide hydrogel accelerates healing of full-thickness wounds in diabetic mice. *ACS Biomaterials Science & Engineering*, *4*(4), 1386–1396.

Castelletto, V., Moulton, C. M., Cheng, G., Hamley, I. W., Hicks, M. R., Rodger, A., & Alemán, C. (2011). Self-assembly of Fmoc-tetrapeptides based on the RGDS cell adhesion motif. *Soft Matter*, *7*(24), 11405–11415.

Chakraborty, P., & Gazit, E. (2018). Amino acid based self-assembled nanostructures: Complex structures from remarkably simple building blocks. *ChemNanoMat*, *4*(8), 730–740.

Chen, S., Wang, Y., Nie, T., Bao, C., Wang, C., Xu, T., & Yang, Y. (2018). An artificial molecular shuttle operates in lipid bilayers for ion transport. *Journal of the American Chemical Society*, *140*(51), 17992–17998.

Chen, Y., Mant, C. T., Farmer, S. W., Hancock, R. E. W., Vasil, M. L., & Hodges, R. S. (2005). Rational design of α-helical antimicrobial peptides with enhanced activities and sSpecificity/therapeutic index*. *Journal of Biological Chemistry*, *280*(13), 12316–12329. Available from https://doi.org/10.1074/jbc.M413406200.

Chen, Y., Wu, J. J., & Huang, L. (2010). Nanoparticles targeted with NGR motif deliver c-myc siRNA and doxorubicin for anticancer therapy. *Molecular Therapy*, *18*(4), 828–834.

Cheng, G., Castelletto, V., Moulton, C. M., Newby, G. E., & Hamley, I. W. (2010). Hydrogelation and self-assembly of Fmoc-tripeptides: Unexpected influence of sequence on self-assembled fibril structure, and hydrogel modulus and anisotropy. *Langmuir: The ACS Journal of Surfaces and Colloids*, *26*(7), 4990–4998.

Cherny, I., & Gazit, E. (2008). Amyloids: Not only pathological agents but also ordered nanomaterials. *Angewandte Chemie International Edition*, *47*(22), 4062–4069.

Ciani, B., Hutchinson, E. G., Sessions, R. B., & Woolfson, D. N. (2002). A designed system for assessing how sequence affects α to β conformational transitions in proteins. *Journal of Biological Chemistry*, *277*(12), 10150–10155.

Colgrave, M. L., Korsinczky, M. J. L., Clark, R. J., Foley, F., & Craik, D. J. (2010). Sunflower trypsin inhibitor-1, proteolytic studies on a trypsin inhibitor peptide and its analogs. *Peptide Science*, *94*(5), 665–672.

Colherinhas, G., Malaspina, T., & Fileti, E. E. (2018). Storing energy in biodegradable electrochemical supercapacitors. *ACS Omega*, *3*(10), 13869–13875.

Collin, J.-P., Gavinã, P., & Sauvage, J.-P. (1996). Electrochemically induced molecular motions in a copper (I) complex pseudorotaxane. *Chemical Communications*, *17*, 2005–2006.

Corti, A., Gasparri, A. M., Ghitti, M., Sacchi, A., Sudati, F., Fiocchi, M., & Valtorta, S. (2017). Glycine N-methylation in NGR-tagged nanocarriers prevents isoaspartate formation and integrin binding without impairing CD13 recognition and tumor homing. *Advanced Functional Mmaterials*, *27*(36), 1701245.

Creasey, R. C. G., Shingaya, Y., & Nakayama, T. (2015). Improved electrical conductance through self-assembly of bioinspired peptides into nanoscale fibers. *Materials Chemistry and Physics*, *158*, 52–59.

Criado-Gonzalez, M., Iqbal, M. H., Carvalho, A., Schmutz, M., Jierry, L., Schaaf, P., & Boulmedais, F. (2020). Surface triggered self-assembly of Fmoc-tripeptide as an antibacterial coating. *Frontiers in Bioengineering and Bbiotechnology*, *8*, 938.

Criado-Gonzalez, M., Wagner, D., Rodon Fores, J., Blanck, C., Schmutz, M., Chaumont, A., ... Combet, J. (2020). Supramolecular hydrogel induced by electrostatic interactions between polycation and phosphorylated-Fmoc-tripeptide. *Chemistry of Materials*, *32*(5), 1946–1956.

Cui, H., Muraoka, T., Cheetham, A. G., & Stupp, S. I. (2009). Self-assembly of giant peptide nanobelts. *Nano Letters*, *9*(3), 945–951.

Cui, R., Zhao, P., Yan, Y., Bao, G., Damirin, A., & Liu, Z. (2021). Outstanding drug-loading/release capacity of hollow Fe-Metal–Organic framework-based microcapsules: A potential multifunctional drug-delivery platform. *Inorganic Chemistry*, *60*(3), 1664–1671.

da Silva, E. R., Walter, M. N. M., Reza, M., Castelletto, V., Ruokolainen, J., Connon, C. J., & Hamley, I. W. (2015). Self-assembled arginine-capped peptide bolaamphiphile nanosheets for cell culture and controlled wettability surfaces. *Biomacromolecules*, *16*(10), 3180–3190.

Dai, B., Li, D., Xi, W., Luo, F., Zhang, X., Zou, M., & Wei, G. (2015). Tunable assembly of amyloid-forming peptides into nanosheets as a retrovirus carrier. *Proceedings of the National Academy of Sciences*, *112*(10), 2996–3001.

Damborsky, J., & Brezovsky, J. (2009). Computational tools for designing and engineering biocatalysts. *Current Opinion in Chemical Biology*, *13*(1), 26–34. Available from https://doi.org/10.1016/j.cbpa.2009.02.021.

D'Andrea, L. D., Iaccarino, G., Fattorusso, R., Sorriento, D., Carannante, C., Capasso, D., & Pedone, C. (2005). Targeting angiogenesis: Structural characterization and biological properties of a de novo engineered VEGF mimicking peptide. *Proceedings of the National Academy of Sciences*, *102*(40), 14215–14220.

Davidson, G. J. E., Sharma, S., & Loeb, S. J. (2010). A [2] rotaxane flip switch driven by coordination geometry. *Angewandte Chemie International Edition*, *49*(29), 4938–4942.

Del Gatto, A., Zaccaro, L., Grieco, P., Novellino, E., Zannetti, A., Del Vecchio, S., & Saviano, M. (2006). Novel and selective αvβ3 receptor peptide antagonist: Design, synthesis, and biological behavior. *Journal of Medicinal Chemistry*, *49*(11), 3416–3420.

Derossi, D., Joliot, A. H., Chassaing, G., & Prochiantz, A. (1994). The third helix of the Antennapedia homeodomain translocates through biological membranes. *Journal of Biological Chemistry*, *269*(14), 10444–10450.

Du, X., Zhou, J., Shi, J., & Xu, B. (2015). Supramolecular hydrogelators and hydrogels: From soft matter to molecular biomaterials. *Chemical Reviews*, *115*(24), 13165–13307.

Echeverria, C., Fernandes, S. N., Godinho, M. H., Borges, J. P., & Soares, P. I. P. (2018). Functional stimuli-responsive gels: Hydrogels and microgels. *Gels*, *4*(2), 54.

Eder, U., Sauer, G., & Wiechert, R. (1971). New type of asymmetric cyclization to optically active steroid CD partial structures. *Angewandte Chemie International Edition in English*, *10*(7), 496–497.

Epstein, J. E., Giersing, B., Mullen, G., Moorthy, V., & Richie, T. L. (2007). Malaria vaccines: Are we getting closer? *Current Opinion in Molecular Therapeutics*, *9*(1), 12–24.

Erwin, S. C., Zu, L., Haftel, M. I., Efros, A. L., Kennedy, T. A., & Norris, D. J. (2005). Doping semiconductor nanocrystals. *Nature*, *436*(7047), 91–94.

Faccone, D., Veliz, O., Corso, A., Noguera, M., Martínez, M., Payes, C., & Maffía, P. C. (2014). Antimicrobial activity of de novo designed cationic peptides against multi-resistant clinical isolates. *European Journal of Medicinal Chemistry*, *71*, 31–35. Available from https://doi.org/10.1016/j.ejmech.2013.10.065.

Fan, Z., Chang, Y., Cui, C., Sun, L., Wang, D. H., Pan, Z., & Zhang, M. (2018). Near infrared fluorescent peptide nanoparticles for enhancing esophageal cancer therapeutic efficacy. *Nature Communications*, *9*(1), 2605. Available from https://doi.org/10.1038/s41467-018-04763-y.

Fan, Z., Sun, L., Huang, Y., Wang, Y., & Zhang, M. (2016). Bioinspired fluorescent dipeptide nanoparticles for targeted cancer cell imaging and real-time monitoring of drug release. *Nature Nanotechnology*, *11*(4), 388–394. Available from https://doi.org/10.1038/nnano.2015.312.

Farha, O. K., Yazaydın, A. Ö., Eryazici, I., Malliakas, C. D., Hauser, B. G., Kanatzidis, M. G., & Hupp, J. T. (2010). De novo synthesis of a metal–organic framework material featuring ultrahigh surface area and gas storage capacities. *Nature Chemistry*, *2*(11), 944–948.

Faul, C. F. J., & Antonietti, M. (2003). Ionic self-assembly: Facile synthesis of supramolecular materials. *Advanced Materials*, *15*(9), 673–683.

Fernandez-Lopez, S., Kim, H.-S., Choi, E. C., Delgado, M., Granja, J. R., Khasanov, A., & Wilcoxen, K. M. (2001). Antibacterial agents based on the cyclic D, L-α-peptide architecture. *Nature*, *412*(6845), 452–455. Available from https://doi.org/10.1038/35086601.

Fichman, G., & Gazit, E. (2014). Self-assembly of short peptides to form hydrogels: Design of building blocks, physical properties and technological applications. *Acta Biomaterialia*, *10*(4), 1671–1682.

Firbas, C., Jilma, B., Tauber, E., Buerger, V., Jelovcan, S., Lingnau, K., & Klade, C. S. (2006). Immunogenicity and safety of a novel therapeutic hepatitis C virus (HCV) peptide vaccine: A randomized, placebo controlled trial for dose optimization in 128 healthy subjects. *Vaccine*, *24*(20), 4343–4353.

Fishwick, C. W. G., Beevers, A. J., Carrick, L. M., Whitehouse, C. D., Aggeli, A., & Boden, N. (2003). Structures of helical β-tapes and twisted ribbons: The role of side-chain interactions on twist and bend behavior. *Nano Letters*, *3*(11), 1475–1479.

Fuentes, J. A., Lebl, T., Slawin, A. M. Z., & Clarke, M. L. (2011). Synthesis of organocatalysts using non-covalent chemistry; understanding the reactivity of ProNap, an enamine-type organocatalyst that can self assemble with complementary co-catalysts. *Chemical Science*, *2* (10), 1997–2005.

Gadepalli, R., Dhawan, B., Sreenivas, V., Kapil, A., Ammini, A. C., & Chaudhry, R. (2006). A clinico-microbiological study of diabetic foot ulcers in an Indian tertiary care hospital. *Diabetes Care*, *29*(8), 1727–1732.

Gahane, A. Y., Ranjan, P., Singh, V., Sharma, R. K., Sinha, N., Sharma, M., & Thakur, A. K. (2018). Fmoc-phenylalanine displays antibacterial activity against Gram-positive bacteria in gel and solution phases. *Soft Matter*, *14*(12), 2234–2244.

Garcia, A. M., Lavendomme, R., Kralj, S., Kurbasic, M., Bellotto, O., Cringoli, M. C., & Marchesan, S. (2020). Self-assembly of an amino acid derivative into an antimicrobial hydrogel biomaterial. *Chemistry–A European Journal*, *26*(8), 1880–1886.

Garcia, J., Arranz-Gibert, P., Sánchez-Navarro, M., Giralt, E., & Teixidó, M. (2020). Peptide shuttle-mediated delivery for brain gene therapies. *Current Topics in Medicinal Chemistry*.

Garcia-Viloca, M., Gao, J., Karplus, M., & Truhlar, D. G. (2004). How enzymes work: Analysis by modern rate theory and computer simulations. *Science (New York, N.Y.)*, *303*(5655), 186–195. Available from https://doi.org/10.1126/science.1088172.

Gaurav, J., & Santhoshkumar, T. R. (2020). Conformationally constrained peptides for drug delivery. *Peptide Science*, *26*(6).

Gawande, P. V., Leung, K. P., & Madhyastha, S. (2014). Antibiofilm and antimicrobial efficacy of DispersinB®-KSL-W peptide-based wound gel against chronic wound infection associated bacteria. *Current Microbiology*, *68*(5), 635–641.

Gentilucci, L., Cardillo, G., Squassabia, F., Tolomelli, A., Spampinato, S., Sparta, A., & Baiula, M. (2007). Inhibition of cancer cell adhesion by heterochiral Pro-containing RGD mimetics. *Bioorganic & Medicinal Chemistry Letters*, *17*(8), 2329–2333.

Ghadiri, M. R., Granja, J. R., Milligan, R. A., McRee, D. E., & Khazanovich, N. (1993). Self-assembling organic nanotubes based on a cyclic peptide architecture. *Nature*, *366*(6453), 324–327.

Ghadiri, M. R., Soares, C., & Choi, C. (1992). A convergent approach to protein design. Metal ion-assisted spontaneous self-assembly of a polypeptide into a triple-helix bundle protein. *Journal of the American Chemical Society*, *114*(3), 825–831.

Görbitz, C. H. (2001). Nanotube formation by hydrophobic dipeptides. *Chemistry–A European Journal*, *7*(23), 5153–5159.

Goyal, R., & Ramakrishnan, V. (2019). Chapter 2—Peptide-based drug delivery systems. In S. S. Mohapatra, S. Ranjan, N. Dasgupta, R. K. Mishra, & S. Thomas (Eds.), *Characterization and Biology of Nanomaterials for Drug Delivery* (pp. 25–45). Elsevier.

Goyal, R., Jerath, G., Akhil, R., Chandrasekharan, A., Puppala, E. R., Ponneganti, S., & Ramakrishnan, V. (2021). Geometry encoded functional programming of tumor homing peptides for targeted drug delivery. *Journal of Controlled Release*, *333*, 16–27.

Goyal, R., Jerath, G., Chandrasekharan, A., Christian, Y., Kumar, T. R., & Ramakrishnan, V. (2021). Molecular hybridization combining tumor homing and penetrating peptide domains for cellular targeting. *Drug Delivery and Translational Research*, 1–8.

Goyal, R., Jerath, G., Chandrasekharan, A., Kumar, T. R. S., & Ramakrishnan, V. (2020). Peptide-based delivery vectors with pre-defined geometrical locks. *RSC Medicinal Chemistry*, *11*(11), 1303–1313.

Grek, C. L., Prasad, G. M., Viswanathan, V., Armstrong, D. G., Gourdie, R. G., & Ghatnekar, G. S. (2015). Topical administration of a connexin43-based peptide augments healing of chronic neuropathic diabetic foot ulcers: A multicenter, randomized trial. *Wound Repair and Regeneration*, *23*(2), 203–212.

Greten, T. F., Forner, A., Korangy, F., N'Kontchou, G., Barget, N., Ayuso, C., & Bruix, J. (2010). A phase II open label trial evaluating safety and efficacy of a telomerase peptide vaccination in patients with advanced hepatocellular carcinoma. *BMC Cancer*, *10*(1), 1–7.

Guler, M. O., & Stupp, S. I. (2007). A self-assembled nanofiber catalyst for ester hydrolysis. *Journal of the American Chemical Society*, *129*(40), 12082–12083.

Guler, M. O., Soukasene, S., Hulvat, J. F., & Stupp, S. I. (2005). Presentation and recognition of biotin on nanofibers formed by branched peptide amphiphiles. *Nano Letters*, *5*(2), 249–252.

Guo, C., Luo, Y., Zhou, R., & Wei, G. (2014). Triphenylalanine peptides self-assemble into nanospheres and nanorods that are different from the nanovesicles and nanotubes formed by diphenylalanine peptides. *Nanoscale*, *6*(5), 2800–2811.

Hamley, I. W., Dehsorkhi, A., & Castelletto, V. (2013). Self-assembled arginine-coated peptide nanosheets in water. *Chemical Communications*, *49*(18), 1850–1852.

Hamley, I. W., Nutt, D. R., Brown, G. D., Miravet, J. F., Escuder, B., & Rodríguez-Llansola, F. (2010). Influence of the solvent on the self-assembly of a modified amyloid beta peptide fragment. II. NMR and computer simulation investigation. *The Journal of Physical Chemistry. B*, *114*(2), 940–951.

Hancock, R. E. W., & Sahl, H.-G. (2006). Antimicrobial and host-defense peptides as new anti-infective therapeutic strategies. *Nature Biotechnology*, *24*(12), 1551–1557. Available from https://doi.org/10.1038/nbt1267.

Harrison, P. (2016). *Quantum wells, wires and dots*. Wiley Online Library.

Hasanzadeh, M., & Shadjou, N. (2017). (Nano)-materials and methods of signal enhancement for genosensing of p53 tumor suppressor protein: Novel research overview. *Materials Science and Engineering: C*, *76*, 1424–1439.

Hauser, C. A., & Zhang, S. (2010). Peptides as biological semiconductors. *Nature*, *468*(7323), 516–517.

Hazam, P. K., Akhil, R., Jerath, G., Saikia, J., & Ramakrishnan, V. (2019). Topological effects on the designability and bactericidal potency of antimicrobial peptides. *Biophysical Chemistry*, *248*, 1–8.

Hazam, P. K., Goyal, R., & Ramakrishnan, V. (2019). Peptide based antimicrobials: Design strategies and therapeutic potential. *Progress in Biophysics and Molecular Biology*, *142*, 10–22. Available from https://doi.org/10.1016/j.pbiomolbio.2018.08.006.

Hazam, P. K., Jerath, G., Chaudhary, N., & Ramakrishnan, V. (2018). Peptido-mimetic approach in the design of syndiotactic antimicrobial peptides. *International Journal of Peptide Research and Therapeutics*, *24*(2), 299–307. Available from https://doi.org/10.1007/s10989-017-9615-3.

Hazam, P. K., Jerath, G., Kumar, A., Chaudhary, N., & Ramakrishnan, V. (2017). Effect of tacticity-derived topological constraints in bactericidal peptides. *Biochimica et Biophysica Acta (BBA) - Biomembranes*, *1859*(8), 1388–1395. Available from https://doi.org/10.1016/j.bbamem.2017.05.002.

Hazam, P. K., Phukan, C., Akhil, R., Singh, A., & Ramakrishnan, V. (2021). Antimicrobial effects of syndiotactic polypeptides. *Scientific Reports*, *11*(1), 1–10.

Hazam, P. K., Singh, A., Chaudhary, N., & Ramakrishnan, V. (2019). Bactericidal potency and extended serum life of stereo-chemically engineered peptides against Mycobacterium. *International Journal of Peptide Research and Therapeutics*, *25*(2), 465–472.

He, L., Sayers, E. J., Watson, P., & Jones, A. T. (2018). Contrasting roles for actin in the cellular uptake of cell penetrating peptide conjugates. *Scientific Reports*, *8*(1), 1–12.

He, Yi, Xie, S., Yang, X., Yuan, R., & Chai, Y. (2015). Electrochemical peptide biosensor based on in situ silver deposition for detection of prostate specific antigen. *ACS Applied Materials & Interfaces*, *7*(24), 13360–13366.

Hill, R. B., & DeGrado, W. F. (1998). Solution structure of α2D, a nativelike de novo designed protein. *Journal of the American Chemical Society*, *120*(6), 1138–1145.

Horcajada, P., Gref, R., Baati, T., Allan, P. K., Maurin, G., Couvreur, P., & Serre, C. (2012). Metal–organic frameworks in biomedicine. *Chemical Reviews*, *112*(2), 1232–1268.

Horcajada, P., Serre, C., Vallet-Regí, M., Sebban, M., Taulelle, F., & Férey, G. (2006). Metal–organic frameworks as efficient materials for drug delivery. *Angewandte chemie*, *118*(36), 6120–6124.

Hu, B., Shen, Y., Adamcik, J., Fischer, P., Schneider, M., Loessner, M. J., & Mezzenga, R. (2018). Polyphenol-binding amyloid fibrils self-assemble into reversible hydrogels with antibacterial activity. *ACS Nano*, *12*(4), 3385–3396.

Hu, X., Liao, M., Gong, H., Zhang, L., Cox, H., Waigh, T. A., & Lu, J. R. (2020). Recent advances in short peptide self-assembly: from rational design to novel applications. *Current Opinion in Colloid & Interface Science*, *45*, 1–13.

Huang, R., Qi, W., Su, R., Zhao, J., & He, Z. (2011). Solvent and surface controlled self-assembly of diphenylalanine peptide: From microtubes to nanofibers. *Soft Matter*, *7*(14), 6418–6421.

Huang, Y., Qiu, Z., Xu, Y., Shi, J., Lin, H., & Zhang, Y. (2011). Supramolecular hydrogels based on short peptides linked with conformational switch. *Organic & Biomolecular Chemistry*, *9*(7), 2149–2155.

Huang, Z., Guan, S., Wang, Y., Shi, G., Cao, L., Gao, Y., & Liu, J. (2013). Self-assembly of amphiphilic peptides into bio-functionalized nanotubes: A novel hydrolase model. *Journal of Materials Chemistry B*, *1*(17), 2297–2304.

Ikeda, M., Tanida, T., Yoshii, T., & Hamachi, I. (2011). Rational molecular design of stimulus-responsive supramolecular hydrogels based on dipeptides. *Advanced Materials*, *23*(25), 2819–2822.

Islam, M. M., Ravichandran, R., Olsen, D., Ljunggren, M. K., Fagerholm, P., Lee, C.-J., & Phopase, J. (2016). Self-assembled collagen-like-peptide implants as alternatives to human donor corneal transplantation. *RSC Advances*, *6*(61), 55745–55749.

Jain, R., & Roy, S. (2020). Tuning the gelation behavior of short laminin derived peptides via solvent mediated self-assembly. *Materials Science and Engineering: C*, *108*, 110483.

Jayawickramarajah, J., & Wilson, A. J. (2017). *Volume 5: Supramolecular medicinal chemistry and chemical biology*. .

Jerath, G., Goyal, R., Trivedi, V., Santhoshkumar, T. R., & Ramakrishnan, V. (2019). Syndiotactic peptides for targeted delivery. *Acta Biomaterialia*, *87*, 130–139.

Johnsson, K., Allemann, R. K., Widmer, H., & Benner, S. A. (1993). Synthesis, structure and activity of artificial, rationally designed catalytic polypeptides. *Nature*, *365*(6446), 530–532.

Kebebe, D., Liu, Y., Wu, Y., Vilakhamxay, M., Liu, Z., & Li, J. (2018). Tumor-targeting delivery of herb-based drugs with cell-penetrating/tumor-targeting peptide-modified nanocarriers. *International Journal of Nanomedicine*, *13*, 1425.

Kim, J., Han, T. H., Kim, Y. -Il, Park, J. S., Choi, J., Churchill, D. G., ... Ihee, H. (2010). Role of water in directing diphenylalanine assembly into nanotubes and nanowires. *Advanced Materials*, *22*(5), 583–587.

Knowles, T. P. J., & Buehler, M. J. (2011). Nanomechanics of functional and pathological amyloid materials. *Nature Nanotechnology*, *6*(8), 469–479.

Koivunen, E., Gay, D. A., & Ruoslahti, E. (1993). Selection of peptides binding to the alpha 5 beta 1 integrin from phage display library. *Journal of Biological Chemistry*, *268*(27), 20205–20210.

Kolundzic, F., Noshi, M. N., Tjandra, M., Movassaghi, M., & Miller, S. J. (2011). Chemoselective and enantioselective oxidation of indoles employing aspartyl peptide catalysts. *Journal of the American Chemical Society*, *133*(23), 9104–9111.

Krone, M. G., Hua, L., Soto, P., Zhou, R., Berne, B. J., & Shea, J.-E. (2008). Role of water in mediating the assembly of Alzheimer amyloid-β Aβ16 – 22 protofilaments. *Journal of the American Chemical Society*, *130*(33), 11066–11072.

Kubiak, P. S., Awhida, S., Hotchen, C., Deng, W., Alston, B., McDonald, T. O., & Cameron, P. J. (2015). Polymerization of low molecular weight hydrogelators to form electrochromic polymers. *Chemical Communications*, *51*(52), 10427–10430.

Kumar, A., & Ramakrishnan, V. (2010). Creating novel protein scripts beyond natural alphabets. *Systems and Synthetic Biology*, *4*(4), 247–256. Available from https://doi.org/10.1007/s11693-011-9068-5.

Kumar, A., Ranbhor, R., Patel, K., Ramakrishnan, V., & Durani, S. (2017). Automated protein design: Landmarks and operational principles. *Progress in Biophysics and Molecular Biology*, *125*, 24–35. Available from https://doi.org/10.1016/j.pbiomolbio.2016.12.002.

Kumar, S. R., Gallazzi, F. A., Ferdani, R., Anderson, C. J., Quinn, T. P., & Deutscher, S. L. (2010). In vitro and in vivo evaluation of 64Cu-radiolabeled KCCYSL peptides for targeting epidermal growth factor receptor-2 in breast carcinomas. *Cancer Biotherapy & Radiopharmaceuticals*, *25*(6), 693–703.

Kumar, V. A., Taylor, N. L., Shi, S., Wang, B. K., Jalan, A. A., Kang, M. K., & Hartgerink, J. D. (2015). Highly angiogenic peptide nanofibers. *ACS Nano*, *9*(1), 860–868.

Kumari, A., & Ahmad, B. (2019). The physical basis of fabrication of amyloid-based hydrogels by lysozyme. *RSC Advances*, *9*(64), 37424–37435.

Kundu, B., Eltohamy, M., Yadavalli, V. K., Kundu, S. C., & Kim, H.-W. (2016). Biomimetic designing of functional silk nanotopography using self-assembly. *ACS Applied Materials & Interfaces*, *8*(42), 28458–28467.

Langel, U. (2002). *Cell-penetrating peptides: Processes and applications*. CRC press.

Laungani, A. C., Slattery, J. M., Krossing, I., & Breit, B. (2008). Supramolecular bidentate ligands by metal-directed in situ formation of antiparallel β-sheet structures and application in asymmetric catalysis. *Chemistry–A European Journal*, *14*(15), 4488–4502.

Laverty, G., McCloskey, A. P., Gilmore, B. F., Jones, D. S., Zhou, J., & Xu, B. (2014). Ultrashort cationic naphthalene-derived self-assembled peptides as antimicrobial nanomaterials. *Biomacromolecules*, *15*(9), 3429–3439.

Lee, J. Y., Farha, O. K., Roberts, J., Scheidt, K. A., Nguyen, S. T., & Hupp, J. T. (2009). Metal–organic framework materials as catalysts. *Chemical Society Reviews*, *38*(5), 1450–1459.

Lehn, J.-M. (1993). Supramolecular chemistry. *Science (New York, N.Y.)*, *260*(5115), 1762–1764.

Li, Q., Jia, Y., Dai, L., Yang, Y., & Li, J. (2015). Controlled rod nanostructured assembly of diphenylalanine and their optical waveguide properties. *ACS Nano*, *9*(3), 2689–2695.

Lin, S., Li, H., Tao, Y., Liu, J., Yuan, W., Chen, Y., & Liu, S. (2020). In vitro and in vivo evaluation of membrane-active flavone amphiphiles: Semisynthetic kaempferol-derived antimicrobials against drug-resistant Gram-positive bacteria. *Journal of Medicinal Chemistry*, *63*(11), 5797–5815. Available from https://doi.org/10.1021/acs.jmedchem.0c00053.

Liu, Q., Wang, J., & Boyd, B. J. (2015). Peptide-based biosensors. *Talanta*, *136*, 114–127.

Liu, X., Fei, J., Wang, A., Cui, W., Zhu, P., & Li, J. (2017). Transformation of dipeptide-based organogels into chiral crystals by cryogenic treatment. *Angewandte Chemie International Edition*, *56*(10), 2660–2663.

Liu, X, Sun, X, & Liang, G. (2020). Peptide-based supramolecular hydrogels for bioimaging applications. Biomaterials Science.

Liu, Y., McNevin, J., Zhao, H., Tebit, D. M., Troyer, R. M., McSweyn, M., & McElrath, M. J. (2007). Evolution of human immunodeficiency virus type 1 cytotoxic T-lymphocyte epitopes: Fitness-balanced escape. *Journal of Virology*, *81*(22), 12179–12188.

Luo, Z., Dai, Y., & Gao, H. (2019). Development and application of hyaluronic acid in tumor targeting drug delivery. *Acta Pharmaceutica Sinica B*, *9*(6), 1099–1112.

Mahalakshmi, R., Sengupta, A., Raghothama, S., Shamala, N., & Balaram, P. (2005). Tryptophan-containing peptide helices: Interactions involving the indole side chain. *The Journal of Peptide Research*, *66*(5), 277–296.

Mahler, A., Reches, M., Rechter, M., Cohen, S., & Gazit, E. (2006). Rigid, self-assembled hydrogel composed of a modified aromatic dipeptide. *Advanced Materials*, *18*(11), 1365–1370.

Malonis, R. J., Lai, J. R., & Vergnolle, O. (2019). Peptide-based vaccines: Current progress and future challenges. *Chemical Reviews*, *120*(6), 3210–3229.

Marchesan, S., Vargiu, A. V., & Styan, K. E. (2015). The Phe-Phe motif for peptide self-assembly in nanomedicine. *Molecules (Basel, Switzerland)*, *20*(11), 19775–19788.

Marsh, E. N. G., & DeGrado, W. F. (2002). Noncovalent self-assembly of a heterotetrameric diiron protein. *Proceedings of the National Academy of Sciences*, *99*(8), 5150–5154.

Martin, A. D., Robinson, A. B., Mason, A. F., Wojciechowski, J. P., & Thordarson, P. (2014). Exceptionally strong hydrogels through self-assembly of an indole-capped dipeptide. *Chemical Communications*, *50*(98), 15541–15544.

Martins, T. D., de Souza, M. I., Cunha, B. B., Takahashi, P. M., Ferreira, F. F., Souza, J. A., & Alves, W. A. (2011). Influence of pH and pyrenyl on the structural and morphological control of peptide nanotubes. *The Journal of Physical Chemistry C*, *115*(16), 7906–7913.

Mas-Moruno, C., Rechenmacher, F., & Kessler, H. (2010). Cilengitide: The first anti-angiogenic small molecule drug candidate. Design, synthesis and clinical evaluation. *Anti-cancer Agents in Medicinal Chemistry (Formerly Current Medicinal Chemistry-Anti-Cancer Agents)*, *10*(10), 753–768.

Masters, B. R. (2013). Molecular fluorescence: Principles and applications. *Journal of Biomedical Optics*, *18*(3), 039901.

McConney, M. E., Anderson, K. D., Brott, L. L., Naik, R. R., & Tsukruk, V. V. (2009). Bioinspired material approaches to sensing. *Advanced Functional Materials*, *19*(16), 2527–2544.

Mei, L., Xu, K., Zhai, Z., He, S., Zhu, T., & Zhong, W. (2019). Doxorubicin-reinforced supramolecular hydrogels of RGD-derived peptide conjugates for pH-responsive drug delivery. *Organic & Biomolecular Chemistry*, *17*(15), 3853–3860.

Meng, N., Ren, X., Santagiuliana, G., Ventura, L., Zhang, H., Wu, J., & Bilotti, E. (2019). Ultrahigh β-phase content poly (vinylidene fluoride) with relaxor-like ferroelectricity for high energy density capacitors. *Nature Communications*, *10*(1), 1–9.

Meyer, D. E., & Chilkoti, A. (2004). Quantification of the effects of chain length and concentration on the thermal behavior of elastin-like polypeptides. *Biomacromolecules*, *5*(3), 846–851.

Migliore, A., Polizzi, N. F., Therien, M. J., & Beratan, D. N. (2014). Biochemistry and theory of proton-coupled electron transfer. *Chemical Reviews*, *114*(7), 3381–3465.

Mitchell, D. J., Steinman, L., Kim, D. T., Fathman, C. G., & Rothbard, J. B. (2000). Polyarginine enters cells more efficiently than other polycationic homopolymers. *The Journal of Peptide Research*, *56*(5), 318–325.

Moorcroft, S. C. T., Roach, L., Jayne, D. G., Ong, Z. Y., & Evans, S. D. (2020). Nanoparticle-loaded hydrogel for the light-activated release and photothermal enhancement of antimicrobial peptides. *ACS Applied Materials & Interfaces*, *12*(22), 24544–24554.

Nair, R. V., Baravkar, S. B., Ingole, T. S., & Sanjayan, G. J. (2014). Synthetic turn mimetics and hairpin nucleators: Quo Vadimus? *Chemical Communications*, *50*(90), 13874–13884. Available from https://doi.org/10.1039/C4CC03114H.

Nanda, J., Biswas, A., & Banerjee, A. (2013). Single amino acid based thixotropic hydrogel formation and pH-dependent morphological change of gel nanofibers. *Soft Matter*, *9*(16), 4198–4208.

Niyonsaba, F., Madera, L., Afacan, N., Okumura, K., Ogawa, H., & Hancock, R. E. W. (2013). The innate defense regulator peptides IDR-HH2, IDR-1002, and IDR-1018 modulate human neutrophil functions. *Journal of Leukocyte Biology*, *94*(1), 159–170.

Novak, P., Schmidt, R., Kontsekova, E., Kovacech, B., Smolek, T., Katina, S., & Dal-Bianco, P. (2018). FUNDAMANT: An interventional 72-week phase 1 follow-up study of AADvac1, an active immunotherapy against tau protein pathology in Alzheimer's disease. *Alzheimer's Research & Therapy*, *10*(1), 1–16.

Oba, M., Fukushima, S., Kanayama, N., Aoyagi, K., Nishiyama, N., Koyama, H., & Kataoka, K. (2007). Cyclic RGD peptide-conjugated polyplex micelles as a targetable gene delivery system directed to cells possessing $\alpha v\beta 3$ and $\alpha v\beta 5$ integrins. *Bioconjugate Chemistry*, *18*(5), 1415–1423.

Orbach, R., Mironi-Harpaz, I., Adler-Abramovich, L., Mossou, E., Mitchell, E. P., Forsyth, V. T., & Seliktar, D. (2012). The rheological and structural properties of Fmoc-peptide-based hydrogels: The effect of aromatic molecular architecture on self-assembly and physical characteristics. *Langmuir: The ACS Journal of Surfaces and Colloids*, *28*(4), 2015–2022.

Panda, J. J., & Chauhan, V. S. (2014). Short peptide based self-assembled nanostructures: Implications in drug delivery and tissue engineering. *Polymer Chemistry*, *5*(15), 4418–4436.

Pandey, G., & Ramakrishnan, V. (2020). Invasive and non-invasive therapies for Alzheimer's disease and other amyloidosis. *Biophysical Reviews*, *12*(5), 1175–1186.

Pandey, G., Morla, S., Nemade, H. B., Kumar, S., & Ramakrishnan, V. (2019). Modulation of aggregation with an electric field; scientific roadmap for a potential non-invasive therapy against tauopathies. *RSC Advances*, *9*(9), 4744–4750.

Pandey, G., Saikia, J., Sasidharan, S., Joshi, D. C., Thota, S., Nemade, H. B., & Ramakrishnan, V. (2017). Modulation of peptide based nano-assemblies with electric and magnetic fields. *Scientific Reports*, *7*(1), 1–9.

Paramonov, S. E., Jun, H.-W., & Hartgerink, J. D. (2006). Self-assembly of peptide – amphiphile nanofibers: The roles of hydrogen bonding and amphiphilic packing. *Journal of the American Chemical Society*, *128*(22), 7291–7298.

Pasqualini, R., Koivunen, E., Kain, R., Lahdenranta, J., Sakamoto, M., Stryhn, A., & Ruoslahti, E. (2000). Aminopeptidase N is a receptor for tumor-homing peptides and a target for inhibiting angiogenesis. *Cancer Research*, *60*(3), 722–727.

Patel, S. G., Sayers, E. J., He, L., Narayan, R., Williams, T. L., Mills, E. M., & Tsai, Y.-H. (2019). Cell-penetrating peptide sequence and modification dependent uptake and subcellular distribution of green florescent protein in different cell lines. *Scientific Reports*, *9*(1), 1–9.

Pelin, J. N. B. D., Gerbelli, B. B., Edwards-Gayle, C. J. C., Aguilar, A. M., Castelletto, V., Hamley, I. W., & Alves, W. A. (2020). Amyloid peptide mixtures: Self-assembly, hydrogelation, nematic ordering, and catalysts in aldol reactions. *Langmuir: The ACS Journal of Surfaces and Colloids*, *36*(11), 2767–2774.

Pires, M. M., Przybyla, D. E., & Chmielewski, J. (2009). A metal–collagen peptide framework for three-dimensional cell culture. *Angewandte Chemie International Edition*, *48*(42), 7813–7817.

Prabhukumar, G., Matsumoto, M., Mulchandani, A., & Chen, W. (2004). Cadmium removal from contaminated soil by tunable biopolymers. *Environmental Science & Technology*, *38* (11), 3148–3152.

Prakash, V., Ranbhor, R., & Ramakrishnan, V. (2020). De novo designed heterochiral blue fluorescent protein. *ACS Omega*, *5*(41), 26382–26388. Available from https://doi.org/10.1021/acsomega.0c02574.

Pujals, S., Fernández-Carneado, J., Ludevid, M. D., & Giralt, E. (2008). D-SAP: A new, noncytotoxic, and fully protease resistant cell-penetrating peptide. *ChemMedChem: Chemistry Enabling Drug Discovery*, *3*(2), 296–301.

Purcell, A. W., McCluskey, J., & Rossjohn, J. (2007). More than one reason to rethink the use of peptides in vaccine design. *Nature Reviews. Drug Discovery*, *6*(5), 404–414.

Qiu, S., Xue, M., & Zhu, G. (2014). Metal–organic framework membranes: From synthesis to separation application. *Chemical Society Reviews*, *43*(16), 6116–6140.

Qiu, Y., Song, B., Pezzato, C., Shen, D., Liu, W., Zhang, L., & Li, W. (2020). A precise polyrotaxane synthesizer. *Science (New York, N.Y.)*, *368*(6496), 1247–1253.

Rajagopal, K., & Schneider, J. P. (2004). Self-assembling peptides and proteins for nanotechnological applications. *Current Opinion in Structural Biology*, *14*(4), 480–486.

Rajbhandary, A., & Nilsson, B. L. (2017). Investigating the effects of peptoid substitutions in self-assembly of Fmoc-diphenylalanine derivatives. *Peptide Science*, *108*(2), e22994.

Ramakrishnan, V., Ranbhor, R., & Durani, S. (2005). Simulated folding in polypeptides of diversified molecular tacticity: Implications for protein folding and de novo design. *Biopolymers: Original Research on Biomolecules*, *78*(2), 96–105.

Ramakrishnan, V., Ranbhor, R., Kumar, A., & Durani, S. (2006). The link between sequence and conformation in protein structures appears to be stereochemically established. *The Journal of Physical Chemistry. B*, *110*(18), 9314–9323.

Ranbhor, R., Kumar, A., Patel, K., Ramakrishnan, V., & Durani, S. (2018). Automated design evolution of stereochemically randomized protein foldamers. *Physical Biology*, *15*(3), 036001.

Ranbhor, R., Kumar, A., Tendulkar, A., Patel, K., Ramakrishnan, V., & Durani, S. (2018). IDeAS: Automated design tool for hetero-chiral protein folds. *Physical Biology*, *15*(6), 066005. Available from https://doi.org/10.1088/1478-3975/aacdc3.

Raza, A., Rasheed, T., Nabeel, F., Hayat, U., Bilal, M., & Iqbal, H. (2019). Endogenous and exogenous stimuli-responsive drug delivery systems for programmed site-specific release. *Molecules (Basel, Switzerland)*, *24*(6), 1117.

Raza, F., Zhu, Y., Chen, L., You, X., Zhang, J., Khan, A., & Wu, J. (2019). Paclitaxel-loaded pH responsive hydrogel based on self-assembled peptides for tumor targeting. *Biomaterials Science*, *7*(5), 2023–2036.

Reja, A., Afrose, S. P., & Das, D. (2020). Aldolase cascade facilitated by self-assembled nanotubes from short peptide amphiphiles. *Angewandte Chemie*, *132*(11), 4359–4364.

Rodríguez-Llansola, F., Escuder, B., & Miravet, J. F. (2009). Remarkable increase in basicity associated with supramolecular gelation. *Organic & Biomolecular Chemistry*, *7*(15), 3091–3094.

Rodríguez-Llansola, F., Escuder, B., Hamley, I. W., Hayes, W., & Miravet, J. F. (2012). Structural and morphological studies of the dipeptide based L-Pro-L-Val organocatalytic gels and their rheological behaviour. *Soft Matter*, *8*(34), 8865–8872.

Rosenberg, S. A., Yang, J. C., Schwartzentruber, D. J., Hwu, P., Marincola, F. M., Topalian, S. L., & Spiess, P. J. (1998). Immunologic and therapeutic evaluation of a synthetic peptide vaccine for the treatment of patients with metastatic melanoma. *Nature Medicine*, *4*(3), 321–327.

Rubert Pérez, C. M., Álvarez, Z., Chen, F., Aytun, T., & Stupp, S. I. (2017). Mimicking the bioactivity of fibroblast growth factor-2 using supramolecular nanoribbons. *ACS Biomaterials Science & Engineering*, *3*(9), 2166–2175.

Ruoslahti, E. (2002). Drug targeting to specific vascular sites. *Drug Discovery Today*, *7*(22), 1138–1143.

Ruseska, I., & Zimmer, A. (2020). Internalization mechanisms of cell-penetrating peptides. *Beilstein Journal of Nanotechnology*, *11*(1), 101–123.

Ryan, D. M., Anderson, S. B., & Nilsson, B. L. (2010). The influence of side-chain halogenation on the self-assembly and hydrogelation of Fmoc-phenylalanine derivatives. *Soft Matter*, *6* (14), 3220–3231.

Ryan, D. M., Doran, T. M., & Nilsson, B. L. (2011). Stabilizing self-assembled Fmoc–F 5–Phe hydrogels by co-assembly with PEG-functionalized monomers. *Chemical Communications*, *47*(1), 475–477.

Saikia, J., Dharmalingam, K., Anandalakshmi, R., Redkar, A. S., Bhat, V. T., & Ramakrishnan, V. (2021). Electric field modulated peptide based hydrogel nanocatalysts. *Soft Matter*, *17* (42), 9725–9735. Available from https://doi.org/10.1039/D1SM00724F.

Saikia, J., Pandey, G., Sasidharan, S., Antony, F., Nemade, H. B., Kumar, S., & Ramakrishnan, V. (2019). Electric field disruption of amyloid aggregation: potential noninvasive therapy for Alzheimer's disease. *ACS Chemical Neuroscience*, *10*(5), 2250–2262.

Sakala, G. P., & Reches, M. (2018). Peptide-based approaches to fight biofouling. *Advanced Materials Interfaces*, *5*(18), 1800073.

Salick, D. A., Kretsinger, J. K., Pochan, D. J., & Schneider, J. P. (2007). Inherent antibacterial activity of a peptide-based β-hairpin hydrogel. *Journal of the American Chemical Society*, *129*(47), 14793–14799.

Samanta, U., Pal, D., & Chakrabarti, P. (1999). Packing of aromatic rings against tryptophan residues in proteins. *Acta Crystallographica Section D: Biological Crystallography*, *55*(8), 1421–1427.

Sasidharan, S, & Ramakrishnan, V. (2022). Aromatic interactions directing peptide nano-assembly.

Sasidharan, S., Hazam, P. K., & Ramakrishnan, V. (2017). Symmetry-directed self-organization in peptide nanoassemblies through aromatic π–π interactions. *The Journal of Physical Chemistry. B*, *121*(2), 404–411.

Sasidharan, S., Shyni, P. C., Chaudhary, N., & Ramakrishnan, V. (2017). Single crystal organic nanoflowers. *Scientific Reports*, *7*(1), 1–8.

Schnitzer, T., Paenurk, E., Trapp, N., Gershoni-Poranne, R., & Wennemers, H. (2021). Peptide–metal frameworks with metal strings guided by dispersion interactions. *Journal of the American Chemical Society*, *143*(2), 644–648.

Sedman, V. L., Chen, X., Allen, S., Roberts, C. J., Korolkov, V. V., & Tendler, S. J. B. (2013). Tuning the mechanical properties of self-assembled mixed-peptide tubes. *Journal of Microscopy*, *249*(3), 165–172.

Seow, W. Y., & Hauser, C. A. E. (2014). Short to ultrashort peptide hydrogels for biomedical uses. *Materials Today*, *17*(8), 381–388.

Sesardic, D. (1993). Synthetic peptide vaccines. *Journal of Medical Microbiology*, *39*(4), 241–242.

Sgolastra, F., Deronde, B. M., Sarapas, J. M., Som, A., & Tew, G. N. (2013). Designing mimics of membrane active proteins. *Accounts of Chemical Research*, *46*(12), 2977–2987. Available from https://doi.org/10.1021/ar400066v.

Shen, C. L., & Murphy, R. M. (1995). Solvent effects on self-assembly of beta-amyloid peptide. *Biophysical Journal*, *69*(2), 640–651. Available from https://doi.org/10.1016/S0006-3495(95)79940-4.

Shin, H., Jo, S., & Mikos, A. G. (2003). Biomimetic materials for tissue engineering. *Biomaterials*, *24*(24), 4353–4364.

Shundo, A., Hoshino, Y., Higuchi, T., Matsumoto, Y., Penaloza Jr, D. P., Matsumoto, K., & Tanaka, K. (2014). Facile microcapsule fabrication by spray deposition of a supramolecular hydrogel. *RSC Advances*, *4*(68), 36097–36100.

Siepi, M., Oliva, R., Petraccone, L., Del Vecchio, P., Ricca, E., Isticato, R., & Leone, L. (2018). Fluorescent peptide dH3w: A sensor for environmental monitoring of mercury (II). *PLoS One*, *13*(10), e0204164.

Singh, D., Bisland, S. K., Kawamura, K., & Gariépy, J. (1999). Peptide-based intracellular shuttle able to facilitate gene transfer in mammalian cells. *Bioconjugate Chemistry*, *10*(5), 745–754.

Skwarczynski, M., & Toth, I. (2016). Peptide-based synthetic vaccines. *Chemical Science*, *7*(2), 842–854.

Sóvágó, I., Kállay, C., & Várnagy, K. (2012). Peptides as complexing agents: Factors influencing the structure and thermodynamic stability of peptide complexes. *Coordination Chemistry Reviews*, *256*(19), 2225–2233. Available from https://doi.org/10.1016/j.ccr.2012.02.026.

Sreekumar, P. G., Li, Z., Wang, W., Spee, C., Hinton, D. R., Kannan, R., & MacKay, J. A. (2018). Intra-vitreal αB crystallin fused to elastin-like polypeptide provides neuroprotection in a mouse model of age-related macular degeneration. *Journal of Controlled Release*, *283*, 94–104.

Steele, J. T., Allen, S. D., & Kaumaya, P. T. P. (2006). *Cancer immunotherapy with rationally designed synthetic peptides. Handbook of Biologically Active Peptides* (pp. 491–498). Elsevier.

Stephanopoulos, N., Ortony, J. H., & Stupp, S. I. (2013). Self-assembly for the synthesis of functional biomaterials. *Acta Materialia*, *61*(3), 912–930.

Stupp, S. I. (2010). Self-assembly and biomaterials. *Nano Letters*, *10*(12), 4783–4786.

Stupp, R., Hegi, M. E., Neyns, B., Goldbrunner, R., Schlegel, U., Clement, P. M., & Dietrich, P.-Y. (2010). Phase I/IIa study of cilengitide and temozolomide with concomitant radiotherapy followed by cilengitide and temozolomide maintenance therapy in patients with newly diagnosed glioblastoma. *Journal of Clinical Oncology*, *28*(16), 2712–2718.

Sugahara, K. N., Teesalu, T., Karmali, P. P., Kotamraju, V. R., Agemy, L., Girard, O. M., & Ruoslahti, E. (2009). Tissue-penetrating delivery of compounds and nanoparticles into tumors. *Cancer Cell*, *16*(6), 510–520.

Sun, H., Zhang, X., Miao, L., Zhao, L., Luo, Q., Xu, J., & Liu, J. (2016). Micelle-induced self-assembling protein nanowires: Versatile supramolecular scaffolds for designing the light-harvesting system. *ACS Nano*, *10*(1), 421–428.

Sun, Z., Wu, S., Peng, Y., Wang, M., Jalalah, M., Al-Assiri, M., & Li, G. (2021). Sensor array for rapid pathogens identification fabricated with peptide-conjugated 2D metal-organic framework nanosheets. *Chemical Engineering Journal*, *405*, 126707.

Svensen, N., Walton, J. G. A., & Bradley, M. (2012). Peptides for cell-selective drug delivery. *Trends in Pharmacological Sciences*, *33*(4), 186–192.

Tahara, K., Lei, S., Adisoejoso, J., De Feyter, S., & Tobe, Y. (2010). Supramolecular surface-confined architectures created by self-assembly of triangular phenylene–ethynylene macrocycles via van der Waals interaction. *Chemical Communications*, *46*(45), 8507–8525.

Tanaka, K., Mori, A., & Inoue, S. (1990). The cyclic dipeptide cyclo [(S)-phenylalanyl-(S)-histidyl] as a catalyst for asymmetric addition of hydrogen cyanide to aldehydes. *The Journal of Organic Chemistry*, *55*(1), 181–185.

Tang, J. D., Mura, C., & Lampe, K. J. (2019). Stimuli-responsive, pentapeptide, nanofiber hydrogel for tissue engineering. *Journal of the American Chemical Society*, *141*(12), 4886–4899.

Tao, K., Fan, Z., Sun, L., Makam, P., Tian, Z., Ruegsegger, M., & Pan, Z. (2018). Quantum confined peptide assemblies with tunable visible to near-infrared spectral range. *Nature Communications*, *9*(1), 1–11.

Tao, K., Levin, A., Adler-Abramovich, L., & Gazit, E. (2016). Fmoc-modified amino acids and short peptides: Simple bio-inspired building blocks for the fabrication of functional materials. *Chemical Society Reviews*, *45*(14), 3935–3953.

Tao, K., O'Donnell, J., Yuan, H., Haq, E. U., Guerin, S., Shimon, L. J. W., & Thompson, D. (2020). Accelerated charge transfer in water-layered peptide assemblies. *Energy & Environmental Science*, *13*(1), 96–101.

Tarabout, C., Roux, S., Gobeaux, F., Fay, N., Pouget, E., Meriadec, C., & Besselievre, F. (2011). Control of peptide nanotube diameter by chemical modifications of an aromatic residue involved in a single close contact. *Proceedings of the National Academy of Sciences*, *108*(19), 7679–7684.

Teesalu, T., Sugahara, K. N., Kotamraju, V. R., & Ruoslahti, E. (2009). C-end rule peptides mediate neuropilin-1-dependent cell, vascular, and tissue penetration. *Proceedings of the National Academy of Sciences*, *106*(38), 16157–16162.

Teplensky, M. H., Fantham, M., Poudel, C., Hockings, C., Lu, M., Guna, A., & Farha, O. K. (2019). A highly porous metal-organic framework system to deliver payloads for gene knockdown. *Chem*, *5*(11), 2926–2941.

Thakuria, R., Nath, N. K., & Saha, B. K. (2019). The nature and applications of π–π interactions: A perspective. *Crystal Growth & Design*, *19*(2), 523–528.

Theunis, C., Crespo-Biel, N., Gafner, V., Pihlgren, M., López-Deber, M. P., Reis, P., & Ndao, D. M. (2013). Efficacy and safety of a liposome-based vaccine against protein Tau, assessed in tau. P301L mice that model tauopathy. *PLoS One*, *8*(8), e72301.

Tran, N. H., Zhang, X., Xin, L., Shan, B., & Li, M. (2017). De novo peptide sequencing by deep learning. *Proceedings of the National Academy of Sciences*, *114*(31), 8247–8252. Available from https://doi.org/10.1073/pnas.1705691114.

Trasatti, J. P., Woo, J., Ladiwala, A., Cramer, S., & Karande, P. (2018). Rational design of peptide affinity ligands for the purification of therapeutic enzymes. *Biotechnology Progress, 34* (4), 987–998.

Tsao, C. T., Chang, C. H., Lin, Y. Y., Wu, M. F., Wang, J.-L., Han, J. L., & Hsieh, K. H. (2010). Antibacterial activity and biocompatibility of a chitosan–γ-poly (glutamic acid) polyelectrolyte complex hydrogel. *Carbohydrate Research, 345*(12), 1774–1780.

Tsonchev, S., Niece, K. L., Schatz, G. C., Ratner, M. A., & Stupp, S. I. (2008). Phase diagram for assembly of biologically-active peptide amphiphiles. *The Journal of Physical Chemistry. B, 112*(2), 441–447.

Urry, D. W., Goodall, M. C., Glickson, J. D., & Mayers, D. F. (1971). The gramicidin A transmembrane channel: Characteristics of head-to-head dimerized π (L, D) helices. *Proceedings of the National Academy of Sciences, 68*(8), 1907–1911.

Vegners, R., Shestakova, I., Kalvinsh, I., Ezzell, R. M., & Janmey, P. A. (1995). Use of a gel-forming dipeptide derivative as a carrier for antigen presentation. *Journal of Peptide Science: An Official Publication of the European Peptide Society, 1*(6), 371–378.

Villalva, J., Nieto-Ortega, B., Melle-Franco, M., & Perez, E. M. (2020). Quasi-barrierless submolecular motion in mechanically interlocked carbon nanotubes. *The Journal of Physical Chemistry C, 124*(28), 15541–15546.

Voutsas, I. F., Gritzapis, A. D., Mahaira, L. G., Salagianni, M., Hofe, E. v, Kallinteris, N. L., & Baxevanis, C. N. (2007). Induction of potent CD4 + T cell-mediated antitumor responses by a helper HER-2/neu peptide linked to the Ii-Key moiety of the invariant chain. *International Journal of Cancer, 121*(9), 2031–2041.

Voyer, N., Roby, J., Deschěnes, D., & Bernier, J. (1995). Design, synthesis, and supramolecular properties of bis-crown ether modified peptides. *Supramolecular Chemistry, 5*(1), 61–69.

Walport, L. J., Obexer, R., & Suga, H. (2017). Strategies for transitioning macrocyclic peptides to cell-permeable drug leads. *Current Opinion in Biotechnology, 48*, 242–250. Available from https://doi.org/10.1016/j.copbio.2017.07.007.

Wang, J., Liu, K., Xing, R., & Yan, X. (2016). Peptide self-assembly: Thermodynamics and kinetics. *Chemical Society Reviews, 45*(20), 5589–5604.

Wang, J., Zhang, J., Wang, J., Fang, G., Liu, J., & Wang, S. (2020). Fluorescent peptide probes for organophosphorus pesticides detection. *Journal of Hazardous Materials, 389*, 122074. Available from https://doi.org/10.1016/j.jhazmat.2020.122074.

Wang, J., Zhao, X., Li, J., Kuang, X., Fan, Y., Wei, G., & Su, Z. (2014). Electrostatic assembly of peptide nanofiber–biomimetic silver nanowires onto graphene for electrochemical sensors. *ACS Macro Letters, 3*(6), 529–533.

Wang, M., Lv, Y., Liu, X., Qi, W., Su, R., & He, Z. (2016). Enhancing the activity of peptide-based artificial hydrolase with catalytic Ser/His/Asp triad and molecular imprinting. *ACS Applied Materials & Interfaces, 8*(22), 14133–14141.

Wang, X.-G., Dong, Z.-Y., Cheng, H., Wan, S.-S., Chen, W.-H., Zou, M.-Z., & Zhang, X.-Z. (2015). A multifunctional metal–organic framework based tumor targeting drug delivery system for cancer therapy. *Nanoscale, 7*(38), 16061–16070.

Wei, S., Xu, P., Yao, Z., Cui, X., Lei, X., Li, L., & Cheng, B. (2021). A composite hydrogel with co-delivery of antimicrobial peptides and platelet-rich plasma to enhance healing of infected wounds in diabetes. *Acta Biomaterialia, 124*, 205–218.

Wheeler, S. E. (2011). Local nature of substituent effects in stacking interactions. *Journal of the American Chemical Society, 133*(26), 10262–10274.

Whitesides, G. M., & Grzybowski, B. (2002). Self-assembly at all scales. *Science (New York, N. Y.), 295*(5564), 2418–2421.

Wiessner, C., Wiederhold, K.-H., Tissot, A. C., Frey, P., Danner, S., Jacobson, L. H., & Reichwald, J. (2011). The second-generation active Aβ immunotherapy CAD106 reduces amyloid accumulation in APP transgenic mice while minimizing potential side effects. *Journal of Neuroscience*, *31*(25), 9323–9331.

Wimley, W. C., & Hristova, K. (2011). Antimicrobial peptides: Successes, challenges and unanswered questions. *The Journal of Membrane Biology*, *239*(1), 27–34. Available from https://doi.org/10.1007/s00232-011-9343-0.

Woolfson, D. N. (2005). The design of coiled-coil structures and assemblies. *Advances in Protein Chemistry*, *70*, 79–112.

Wu, L., Lin, B., Yang, H., Chen, J., Mao, Z., Wang, W., & Gao, C. (2019). Enzyme-responsive multifunctional peptide coating of gold nanorods improves tumor targeting and photothermal therapy efficacy. *Acta Biomaterialia*, *86*, 363–372.

Xiao, S., Xiao, S., & Gräter, F. (2013). Dissecting the structural determinants for the difference in mechanical stability of silk and amyloid beta-sheet stacks. *Physical Chemistry Chemical Physics*, *15*(22), 8765–8771.

Xiao, Y., Reis, L. A., Feric, N., Knee, E. J., Gu, J., Cao, S., & McGuigan, A. P. (2016). Diabetic wound regeneration using peptide-modified hydrogels to target re-epithelialization. *Proceedings of the National Academy of Sciences*, *113*(40), E5792–E5801.

Xu, H., Wang, Y., Ge, X., Han, S., Wang, S., Zhou, P., & Lu, J. R. (2010). Twisted nanotubes formed from ultrashort amphiphilic peptide I3K and their templating for the fabrication of silica nanotubes. *Chemistry of Materials*, *22*(18), 5165–5173.

Yadav, S., Sharma, A. K., & Kumar, P. (2020). Nanoscale self-assembly for therapeutic delivery. *Frontiers in Bioengineering and Biotechnology*, *8*, 127.

Yamashita, M. M., Wesson, L., Eisenman, G., & Eisenberg, D. (1990). Where metal ions bind in proteins. *Proceedings of the National Academy of Sciences*, *87*(15), 5648–5652.

Yang, X., Guo, J.-L., Han, J., Si, R.-J., Liu, P.-P., Zhang, Z.-R., & Zhang, J. (2020). Chitosan hydrogel encapsulated with LL-37 peptide promotes deep tissue injury healing in a mouse model. *Military Medical Research*, *7*, 1–10.

Yardeni, J. L., Amit, M., Ashkenasy, G., & Ashkenasy, N. (2016). Sequence dependent proton conduction in self-assembled peptide nanostructures. *Nanoscale*, *8*(4), 2358–2366.

Yokoi, H., Kinoshita, T., & Zhang, S. (2005). Dynamic reassembly of peptide RADA16 nanofiber scaffold. *Proceedings of the National Academy of Sciences*, *102*(24), 8414–8419.

Yoshii, T., Onogi, S., Shigemitsu, H., & Hamachi, I. (2015). Chemically reactive supramolecular hydrogel coupled with a signal amplification system for enhanced analyte sensitivity. *Journal of the American Chemical Society*, *137*(9), 3360–3365.

Zannetti, A., Del Vecchio, S., Iommelli, F., Del Gatto, A., De Luca, S., Zaccaro, L., & Speranza, A. (2009). Imaging of αvβ3 expression by a bifunctional chimeric RGD peptide not cross-reacting with αvβ5. *Clinical Cancer Research*, *15*(16), 5224–5233.

Zhai, C., Schreiber, C. L., Padilla-Coley, S., Oliver, A. G., & Smith, B. D. (2020). Fluorescent self-threaded peptide probes for biological imaging. *Angewandte Chemie International Edition*, *59*(52), 23740–23747. Available from https://doi.org/10.1002/anie.202009599.

Zhang, H., Lou, S., & Yu, Z. (2019). Polar − π interactions promote self-assembly of dipeptides into laminated nanofibers. *Langmuir: The ACS Journal of Surfaces and Colloids*, *35*(13), 4710–4717.

Zhang, S., & Rich, A. (1997). Direct conversion of an oligopeptide from a β-sheet to an α-helix: A model for amyloid formation. *Proceedings of the National Academy of Sciences*, *94*(1), 23–28.

Zhang, S., Holmes, T., Lockshin, C., & Rich, A. (1993). Spontaneous assembly of a self-complementary oligopeptide to form a stable macroscopic membrane. *Proceedings of the National Academy of Sciences*, *90*(8), 3334–3338.

Zhang, Y., Gu, H., Yang, Z., & Xu, B. (2003). Supramolecular hydrogels respond to ligand – receptor interaction. *Journal of the American Chemical Society*, *125*(45), 13680–13681.

Zhang, Y., Ouyang, H., Lim, C. T., Ramakrishna, S., & Huang, Z. -M. (2005). Electrospinning of gelatin fibers and gelatin/PCL composite fibrous scaffolds. *Journal of Biomedical Materials Research Part B: Applied Biomaterials: An Official Journal of the Society for Biomaterials, The Japanese Society for Biomaterials, and The Australian Society for Biomaterials and the Korean Society for Biomaterials*, *72*(1), 156–165.

Zhao, X., Pan, F., Xu, H., Yaseen, M., Shan, H., Hauser, C. A. E., & Lu, J. R. (2010). Molecular self-assembly and applications of designer peptide amphiphiles. *Chemical Society Reviews*, *39*(9), 3480–3498.

Zhao, Y., Du, X., Jiang, L., Luo, H., Wang, F., Wang, J., & Wang, X. (2020). Glucose oxidase-loaded antimicrobial peptide hydrogels: Potential dressings for diabetic wound. *Journal of Nanoscience and Nanotechnology*, *20*(4), 2087–2094.

Ziserman, L., Lee, H.-Y., Raghavan, S. R., Mor, A., & Danino, D. (2011). Unraveling the mechanism of nanotube formation by chiral self-assembly of amphiphiles. *Journal of the American Chemical Society*, *133*(8), 2511–2517.

Zorzi, A., Deyle, K., & Heinis, C. (2017). Cyclic peptide therapeutics: Past, present and future. *Current Opinion in Chemical Biology*, *38*, 24–29.

Chapter 9

Patents in peptide science

Mouli Sarkar[1] and Ranjit Ranbhor[2]
[1]Department of Biosciences and Bioengineering, IIT Guwahati, Guwahati, Assam, India,
[2]Pergament & Cepeda LLP, Florham Park, NJ, United States

9.1 Introduction

Biotechnology has a positive impact on the globe. As a result, patenting is critical, and it gives inventors exclusive rights to create, use, and sell their discoveries for the next 20 years.

Without patenting in biotechnology, compounds would be treated more like trade secrets, with no one other than the proprietor knowing how something is manufactured or done. This way of thinking prevents technological breakthroughs and, as a result, no significant improvements in living.

Patents, on the other hand, are notoriously difficult to understand. Even if an individual is awarded one on their own, competitors of larger organizations frequently find ways to get around infringement laws and profit from someone else's concept. As a result, it is critical to enlist the assistance of a knowledgeable patent attorney. These are experts that have spent years studying legalese and are well-versed in patent application and related processes. Thus being comprehensive from the start is the greatest option.

9.2 Basics of protein and peptide structure

Peptides are the building components of proteins, and the functioning of a protein is determined by its unique folding. The native state of a folded protein is the most accurate representation of a protein's function. There are reports of primary, secondary, and tertiary protein structures. There are various repositories where information about protein structures can be found.

Although data sets relating to protein structure information are not patentable in and of themselves, they can be indirectly preserved by integrating them with a specific complex or application. Because of the storage of the appropriate data set in a computer-readable memory, it is achievable to secure the tertiary conformation of a protein.

De Novo Peptide Design. DOI: https://doi.org/10.1016/B978-0-323-99917-5.00002-0

Patentable uses include (1) altering amino acid residues in the active region of a protein to improve bioactivity, (2) changing the stability of a protein by inserting or removing intramolecular links that do not influence its function, (3) altering a protein's surface accessible residues to reduce immunogenicity, and (4) improving the solubility of a protein by changing the solvent-accessible regions. A number of recently issued patents show the many types of protection available in the United States for tertiary structures and their applications (Meyers, Turano, Greenhalgh, & Waller, 2000).

9.2.1 Patents

- **US5856116**: Interleukin-1 beta, an inactive precursor of the proinflammatory cytokine resides in Interleukin-1-beta converting enzyme (ICE). The invention identified human ICE crystal structure using X-ray diffraction, combined with an inhibitor. The active site involves both the 20 and 10 kDa subunits. The structural coordinates of the enzyme could be employed to create new types of ICE inhibitors (Wilson, Griffith, Kim, & Livingston, 1999).
- **US5989169**: The present discovery refers to a technique for producing a Termamyl-like amylase, which is a variation of a parent Termamyl-like amylase that has at least one changed property and amylase activity when compared to the parent amylase. This method includes identifying at least one amino acid residue or structural element of the parent Termamyl-like amylase structure by analyzing the structure of the parent Termamyl-like amylase (Svendsen, Bisgård-Frantzen, & Borchert, 1999).
- **US5835382**: Computer-assisted methods for discovering compounds that will attach to the erythropoietin (EPO) receptor and operate as an EPO mimic are included in the invention. In one or more in vitro or in vivo biological assessments of EPO activity, preferred EPO mimetics were identified using the technology of the invention function as agonists of the EPO receptor (I. A. Wilson, Livnah, Stura, Johnson, & Jolliffe, 1998).
- **US6063636**: In a patient, an approach to detect accumulated proteins in the body fluid from the patient, preparing the fluid, and introducing the prepared fluid to size-exclusion chromatography in order to establish an excreted fluid, and analyzing the excreted fluid to identify biomolecules with a predetermined molecular weight (Stevens, Myatt, & Solomon, 2000).
- **US5966712**: The database contains genomic libraries for a variety of species, each of which contains multiple genomic sequences, at least some of which represent open reading frames located along a contiguous sequence on each of the organism's genomes, and a user interface that can recognize a selection of two or more genomic libraries for comparison. A user interface is also included in the system, which can take and

display a list of one or more probe open reading frames for use in determining homologous connections between those probe open reading frames and the open reading frames in genomic libraries (Sabatini et al., 1999).

9.3 Modeling and simulation of peptides

Conformations of proteins and protein complexes are important not just for a basic understanding of molecular biology underpinnings of principles of life, but also for inspection and creating unique medications for therapeutic uses. The two primary types of Protein Structure Prediction (PSP) techniques that have been developed are template-based modeling (TBM) and predicting de novo structure. Using preexisting protein structures as templates, TBM predicts the native structure of a protein target based on homology modeling or threading alignments. De novo PSP (template-free modeling) can be employed to predict unique protein folds in the absence of an appropriate template. Predominantly used as well as successful ones rely on combining associated structural fragments having specific potential energy (scoring) functions acquired by employing protein structure mining. The predicted models of protein can have a huge spectrum of accuracies, making them applicable for a wide range of uses. All of the aforementioned well-known PSP techniques are knowledge-based, as they rely on a database of sequences and structures. Two of the most extensively utilized methods for conformational search are Monte Carlo (MC) and molecular dynamics (MD). Using either atomistic or coarse-grained models, the MC method has been used to investigate peptide folding and protein folding. It has also been used to estimate protein loop shapes. Although MC approaches can be as precise as MD methods, in theory, they may be less efficient when applying fine-grained model, particularly when there is a significant number of explicit solvent molecules (Geng, Chen, Ye, & Jiang, 2019).

Peptide and protein isolation and characterization studies enable the collection of data sets on their structures and properties/activities. Quantitative structure activity relationship modeling is a technique for determining mathematical expressions for such connections, which may subsequently be used to estimate the activities of related compounds and anticipate high-activity structures (Pripp, Isaksson, Stepaniak, Sørhaug, & Ardö, 2005).

9.3.1 Patents

- **US5612895A**: One method of rational drug design is to simulate polypeptides in a way that estimates the most probable two-dimensional and/or three-dimensional structure of a polypeptide, such as an oligopeptide, without making any assumptions about the conformation of the fundamental primary or secondary structure. The method comprises simulating a real-size basic structure in an aqueous environment, isobarically and isothermally reducing the size of the polypeptide, and then expanding the modeled polypeptide to its real size over time (Balaji & Singh, 1997).

- **EP1163639A1**: The invention provides a method for constructing three-dimensional structure of a protein using well-known, insufficiently encoded secondary structure restrictions and scanty knowledge regarding appropriate side chain interactions. The method is based on reduced modeling of protein structure and dynamics in which the protein is defined by side chain mass centers rather than alpha-carbons. The model implicitly simulates long and the short stacking requirements, hydrogen bonding cooperativity, and an overall average force potential characterizing hydrophobic interaction. For successful fold construction, the approach requires fewer tertiary constraints; on average, one for every seven residues. The method can be used to build a model depending upon a variety of derived structural constraints from several studies (Skolnick & Kolinski, 2001).

9.4 Glucagon-like peptide-1 analogs

The endogenous incretin glucagon-like peptide-1 (GLP-1) is important for glucose homeostasis. Intestinal L-cells release GLP-1, a 30-amino acid peptide, in response to dietary intake (Fig. 9.1). Preproglucagon, a precursor to numerous peptides associated with glucagon, undergoes posttranslational processing to create it. It comes in two forms: GLP-1(736)-NH2 and GLP-1 (737), and it works by activating the GLP-1 receptor (GLP1R), which is part of G-protein-coupled receptors (Manandhar & Ahn, 2015). While GLP1R is present in a variety of organs, including the heart, brain, pancreas, kidney, and GI tract, research has indicated that it is specifically expressed in the pancreas (Manandhar & Ahn, 2015).

9.4.1 Patents

- **WO2015000942**: The invention concerns peptides similar to GLP-1 that is C-terminally expanded GLP-1 mimics. The derivatives are made up of two side chains, one of which corresponds to position 42 and the other to 18, 23, 27, 31, 36, or 38 positions when compared to GLP-1 (7–37). A C19, C20, or C22 diacid protracting moiety and a linker are included in the side chains. The invention also encompasses intermediate products, such as novel GLP-1 analogs incorporated in the invention's derivatives, as well as pharmaceutical formulations and medical applications. The derivatives have unusually long half-lives while still being strong, making them potentially suitable for administration once a month (Reedtz-Runge, Kofoed, Tornoe, & Sauerberg, 2015).

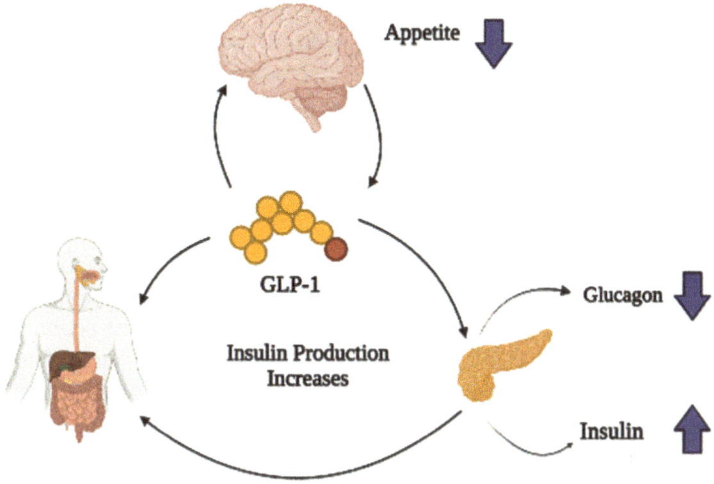

FIGURE 9.1 Analog of glucagon-like peptide-1 (GLP-1) created to treat type 2 diabetes.

- **WO2019060660**: Long-acting coagonists peptides of the glucagon and GLP-1 receptors are described in this invention (Palani et al., 2019).
- **WO2019016306**: Compounds beneficial in the treatment of diabetes, weight loss, and/or cardiovascular risk reduction is disclosed in the application. Because the compounds are bi-functional, they can be used as straightforward treatment for patients who benefit from both GLP-1 receptor agonist and PCSK9 inhibitor therapy (Tornøe, Kodal, Reedtz-Runge, & Lykke, 2019).
- **WO2016198628**: The current application relates to exendin-4 derivatives and their medicinal implementations, for example, the treatment of metabolic syndrome disorders such as obesity and diabetes, as well as calorie restriction (Kadereit et al., 2016b).
- **WO2017153575**: GLP-1/glucagon agonist peptides are administered to patients to prevent and treat obesity and diabetes (Bednarek, Jermutus, Ambery, & Petrone, 2017).
- **WO2017178829**: Amino acid alterations at one or more of positions 1 − 5 are supplied in peptide hormone analogs of formulae (I) − (IV). Pharmaceutical compositions containing said analogs, as well as methods of using said analogs to treat obesity and diabetes, are also disclosed (Bloom & Jones, 2017).
- **US20180155406**: The current invention refers to trigonal GLP-1/glucagon/GIP receptor agonists and their medicinal applications, such as treating metabolic syndrome disorders such as obesity and diabetes, as well as reducing excessive calorie intake (Bossart et al., 2018).
- **WO2018178796**: The Patent Cooperation Treaty (PCT) application relates to making KEX-1 protease sensitive polypeptides using yeast

strain, which utilizes nucleic acids that encode precursor forms of the polypeptides that contain, at their C-termini, at least one additional amino acid each of which is susceptible to degradation by carboxypeptidase B, and wherein the C-terminal amino acid of the precursor is resistant to degradation by KEX1. Precursor polypeptides that are biologically inactive may then be converted to the mature, biologically active polypeptides via cleavage of at least one C-terminal amino acid using carboxypeptidase B (Thennati et al., 2018).

- **WO2019125929**: Incretin analogs with action at the GIP, GLP-1, and glucagon receptors are reported. The structural properties of the incretin analogs result in balanced activity and a longer duration of action at each of these receptors. Methods for treating diabetes, dyslipidemia, fatty liver disease, metabolic syndrome, nonalcoholic steatohepatitis, and obesity are also provided (Alsina-Fernandez, Coskun, Guo, & Qu, 2019).

9.5 Peptide-based antibiotics

Enhancing the antibacterial effect of medications is a crucial technique for combating antimicrobial resistance. High-throughput screening and cloning of numerous pathogens' genomes, combined with in silico methods, facilitated the discovery of alternative cellular pathways as a reservoir of potentially new antimicrobial targets thought to be important for pathogen metabolism and reproduction (Annunziato & Costantino, 2020). Antibiotic peptides are a wide family of molecules that are found in all forms of life and are thought to be promising compounds for combating antimicrobial resistance. These AMPs have been shown to have several advantages over standard antibiotics, including a wider range of antibacterial activity and the ability to overcome bacterial drug resistance (Annunziato & Costantino, 2020). Gramicidin, derived from *Bacillus brevis*, which proved active against a wide spectrum of Gram-positive and Gram-negative bacteria, was the first natural peptide reported in 1939 (Annunziato & Costantino, 2020) (Fig. 9.2).

9.5.1 Patents

- **US9707282B2**: The current invention concerns antibacterial and immunomodulatory host defense peptides in general, and more US patent documents particularly (Hancock, Hilpert, Cherkasov, & Fjell, 2017).
- **US10077293**: The current invention concerns an antimicrobial peptide analog obtained from abalone, as well as an antimicrobial pharmaceutical formulation including it. The antimicrobial peptide analog according to the present disclosure is based on hdMolluscidin, a peptide generated from the gill of abalone that has been reduced in number of amino acids to make it commercially viable. Despite having fewer amino acids, the proposed peptide analog has excellent antibacterial action, good

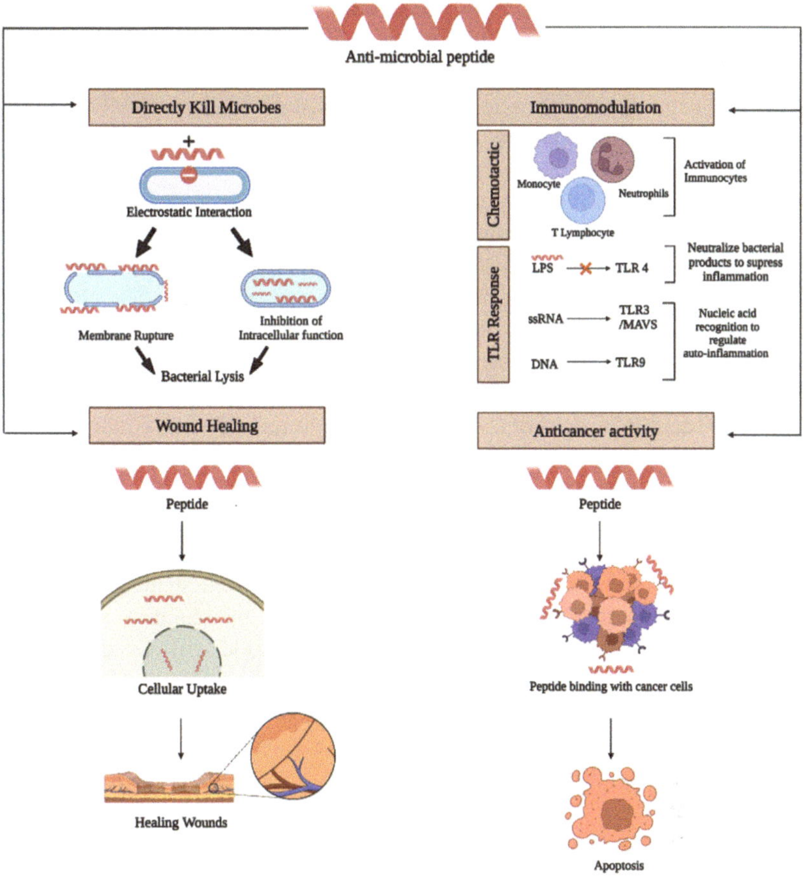

FIGURE 9.2 Diagrammatic representation of the activity of antimicrobial peptides.

membrane permeability, and minimal hemolytic activity (Seo, Lee, Cho, & Lee, 2018).

- **US2015/0344524**: This invention is about antibiotic peptides that have been modified for use in medicine. Compositions and methods for eliminating microorganisms such as bacteria, viruses, or fungi, as well as methods for treating microbial illnesses, are also included in the invention. The goal of the idea is to create new antibiotic peptides with improved antibiotic action and a broader spectrum of activity against various bacteria, especially Gram-positive bacteria like *Staphylococcus aureus* (Hoffmann, Knappe, Hilpert, Mikut, & Ruden, 2015).
- **US2016/0376306 A1**: The invention relates to antibacterial compositions that are therapeutically viable and based on extremely short cationic peptides that mimic host defense (HDCPs). As new antibacterial drugs, the

invention relates to template-based N-terminal modified di-peptidomimetics with or without alterations in the polyamine backbone. Under in vitro settings, the compounds described in the current invention do not acquire resistance to methicillin-resistant *S. aureus* (MRSA), and so may be employed as topical treatments or in other applications (Santosh, Prasad, & Seema, 2016).

- **WO2019/027366 A1**: The current invention pertains to peptides comprising guanidine and halogenated biphenyl moieties of the formula (I), as well as salts and solvates of each of these peptides, procedures for preparing them, compositions containing them, and their uses. The chemicals have been discovered to have a strong microbicide activity and are suitable for resisting bacteria, such as MRSA strains, at extremely low concentrations (Chia, 2019).
- **WO2016/150576 A1**: Beta-hairpin peptidomimetics with Gram-negative antibacterial activity to inhibit the growth or kill microorganisms such as *Klebsiella pneumonia*, *Acinetobacter baumannii*, *Escherichia coli*, and *Pseudomonas aeruginosa* are provided by the present invention. Furthermore, the current invention provides an efficient synthetic procedure that allows these compounds to be synthesized in parallel library style if required. Furthermore, the invention's beta-hairpin peptidomimetics have increased efficacy, reduced red blood cell hemolysis, and reduced or no cytotoxicity. They can be used to treat or prevent illnesses, as well as to disinfect foods, cosmetics, pharmaceuticals, and other nutrient-containing items (Obrecht, Luther, Bernardini, Zbinden, & Lederer, 2016).

9.6 Drug delivery vehicles

Pharmaceutical corporations are becoming interested in small molecular, oligonucleotide, and protein therapies. The various drug candidates' intrinsic properties, however, hindered their therapeutic efficacy. Drugs containing oligonucleotides and proteins are tough to enter the cell membrane and are susceptible to protease and nuclease degradation. As a result, building a robust and to develop an effective mechanism to address these drug delivery constraints. An optimal drug delivery system protects drug molecules from disintegration, prolongs circulation in vivo, targets the area to decrease toxic side effects for cells, regulates the release of the drug, and efficiently removes drug molecules (Yang, An, & Wang, 2021).

Peptides are polymers, comprised of monomeric amino acids connected by amide bonds that are abundant in biological systems and play a vital role in biological functions. As a result, it is biodegradable and biocompatible by nature, formulation an ideal building component for biomaterials. In biology, 20 natural amino acids are used to make peptides. As fundamental assembly units, these peptide sequences with specified characteristics are able to self-assemble into various sizes and morphologies nanomaterials in several dimensions, as well as with various functions and stable structures. The qualities are responsible for peptide nanostructures' unique scaffolds with excellent delivery potentiality, which have the potential for nanoscale drug delivery system design from the bottom up (Yang et al., 2021).

Pharmaceutical companies are increasingly turning their attention to peptide vaccines and treatments (Storz et al., 2012). The functions of each peptide category, including peptides, target tumors, peptides that have the ability to penetrate cells, peptides targeting to subcellular organelles, and peptides able to penetrate different tumors, and an exhibition of four different types of peptide-functionalized delivery vehicles were presented (Tang, Wei, & Yu, 2021) (Fig. 9.3). Tumor homing peptides are versatile biomolecules that have been used to develop peptide-based drug delivery systems in a range of cancers (Goyal et al., 2021).

9.6.1 Patents

- **US20080207497**: The current invention relates to cancer prevention, treatment, and diagnosis, particularly carcinomas such as ovarian carcinoma. Peptides, polypeptides, and polynucleotides are among the peptides, polypeptides, and polynucleotides described in the invention (Ramakrishna, Ross, Philip, & Keller, 2002).
- **WO200246416**: The current invention relates to cancer prevention, treatment, and diagnosis, particularly carcinomas such as ovarian carcinoma. Peptides, polypeptides, and polynucleotides which are disclosed in the

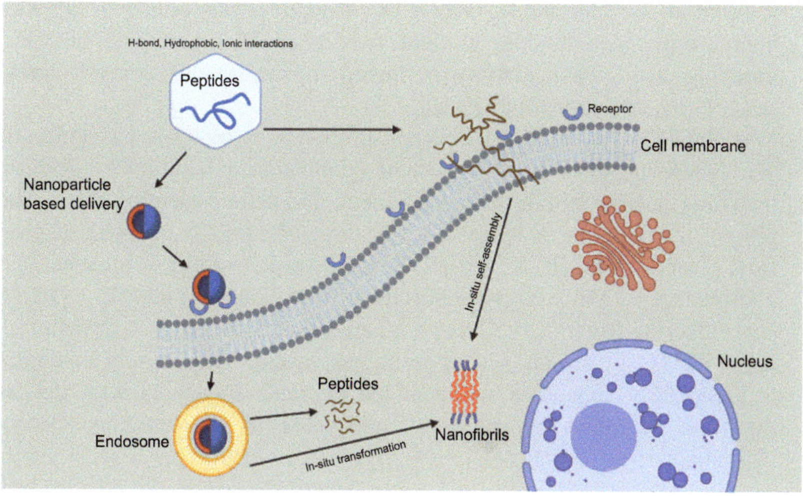

FIGURE 9.3 Schematic representation of peptide-based drug delivery system.

invention can act as a stimulant of cytotoxic T lymphocyte (CTL) response for cancer (Ramakrishna, Ross, Philip, & Keller, 2002).

- **WO2008070047**: The current invention pertains to cancer stem cell prevention, treatment, and diagnosis formulations and techniques. Peptides, polypeptides, and polynucleotides are disclosed that can be employed to activate a CTL response against cancer cells, particularly cancer stem cells (Philip & Keller, 2008b).
- **WO2008088583**: The current invention is concerned with formulations and techniques for the treatment, and diagnostics of several types of cancer, particularly carcinomas such as breast carcinoma. Peptides, polypeptides, and polynucleotides that can be employed to activate a CTL reaction against breast cancer are disclosed in this invention (Philip & Keller, 2008a).
- **WO2004052917**: At least one epitope or analog from p53, HER2/neu, CEA, MAGE3, or MAGE2 is included in a peptide or composition (Keogh, Southwood, Fikes, & Sette, 2002).
- **WO2002061435**: Methods for developing vaccinations that are efficacious in people who have the A2-supertype allele are outlined. Single amino acid substitution analogs of known peptides bind to A2-supertype and comprehensive peptide libraries were employed to characterize the specificities of peptide binding to A2-supertype molecules. During discoveries of each molecule, in order to have its own preferences, there were significant commonalities in specificity. A2-supertype molecules tolerated the hydrophobic and aliphatic residues L, I, V, M, A, T, and Q in position 2 of peptide ligands. At the C-terminus, T, V, A, M, I, and L were tolerated. Allele-specific preferences emerged from secondary effects on peptide binding, but similar features were uncovered and used to build an A2-supermotif. Over 70% of peptides that bind A*0201 with high affinity also bind to at least two other A2-supertype molecules, implying that cross-reactivity is linked to similar properties (Sidney, Sette, Grey, & Southwood, 2002).
- **WO0006723**: The subject of current discovery is tumor-associated antigen (TAA) peptides as well as their application as antitumor vaccines, including polynucleotides encoding them, and cells presenting them. The present invention is more specifically directed to TAA peptides derived from Uroplakin Ia, Ib, II, and III. Prostate-specific antigen, prostate acid phosphatase, and prostate-specific membrane antigen, BA-46 (Lactadherin), Mucin (MUC-1), and Teratocarcinoma-derived growth factor (CRIPTO-1), are applied as antitumor vaccine. The current invention is particularly concerned with antigen peptides associated with tumors that are presented by HLA-A2 molecules to the immune system (Eisenbach et al., 2000).

9.7 Protein aggregation

In vivo, protein/peptide aggregation frequently occurs at the membrane–cytochylema contact and behaves differently than in bulk solution. Blood clotting, the ineffectiveness of proteinic medicines, and the formation of amyloid plaques are all examples of protein aggregation. Protein aggregation can be beneficial in some situations. Blood coagulation and wound healing, for example, are dependent on the aggregation of a certain protein. Protein aggregation, on the other hand, is usually associated with undesirable biological processes or even fatal disorders such as cataract development, sickle cell disease, and arterial thrombosis (Fig. 9.4). Furthermore, neurodegenerative diseases such as Alzheimer's disease (AD), Huntington's disease, Parkinson's disease (PD), amyotrophic lateral sclerosis (ALS), and prion diseases are likely the most well-known diseases related to protein aggregation-induced disorders (Lu, Tang, Xiong, Qing, & Sun, 2016). The aggregation of improperly phosphorylated microtubule-associated protein tau into paired helical filaments and neurofibrillary tangles (NFTs) is defined by tauopathies, a group of progressive neuropathological disorders (Pandey, Morla, Kumar, & Ramakrishnan, 2020).

9.7.1 Patents

- **US8268628B1**: Without the use of probes or additives, the discovery proposes a technique for assessing aggregation in protein, peptide, or peptoid formulations. The method combines FTIR spectroscopy with two-dimensional correlation analysis, allowing for the detection of aggregates,

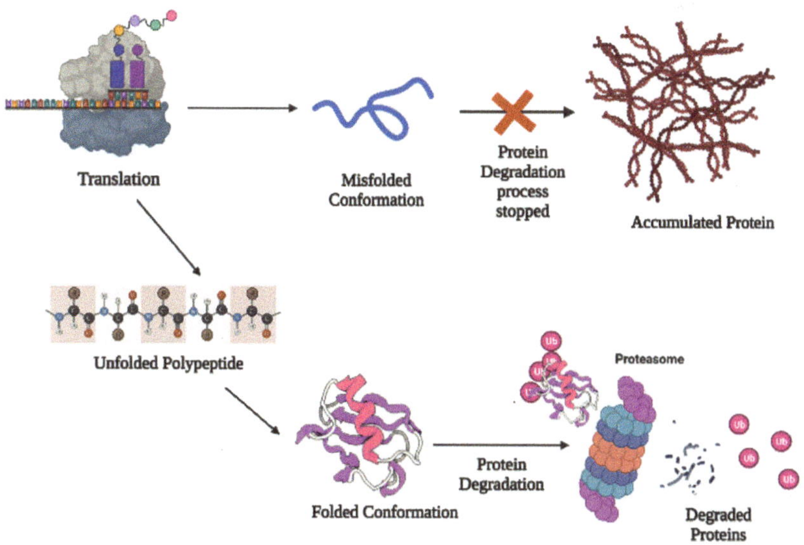

FIGURE 9.4 Several conformational states of a protein in a cell.

the identification of the mechanism of aggregation, and the modification of the protein's pipeline production process to produce viable protein once again (Pastrana-Rios & Lloro-Manzano, 2012).

9.8 Peptide nanomaterials

Molecular self-assembly is a pervasive phenomenon in nature that describes the spontaneous connection of individual molecules under thermodynamic conditions into a well-defined and rather stable supramolecule through non-covalent interactions (Li, Lu, Zhang, Hu, & Li, 2022). Various supramolecular structures are formed by the symmetric self-assembly of proteins. Viruses, bacterial microcompartments, and eukaryotic vaults are examples of such structures. Synthetic biology is introducing peptide/protein-based nanoparticles for a variety of medicinal uses, mostly as drug targeting and delivery systems or vaccinations (Doll, Dey, & Burkhard, 2015). Molecules are sustained in a stable, low-energy state by their interactions (Li et al., 2022). Nanotubes, nanofibers, hydrogels, and nanovesicles are some of the nanostructures that result, and they all have different physico-chemical and mechanical properties (Pentlavalli, Coulter, & Laverty, 2020) (Fig. 9.5).

9.8.1 Patents

- **US7745708B2**: Compounds, formulations, and procedures involving peptide amphiphiles for self-assembly or the production of nanofibrous

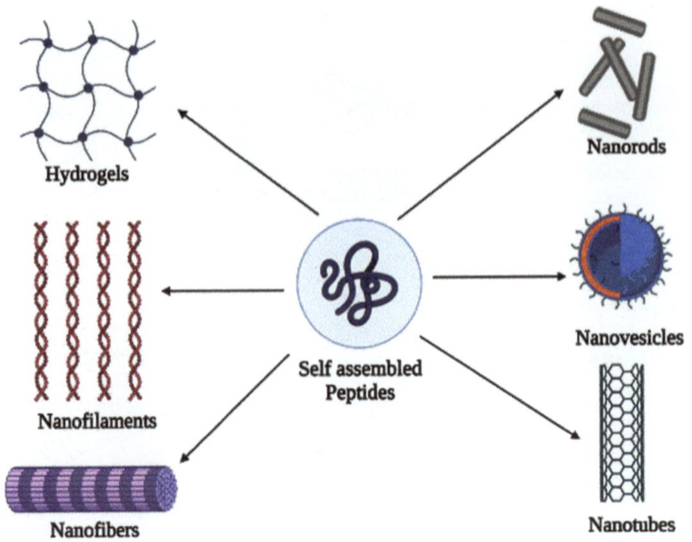

FIGURE 9.5 Self-assembly of peptides and formation of several nanostructures.

networks in neutral or physiological circumstances (Stupp, Hartgerink, & Beniash, 2010).

- **US7390526B2**: Biomimetic composites with peptide amphiphiles and surface-modified substrates, as well as attachment mechanisms (Stupp, Spoerke, Anthony, & Niece, 2008).
- **243/KOL/2015**: In an X-ray diffractogram obtained with X-ray radiation of wavelength 1.5418 A°, the present invention relates to a crystalline dihistidine nanostructure with peaks at 2 theta values of about 11.6 A° and 14.2 A°. A method for the manufacture of crystalline dihistidine nanostructure is also included in the invention.
- **US9486409B2**: Nanoparticle compositions including one or more peptides are provided by the present invention. The present invention facilitates the transdermal distribution of such peptides without peptide alteration, chemical or mechanical abrasion, or skin disturbance (Edelson & Kotyla, 2016).

9.9 Peptide-based hydrogels

Hydrogels are the current biomedical engineering achievements. They are water-retentive, flexible, jelly-like materials. Such hydrogels have a molecular-level fibrous network capable of trapping water and other water-soluble substances (Datta & Chaudhary, 2021). Short to ultra-short peptides have emerged as minimalistic building blocks, resulting in the identification of new functional capabilities as a result of their self-assembly (Saikia et al., 2021). Regenerative medicine, sustained drug delivery, dressing wound, vaccine adjuvants, and tissue engineering are all possible applications for these materials. Peptides and peptide-based materials have received special attention among the several molecules that might cause hydrogelation (Fig. 9.6). Peptides have several advantages over other polymers, including biodegradability, biocompatibility, the ability to create peptide/nonpeptide hybrids, and the ability to tune properties by perturbing amino acids (Datta & Chaudhary, 2021).

9.9.1 Patents

- **US8835395B2**: The present invention introduces a new class of peptides that can be employed to make hydrogels. The peptides are short (30 amino acids or less) and contain hydrophilic and hydrophobic segments connected by a turning segment. The hydrogels are created by adjusting the pH of a peptide solution or adding a source of ions to the solution. Shear thinning hydrogels with high storage moduli and strong recovery rates after destruction are the result. They are used in a variety of medicinal fields, including tissue engineering (Sun & Huang, 2014).

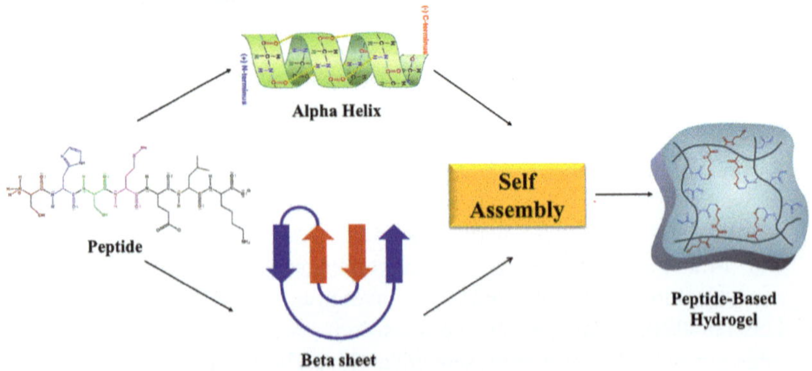

FIGURE 9.6 Schematic representation of peptide-based hydrogels.

- **US20040242469A1**: Compositions containing self-assembling peptide hydrogels and cardiovascular system cells, particularly endothelial cells, cultured in or on the gels are described in the invention. Endothelial cells create capillary-like structures according to some versions of studies. The endothelial cell – gel matrix acts as a prevascularized scaffold for the growth of other cell types. Smooth muscle cells and/or fibroblasts are used to form mature blood arteries in or on the scaffold. The invention's compositions can be utilized for cell-culture and administering to a person to diagnose several diseases, such as cardiac malfunction or injury (Lee, Kamm, Narmoneva, & Zhang, 2004).
- **WO2004007683A2**: Cells, such as progenitor cells, stem cells, and their progeny, are contained within self-assembling three-dimensional peptide hydrogel frameworks in the present invention (scaffolds). A nanoscale environment is created by the scaffolds. As a result, the invention creates a nanoscale structure that encases cells. Progenitor cells and/or their offspring can differentiate or transdifferentiate inside the structures, according to some embodiments of the invention. The invention entails implanting peptide hydrogel structures containing cells or cells retrieved after growing in peptide hydrogels into the body to cure disease (Semino, Shen, Sherley, & Zhang, 2004).
- **US9994615B2**: The invention provides a glucose-binding amphiphilic peptide hydrogel insulin delivery device that is responsive to glucose concentrations under physiological conditions. Insulin is encased in a glucose binding hydrogel consisting of self-assembling amphiphilic peptides with a hydrophobic domain with a beta-sheet generating region and a charged hydrophilic domain with a glucose binding segment. The formulations are designed to release insulin in response to blood glucose levels, keeping patients' blood glucose levels in a healthy range and preventing hyper- and hypoglycemia (Langer, Anderson, Gu, & Aimetti, 2018).

9.10 Therapeutic interventions against protein/peptide aggregation diseases

The aggregation of certain proteins is thought to be the cause of various degenerative diseases known as amyloid disorders. Protein aggregation and deposition in affected patient tissues occur mostly at the aggregation-prone protein's typical extracellular or intracellular site. However, evidence for the existence of intracellular and extracellular aggregates in nearly all aggregation-related degenerative illnesses is growing. The assumption that human amyloid disorders are caused by protein aggregation or amyloidogenesis is supported by strong genetic, pharmacologic, biochemical, and pathologic evidence.

There are several strategies to target protein aggregation in degenerative diseases that have been reported. Surgically administered gene therapy was used in the "protein reduction" strategy. The aggregation of primarily mutant transthyretin (TTR) and, to a lesser extent, wild-type TTR causes familial TTR amyloid disorders. The "protein stabilization" or "kinetic stabilization" strategy is an alternative approach to preventing the harmful effects of misfolding and aggregation of proteins. This kinetic stabilizing method demonstrates that aggregation, or amyloidogenesis, is the cause of these diseases. A "protein quality control technique" focuses on altering intracellular and/or extracellular proteostasis processes involved in protein folding, trafficking, and/or degradation to reduce disease-associated protein aggregation and maybe other biological proteins (Eisele et al., 2015) (Fig. 9.7).

9.10.1 Patents

- **US6946135B2**: Compositions and techniques for treating amyloidogenic disorders are described in the invention. A drug that generates a favorable

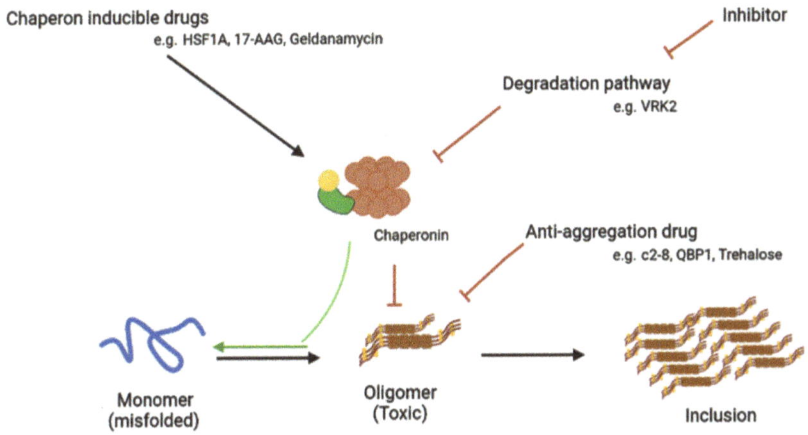

FIGURE 9.7 Schematic representation of strategies of therapeutic interventions against protein aggregation disorders.

immune response against an amyloid deposit in the patient is used in these procedures. The techniques are particularly beneficial for AD prevention and therapy. A peptide or an antibody to it can be used in such techniques (Schenk, 2005).

- **US20100113613A1**: The study focuses on cyclohexane polyalcohol compounds that have favorable pharmacokinetic profiles in the treatment of disorders involving protein folding and aggregation, and amyloid production, accumulation, or persistence. Some aspects of the invention provide an active ingredient for a cyclohexane polyalcohol compound appropriate for administration in order to include a desired therapeutic concentration of the compound in plasma, brain, and/or cerebral spinal fluid, as well as a pharmaceutically acceptable carrier, diluent, or excipient (McLaurin & Cruz, 2007).
- **WO2017147044A1**: Pharmaceutical compositions containing judiciously chosen peptide analogs of GILZ's p65-TAD interacting region are revealed for selectively sequestering activated p65. According to structural and functional studies, selective GILZ analogs (GA) bind p65-TAD with optimum affinity, have a half-minimal lethal dose comparable to established peptide drugs, and minimize A1−42-induced cytotoxicity. For the treatment of neurodegenerative diseases such as AD, PD, multiple sclerosis, and ALS, the present disclosure also includes uses and techniques of using pharmaceutical combinations, and strategies of using pharmaceutical formulations (Srinivasan & Lahiri, 2017).
- **US10894822B2**: The innovation pertains to methods to treat AD and other tauopathies by administering antibodies that are specific for abnormal forms of tau protein, have no binding and/or reactivity to normal tau protein, and are administered under conditions and in amounts that are effective in preventing or treating AD or other tauopathies (Chain, 2021a, 2021a).

9.11 Computational tools and algorithms in peptide design

Because each peptide residue can be identified from natural amino acids, determining the affinity of a particular peptide sequence for a specific protein surface is a difficult task. A reasonable mechanism for predicting particular and selective peptide sequences is needed. The required methodology should be fast and capable of designing peptides for specific places on a set of proteins. It is desirable to have a computational tool that is both rapid and global (Unal, Gursoy, & Erman, 2010). Peptides are the most adaptable modulators for dealing with protein−protein interactions' (PPI) large surface area. They are simple in construction, resemble proteins, and can be adjusted to improve stability, bioavailability, and binding strength. Peptide design can be divided into two approaches: sequence-based design and structure-based design. Peptide stability, toxicity, immunogenicity, and antibody specificity are all factors to consider when creating sequence-based peptides. In this context, *in silico* techniques for creating new peptides with therapeutic qualities ranging from cell-penetrating to antimicrobial, antiparasitic, anticancer, and antihypertension capabilities have been established (Hashemi et al., 2021).

9.11.1 Patents

- **WO2016005969A1**: The present invention includes a method for creating a library of amino acid sequences with a common structural fold, as well as a method for developing and choosing an amino acid sequence with the desired affinity to a molecular surface of interest of a molecular entity utilizing the library. The methods are based on a random sample of backbone conformations and amino acid conservation patterns in experimentally available protein structures with the same structural fold (Fleishman, Lapidoth, Pszolla, & Norn, 2016).

References

Alsina-Fernandez, J., Coskun, T., Guo, L., & Qu, H. (2019). Incretin analogs and uses thereof. In: *World Intellectual Property Organization patent no. WO2019125929A1*. https://patents.google.com/patent/WO2019125929A1/en?oq = WO2019125929.

Annunziato, G., & Costantino, G. (2020). Antimicrobial peptides (AMPs): a patent review (2015–2020). *Expert Opinion on Therapeutic Patents*, *30*(12), 931–947. Available from https://doi.org/10.1080/13543776.2020.1851679.

Balaji, V. N., & Singh, C. U. (1997). Method of rational drug design based on ab initio computer simulation of conformational features of peptides. In: *United States Patent No. US5612895A*. https://patents.google.com/patent/US5612895A/en.

Bednarek, M. A., Jermutus, L. U., Ambery, P., & Petrone, M. (2017). Glucagon and glp-1 co-agonists for the treatment of obesity. In: *World Intellectual Property Organization patent no. WO2017153575A1*. https://patents.google.com/patent/WO2017153575A1/en?oq = WO2017153575.

Bloom, S. R., & Jones, B. J. (2017). Peptide analogs. In: *World Intellectual Property Organization patent no. WO2017178829A1*. https://patents.google.com/patent/WO2017178829A1/en?oq = WO2017178829.

Bossart, M., Evers, A., Haack, T., Kadereit, D., Lorenz, K., Wagner, M., ... Lorenz, M. (2018). Compounds as peptidic trigonal glp1/glucagon/gip receptor agonists. In: *United States Patent No. US20180155406A1*. https://patents.google.com/patent/US20180155406A1/en?oq = US20180155406.

Chain, D. G. (2021a). Treatment of tauopathies. In: *Canada Patent No. CA2826286C*. https://patents.google.com/patent/CA2826286C/en?q = protein%2fpeptide + aggregation + patent&oq = protein%2fpeptide + aggregation + patent.

Chain, D. G. (2021b). Treatment of tauopathies. In: *US Patent 10894822*. https://patents.google.com/patent/US10894822B2/en.

Chia, B. C. S. (2019). Antimicrobial peptidomimetics. In: *World Intellectual Property Organization patent no. WO2019027366A1*. https://patents.google.com/patent/WO2019027366A1/en?oq = WO2019%2f027366 + A1.

Datta, D., & Chaudhary, N. (2021). Chapter 8—Peptide-based hydrogels for biomedical applications. In Y. Hasija (Ed.), *Translational biotechnology* (pp. 203–232). Academic Press. Available from https://doi.org/10.1016/B978-0-12-821972-0.00003-4.

Doll, T. A. P. F., Dey, R., & Burkhard, P. (2015). Design and optimization of peptide nanoparticles. *Journal of Nanobiotechnology, 13*(1), 73. Available from https://doi.org/10.1186/s12951-015-0119-z.

Edelson, J., & Kotyla, T. (2016). Peptide nanoparticles and uses therefor. In: *United States Patent No. US9486409B2*. https://patents.google.com/patent/US9486409B2/en.

Eisele, Y. S., Monteiro, C., Fearns, C., Encalada, S. E., Wiseman, R. L., Powers, E. T., & Kelly, J. W. (2015). Targeting protein aggregation for the treatment of degenerative diseases. *Nature Reviews. Drug Discovery, 14*(11), 759. Available from https://doi.org/10.1038/nrd4593.

Eisenbach, L., Carmon, L., Tirosh, B., Bar-Haim, E., Paz, A., Fridkin, M., & Fitzer-Attas, C. (2000). Tumor associated antigen peptides and use of same as anti-tumor vaccines. In: *World Intellectual Property Organization patent no. WO2000006723A1*. https://patents.google.com/patent/WO2000006723A1/en?oq = WO0006723.

Fleishman, S., Lapidoth, G., Pszolla, M.G., & Norn, C. (2016). Method of computational protein design. In: *World Intellectual Property Organization patent no. WO2016005969A1*. https://patents.google.com/patent/WO2016005969A1/en.

Geng, H., Chen, F., Ye, J., & Jiang, F. (2019). Applications of molecular dynamics simulation in structure prediction of peptides and proteins. *Computational and Structural Biotechnology Journal, 17*, 1162–1170. Available from https://doi.org/10.1016/j.csbj.2019.07.010.

Goyal, R., Jerath, G., Akhil, R., Chandrasekharan, A., Puppala, E. R., Ponneganti, S., ... Ramakrishnan, V. (2021). Geometry encoded functional programming of tumor homing peptides for targeted drug delivery. *Journal of Controlled Release, 333*, 16–27. Available from https://doi.org/10.1016/j.jconrel.2021.03.010.

Hancock, R. E. W., Hilpert, K., Cherkasov, A., & Fjell, C. (2017). Small cationic antimicrobial peptides. In: *United States Patent No. US9707282B2*. https://patents.google.com/patent/US9707282B2/en.

Hashemi, Z. S., Zarei, M., Fath, M. K., Ganji, M., Farahani, M. S., Afsharnouri, F., ... Khalili, S. (2021). In silico approaches for the design and optimization of interfering peptides against protein – protein interactions. *Frontiers in Molecular Biosciences, 8*, 669431. Available from https://doi.org/10.3389/fmolb.2021.669431.

Hoffmann, R., Knappe, D., Hilpert, K., Mikut, R., & Ruden, S. (2015). Modified apidaecin derivatives as antibiotic peptides. In: *United States Patent No. US20150344524A1*. https://patents.google.com/patent/US20150344524A1/en?oq = US2015%2f0344524.

Kadereit, D., Wagner, M., Bossart, M., Lorenz, K., Haack, T., & Evers, A. (2016b). Non-acylated exendin-4 derivatives as dual glp-1/glucagon receptor agonists. In: *World Intellectual Property Organization patent no. WO2016198628A1*. https://patents.google.com/patent/WO2016198628A1/en?oq = WO2016198628.

Keogh, E. A., Southwood, S., Fikes, J. D., & Sette, A. (2002). Hla-a1, a2 -a3, -a24, -b7, and -b44 tumor associated antigen peptides and compositions. In: United States Patent No. WO2004052917.

Langer, R. S., Anderson, D. G., Gu, Z., & Aimetti, A. A. (2018). Self-regulated peptide hydrogel for insulin delivery. In: *United States Patent No. US9994615B2*. https://patents.google.com/patent/US9994615B2/en?q = peptide + hydrogels&oq = peptide + hydrogels&page = 4.

Lee, R., Kamm, R., Narmoneva, D., & Zhang, S. (2004). Angiogenesis and cardiac tissue engineering with peptide hydrogels and related compositions and methods of use thereof. In:

United States Patent No. US20040242469A1. https://patents.google.com/patent/US20040242469A1/en?q = peptide + hydrogels&oq = peptide + hydrogels.

Li, T., Lu, X.-M., Zhang, M.-R., Hu, K., & Li, Z. (2022). Peptide-based nanomaterials: Self-assembly, properties, and applications. *Bioactive Materials*, *11*, 268–282. Available from https://doi.org/10.1016/j.bioactmat.2021.09.029.

Lu, Q., Tang, Q., Xiong, Y., Qing, G., & Sun, T. (2016). Protein/peptide aggregation and amyloidosis on biointerfaces. *Materials*, *9*(9), 740. Available from https://doi.org/10.3390/ma9090740.

Manandhar, B., & Ahn, J.-M. (2015). Glucagon-like peptide-1 (GLP-1) analogs: recent advances, new possibilities, and therapeutic implications. *Journal of Medicinal Chemistry*, *58*(3), 1020–1037. Available from https://doi.org/10.1021/jm500810s.

McLaurin, J., & Cruz, A. (2007). Cyclohexane polyalcohol formulation for treatment of disorders of protein aggregation. In: United States Patent No. US20100113613A1.

Meyers, T. C., Turano, T. A., Greenhalgh, D. A., & Waller, P. R. (2000). Patent protection for protein structures and databases. *Nature Structural Biology*, *7 Suppl*, 950–952. Available from https://doi.org/10.1038/80723.

Obrecht, D., Luther, A., Bernardini, F., Zbinden, P., & Lederer, A. (2016). Beta-hairpin peptidomimetics. In: *World Intellectual Property Organization patent no. WO2016150576A1.* https://patents.google.com/patent/WO2016150576A1/en?oq = WO2016%2f150576 + A1.

Palani, A., Deng, Q., Huang, C., Zhu, Y., Bianchi, E., & Orvieto, F. (2019). Co-agonists with extended action of glucagon and GLP-1 receptors. In: *World Intellectual Property Organization patent no. WO2019060660A1.* https://patents.google.com/patent/WO2019060660A1/en?oq = WO2019060660.

Pandey, G., Morla, S., Kumar, S., & Ramakrishnan, V. (2020). Modulation of tau protein aggregation using 'Trojan' sequences. *Biochimica et Biophysica Acta (BBA) - General Subjects*, *1864*(7), 129569. Available from https://doi.org/10.1016/j.bbagen.2020.129569.

Pastrana-Rios, B., & Lloro-Manzano, I. (2012). Method for determination of protein, peptide or peptoid aggregation, stability, and viability and system using the same. In: *United States Patent No. US8268628B1.* https://patents.google.com/patent/US8268628B1/en.

Pentlavalli, S., Coulter, S., & Laverty, G. (2020). Peptide nanomaterials for drug delivery applications. *Current Protein & Peptide Science*, *21*(4), 401–412. Available from https://doi.org/10.2174/1389203721666200101091834.

Philip, R., & Keller, L. (2008a). Cytotoxic t-lymphocyte-inducing immunogens for prevention, treatment, and diagnosis of cancer. In: *World Intellectual Property Organization patent no. WO2008088583A2.* https://patents.google.com/patent/WO2008088583A2/en?oq = WO2008088583.

Philip, R., & Keller, L. H. (2008b). Immunogens in cancer stem cells. In: *World Intellectual Property Organization patent no. WO2008070047A3.* https://patents.google.com/patent/WO2008070047A3/en?oq = WO2008070047.

Pripp, A. H., Isaksson, T., Stepaniak, L., Sørhaug, T., & Ardö, Y. (2005). Quantitative structure activity relationship modelling of peptides and proteins as a tool in food science. *Trends in Food Science & Technology*, *16*(11), 484–494. Available from https://doi.org/10.1016/j.tifs.2005.07.003.

Ramakrishna, V., Ross, M. M., Philip, R., & Keller, L. H. (2002). Cytotoxic t-lymphocyte-inducing immunogens for prevention, treatment, and diagnosis of cancer. In: United States Patent No. US20080207497A1. https://patents.google.com/patent/US20080207497A1/en?oq = US20080207497.

Reedtz-Runge, S., Kofoed, J., Tornoe, C. W., & Sauerberg, P. (2015). Derivatives of glp-1 like peptides, and uses thereof. In: *World Intellectual Property Organization patent no. WO2015000942A1.* https://patents.google.com/patent/WO2015000942A1/en?oq = WO2015000942.

Sabatini, C. E., Heath, J. D., Covitz, P. A., Klingler, T. M., Russo, F. D., & Berry, S. F. (1999). Database and system for storing, comparing, and displaying genomic information. In: *United States Patent No. US5966712A*. https://patents.google.com/patent/US5966712A/en?oq = U.S. + Patent + No. + 5%2c966%2c712%2c.

Saikia, J., Dharmalingam, K., Anandalakshmi, R., Redkar, A. S., Bhat, V. T., & Ramakrishnan, V. (2021). Electric field modulated peptide based hydrogel nanocatalysts. *Soft Matter*, *17* (42), 9725–9735. Available from https://doi.org/10.1039/D1SM00724F.

Santosh, P., Prasad, D. R., & Seema, J. (2016). N-terminally modified linear and branched polyamine conjugated peptidomimetics as antimicrobials agents. In: *United States Patent No. US20160376306A1*. https://patents.google.com/patent/US20160376306A1/en?oq = US2016%2f0376306 + A1.

Schenk, D. B. (2005). Prevention and treatment of amyloidogenic disease. In: *United States Patent No. US6946135B2*. https://patents.google.com/patent/US6946135/en.

Semino, C., Shen, C., Sherley, J., & Zhang, S. (2004). Cellular reprogramming in peptide hydrogel and uses thereof. In: *World Intellectual Property Organization patent no. WO2004007683A2*. https://patents.google.com/patent/WO2004007683A2/en?q = peptide + hydrogels&oq = peptide + hydrogels.

Seo, J.-K., Lee, K.-Y., Cho, S., & Lee, I.-A. (2018). Antimicrobial peptide analogues derived from abalone (*Haliotis discus*) and antimicrobial pharmaceutical composition containing the same. In: *United States Patent No. US10077293B2*. https://patents.google.com/patent/US10077293B2/en?oq = US10077293.

Sidney, J., Sette, A., Grey, H. M., & Southwood, S. (2002). Subunit vaccines with a2 supermotifs. WIPO (PCT).

Skolnick, J., & Kolinski, A. (2001). Protein modeling tools. *European Patent Office*, EP1163639A1.

Srinivasan, M., & Lahiri, D. (2017). Peptides and methods for treatment of neurodegenerative diseases. In: *World Intellectual Property Organization patent no. WO2017147044A1*. https://patents.google.com/patent/WO2017147044A1/en.

Stevens, F. J., Myatt, E. A., & Solomon, A. (2000). Method for detecting and diagnosing disease caused by pathological protein aggregation. In: *United States Patent No. US6063636A*. https://patents.google.com/patent/US6063636A/en?oq = U.S. + Patent + No. + 6%2c063%2c636.

Stupp, S. I., Hartgerink, J. D., & Beniash, E. (2010). Self-assembly of peptide-amphiphile nanofibers under physiological conditions. In: *United States Patent No. US7745708B2*. https://patents.google.com/patent/US7745708/en?oq = US7745708.

Stupp, S. I., Spoerke, E. D., Anthony, S. G., & Niece, K. L. (2008). Methods and materials for nanocrystalline surface coatings and attachment of peptide amphiphile nanofibers thereon. In: *United States Patent No. US7390526B2*. https://patents.google.com/patent/US7390526/en.

Sun, X. S., & Huang, H. (2014). Protein peptide hydrogels. In: *United States Patent No. US8835395B2*. https://patents.google.com/patent/US8835395B2/en.

Svendsen, A., Bisgård-Frantzen, H., & Borchert, T. V. (1999). α-Amylase mutants. In: *United States Patent No. US5989169A*. https://patents.google.com/patent/US5989169A/en?oq = Patent + No. + 5%2c989%2c169.

Tang, S.-Y., Wei, H., & Yu, C.-Y. (2021). Peptide-functionalized delivery vehicles for enhanced cancer therapy. *International Journal of Pharmaceutics*, *593*, 120141. Available from https://doi.org/10.1016/j.ijpharm.2020.120141.

Thennati, R., Singh, S. K., Chaturvedi, N., Nage, N. B., Ranbhor, R. S., Ravishankara, M., & Bhagat, N. G. (2018). Method for making kex1 protease sensitive polypeptides using yeast strain. In: *World Intellectual Property Organization patent no. WO2018178796A1*. https://patents.google.com/patent/WO2018178796A1/en?oq = WO2018178796.

Tornøe, C. W., Kodal, A. L. B., Reedtz-Runge, S., & Lykke, L. (2019). Bifunctional compounds. In: *World Intellectual Property Organization patent no. WO2019016306A1*. https://patents.google.com/patent/WO2019016306A1/en?oq = WO2019016306.

Unal, E. B., Gursoy, A., & Erman, B. (2010). VitAL: Viterbi algorithm for de novo peptide design. *PLoS One*, *5*(6), e10926. Available from https://doi.org/10.1371/journal.pone.0010926.

Wilson, I. A., Livnah, O., Stura, E. A., Johnson, D. L., & Jolliffe, L. K. (1998). Small molecule mimetics of erythropoietin. In: *United States Patent No. US5835382A*. https://patents.google.com/patent/US5835382A/en?oq = U.S. + Patent + No. + 5%2c835%2c382.

Wilson, K. P., Griffith, J. P., Kim, E. E., & Livingston, D. J. (1999). Crystal structure and mutants of interleukin-1 beta converting enzyme. In: *United States Patent No. US5856116A*. https://patents.google.com/patent/US5856116A/en?oq = U.S. + Patent + No. + 5%2c856%2c116.

Yang, J., An, H.-W., & Wang, H. (2021). Self-assembled peptide drug delivery systems. *ACS Applied Bio Materials*, *4*(1), 24–46. Available from https://doi.org/10.1021/acsabm.0c00707.

Index

Note: Page numbers followed by "*f*" and "*t*" refer to figures and tables, respectively.

R

S

T

V

FSC
www.fsc.org
MIX
Papier | Fördert
gute Waldnutzung
FSC® C083411

Kolibri 360 GmbH
Ferdinand-Jühlke-Straße 7
99095 Erfurt, Deutschland
produktsicherheit@kolibri360.de